The development of the shapes of living organisms and their parts is a field of science in which there are no generally accepted theoretical principles. What form these principles are likely to take, when they emerge, is a subject in which there is a wide gulf of disagreement between physical scientists and experimental biologists.

This book contains both an extensive philosophical commentary on this dichotomy in views and an exposition of the type of theory most favoured by physical scientists. In this theory, living form is a manifestation of the dynamics of chemical change and physical transport or other physics of spatial communication. The reaction-diffusion theory, as initiated by Turing in 1952 and since elaborated by Prigogine and by Gierer and Meinhardt and others, is discussed in detail at a level that requires a good knowledge of a first course in calculus, but no more than that. In some respects this book takes up the theme that "the *things* which we see in the cell are less important than the *actions* which we recognize in the cell," which was a major theme of D'Arcy W. Thompson's classic 1917 work *On Growth and Form*. The rapid growth of the field of molecular biology has tended to overshadow the increase in our understanding of the nature of these kinetic processes. This book seeks to reawaken interest in dynamics in the hope that a better balance between the importance of things and the importance of actions may gradually emerge in the field of biology in the twenty-first century.

DEVELOPMENTAL AND CELL BIOLOGY SERIES 28
EDITORS
P. W. BARLOW   D. BRAY   P. B. GREEN   J. M. W. SLACK

KINETIC THEORY OF LIVING PATTERN

Developmental and cell biology series

SERIES EDITORS

Dr P. W. Barlow, *Long Ashton Research Station, Bristol*
Dr D. Bray, *King's College, London*
Dr P. B. Green, *Dept of Biology, Stanford University*
Dr J. M. W. Slack, *ICRF Developmental Biology Unit, Oxford*

The aim of the series is to present relatively short critical accounts of areas of developmental and cell biology where sufficient information has accumulated to allow a considered distillation of the subject. The fine structure of the cells, embryology, morphology, physiology, genetics, biochemistry and biophysics are subjects within the scope of the series. The books are intended to interest and instruct advanced undergraduates and graduate students and to make an important contribution to teaching cell and developmental biology. At the same time, they should be of value to biologists who, while not working directly in the area of a particular volume's subject matter, wish to keep abreast of developments relative to their particular interests.

BOOKS IN THE SERIES

R. Maksymowych *Analysis of leaf development*
L. Roberts *Cytodifferentiation in plants: xylogenesis as a model system*
P. Sengel *Morphogenesis of skin*
A. McLaren *Mammalian chimaeras*
E. Roosen-Runge *The process of spermatogenesis in animals*
F. D'Amato *Nuclear cytology in relation to development*
P. Nieuwkoop & L. Sutasurya *Primordial germ cells in the chordates*
J. Vasiliev & I. Gelfand *Neoplastic and normal cells in culture*
R. Chaleff *Genetics of higher plants*
P. Nieuwkoop & L. Sutasurya *Primordial germ cells in the invertebrates*
K. Sauer *The biology of* Physarum
N. Le Douarin *The neural crest*
M. H. Kaufman *Early mammalian development: parthenogenic studies*
V. Y. Brodsky & I. V. Uryvaeva *Genome multiplication in growth and development*
P. Nieuwkoop, A. G. Johnen & B. Albers *The epigenetic nature of early chordate development*
V. Raghavan *Embryogenesis in angiosperms: a development and experimental study*
C. J. Epstein *The consequences of chromosome imbalance: principles, mechanisms, and models*
L. Saxen *Organogenesis of the kidney*
V. Raghaven *Developmental biology of fern gametophytes*
R. Maksymowych *Analysis of growth and development in* Xanthium
B. John *Meiosis*
J. Bard *Morphogenesis: the cellular and molecular processes of developmental anatomy*
R. Wall *This side up: spatial determination in the early development of animals*
T. Sachs *Pattern formation in plant tissues*
J. M. W. Slack *From egg to embryo: regional specification in early develoment*
A. I. Farbman *Cell biology of olfaction*

KINETIC THEORY OF LIVING PATTERN

LIONEL G. HARRISON
University of British Columbia

CAMBRIDGE
UNIVERSITY PRESS

CAMBRIDGE UNIVERSITY PRESS
Cambridge, New York, Melbourne, Madrid, Cape Town, Singapore, São Paulo

Cambridge University Press
The Edinburgh Building, Cambridge CB2 2RU, UK

Published in the United States of America by Cambridge University Press, New York

www.cambridge.org
Information on this title: www.cambridge.org/9780521306911

First published 1993
This digitally printed first paperback version 2005

A catalogue record for this publication is available from the British Library

Library of Congress Cataloguing in Publication data
Harrison, Lionel G.
Kinetic theory of living pattern / Lionel G. Harrison.
   p.   cm. – (Developmental and cell biology series : 28)
Includes bibliographical references and index.
ISBN 0–521–30691–4
1. Biology–Philosophy. 2. Matter, Kinetic theory of.
3. Biology–Mathematical models.   I. Title.   II. Series.
QH331.H368   1992
574´.01–dc20   92-12578
                                                        CIP

ISBN-13 978-0-521-30691-1 hardback
ISBN-10 0-521-30691-4 hardback

ISBN-13 978-0-521-01991-0 paperback
ISBN-10 0-521-01991-5 paperback

To the memory of Joyce and Julian

Contents

Preface   *page* xiii

Part I: Macroscopics without mathematics   1

1 Introduction   3
   1.1 Philosophy: lumping, splitting, abstraction, and reality   3
      1.1.1 Lumping and splitting   4
      1.1.2 Abstraction *is* reality   6
      1.1.3 Macroscopics and the structure of processes 8
   1.2 Strategies of research   10
      1.2.1 On starting to build a bridge from both ends 10
      1.2.2 Making the join: bridges versus brains   11
      1.2.3 Visions of the invisible   16
      1.2.4 Preconceptions and experimental programs   17
   1.3 Mixing and unmixing: molecules versus large objects   21
      1.3.1 Molecules and individual identity: classical
             physical chemistry versus modern molecular
             biology   21
      1.3.2 Arrangements, aggregations, amplifiers, automata 24
      1.3.3 The kinetic preconception   28

2 Morphogen: one word for at least two concepts   31
   2.1 Type I morphogens: Wolpert positional signallers   33
   2.2 Activators versus morphogens   37
   2.3 Inhibitors, such as chalones, versus morphogens   46
   2.4 Type II morphogens: Turing morphogen pairs   47
   2.5 The two-morphogen interaction   49
   2.6 Activation and inhibition versus activated and
      inhibited regions   53

3 Pictorial reasoning in kinetic theory of pattern and form   56
   3.1 The fit of pattern to boundaries and the dynamics
      of pattern growth   58
      3.1.1 The fit of pattern to boundaries   59
      3.1.2 The dynamics of pattern growth   63

3.1.3   The guests of Procrustes: responses to chopping
and stretching   65
3.2   Modes of cell division in plants   67
3.2.1   One dimension: filaments   67
3.2.2   Location of determinative events within the cell   72
3.2.3   Division sequence of the *Azolla* root meristem   74
3.3   Animal development: response to damage and grafting 80
3.3.1   Teratogenesis and nonlinearities in the
Gierer–Meinhardt model   80
3.3.2   "Firing" a persistent gradient: grafting
behaviour of *Hydra*   84

4   Structure, equilibrium, kinetics   89
4.1   Definitions of the categories   90
4.1.1   Structure   90
4.1.2   Equilibrium   91
4.1.3   Kinetics   92
4.1.4   Mechanochemistry   94
4.1.5   Semantics of the term "field theory,"
and electric fields   96
4.2   Developmental control of the shapes of crystals 101
4.2.1   Structural aspect: just a few symmetry elements 101
4.2.2   Equilibrium shapes: surface free energy
and Wulff's theorem   102
4.2.3   Kinetic aspect: the diverse shapes of snow crystals 104
4.2.4   Biological control: the echinoid spicule   106
4.3   Division of plant cells: Is control kinetic,
thermodynamic, or mechanical?   108
4.4   Animal morphogenesis: rearrangement, deformation, and
proliferation of cells   110
4.4.1   Some phenomena, and the cell-as-molecule
concept   110
4.4.2   Equilibrium aspects: differential adhesion   118
4.4.3   From equilibrium to kinetics: incompleteness of
adhesive-gradient theory   123
4.4.4   Kinetic aspects involving mechanical forces   129
4.5   A few problems (without solutions)   137

Part II: Pattern-forming processes   143

5   The making and breaking of symmetry   145
5.1   Symmetry is in the spline of the beholder   145
5.2   Open and closed traverses: the accuracy of
self-organization   149

5.3   The simplest reaction-diffusion mechanism:
      optical resolution   154
     5.3.1   The mechanism of W. H. Mills   157
     5.3.2   A model without mathematics   158
5.4   Asymmetry begets asymmetry   167
     5.4.1   Trivial and significant antecedents   167
     5.4.2   Life on the planet Gedanken   170
5.5   The paradoxical nature of symmetry   172

6 Matters needing mathematics: an introduction   173
6.1   The language of rates, and the need for it   173
6.2   Differential equations, diffusion, and a Cheshire Cat   175
     6.2.1   A Cheshire Cat   175
     6.2.2   Differential equations   177
     6.2.3   Diffusion   178
     6.2.4   The Cheshire Cat, with mathematics,
         and three entities   181
6.3   Reaction-diffusion and growth of pattern: departure
     from uniformity, both ways   183
     6.3.1   Linearization about the spatially uniform steady
         state: a simple example   184
     6.3.2   A brief comment on the Brusselator   186
     6.3.3   The mechanism for optical resolution   187
     6.3.4   Reaction-diffusion: rate versus wavelength
         for a single morphogen   188
     6.3.5   Must the self-enhancement involve a squared
         concentration?   190
6.4   Thermodynamics, thresholds, bifurcations, and
     catastrophes   192
     6.4.1   Threshold flow rate: kinetic analysis of the
         optical-resolution model   193
     6.4.2   The same threshold condition with more entropy
         and less algebra   198
     6.4.3   An assortment of jargon: bifurcation, instability,
         catastrophe   201
6.5   Problems illustrating principles   209
6.6   Brief indications of solutions to problems   215

7 Kinetic models for stable pattern: an introduction   223
7.1   Turing's model without equations   223
     7.1.1   Maynard Smith's illustration   223
     7.1.2   Waves in phase as the starting point   225
7.2   Turing's equations and the growth or decay of a
     sine-wave pattern   229

7.3   Turing's conditions   233
    7.3.1   The conditions from a computer programmer's
        viewpoint: Lacalli's (k_1', k_4') space   235
    7.3.2   The conditions from a chemical kineticist's
        viewpoint: the Brusselator as an example   243

Part III: Bringing experiment and theory together   247

8   Classifications   251
  8.1   Beginnings of a classification of developmental theories   252
  8.2   The idiosyncrasies of some reaction-diffusion models   252
    8.2.1   Nonlinearity and the history of reaction-diffusion
        models   252
    8.2.2   Beginnings of a classification of reaction-diffusion
        models   256
    8.2.3   When can *dynamics* be classified as chiral?   261

9   Nonlinear reaction-diffusion models   264
  9.1   The Brusselator   265
    9.1.1   Chemical nature of the model (1): elementary two-
        intermediate schemes   265
    9.1.2   Chemical nature of the model (2): the Brusselator
        itself   268
    9.1.3   Pattern localization: its control by reactant
        concentrations   272
    9.1.4   The joining of models in sequence   275
    9.1.5   The "adaptable" character of the Brusselator   278
  9.2   The hyperchirality model   283
    9.2.1   Big hands from little hands   283
    9.2.2   Dynamics of the model   285
    9.2.3   A structural model, and a wider dynamic
        significance   289
  9.3   Brief comments on other models   291
    9.3.1   The Gierer–Meinhardt model   291
    9.3.2   Murray's model   294
    9.3.3   The Sachs–Mitchison model, and an
        acknowledgment to Rashevsky   296

10   Approaching agreement?   298
  10.1   *Acetabularia* and some desmids   299
    10.1.1   Choice of organism and of developmental event
        to study   300

10.1.2	Predictions, and assessment of the significance of results	302
10.1.3	Morphogens and mechanisms	305
10.1.4	Three feedback loops	307
10.1.5	Branching tip growth in some desmids	310
10.2	*Drosophila* segmentation	314
10.2.1	Communication versus "no crosstalk"	315
10.2.2	Carving out a theory	322
10.3	From slime moulds to salamanders	323
10.3.1	A morphogen pair in *Dictyostelium discoideum?*	324
10.3.2	Stages of Turing kinetics in *Polysphondylium pallidum*	325
10.3.3	Complex kinetics in the mesoderm of *Ambystoma mexicanum*	328
10.3.4	Turing patterns in nonliving chemical systems	332
10.4	Measuring, counting, regulation: *Acetabularia* versus *Drosophila* versus *Dictyostelium*	334
10.5	Confirmed predictions of kinetic theory	337
References		339
Index		351

Preface

Once upon a time I was a physical chemist working chiefly on the reactivity of sodium chloride with rather simple gases, and never on a complex organism, not even anything with a carbon atom in it. This book is the product of a change of field which started twenty years ago. This Preface must serve both to acknowledge those people who were most important in bringing about that change and to argue the case for the possible utility of a classical physical chemist in biological theory.

In 1971 my colleague Dr. R. E. Pincock gave a seminar on an instance of spontaneous optical resolution which he had discovered: Supercooled liquid 1,1'-binaphthyl in a sealed vial, when induced to crystallize suddenly, usually gave an asymmetric product. In some experiments that product contained more left-handed crystals than right-handed ones; in others, the reverse. He claimed that this was closer to truly spontaneous resolution than was Pasteur's crystallization of sodium ammonium tartrate, in which asymmetry was finally achieved only through the intervention of a biological organism, namely, Pasteur himself sorting the crystals into two piles.

Pincock's report presented two problems: first, the philosophical meaning of "spontaneity." How can asymmetry arise with no apparent antecedent, seeming to defy the precept that asymmetry begets asymmetry? Second, what kind of mechanism can one envisage for this phenomenon? Given this provocation, I published (Harrison, 1973) a speculative kinetic mechanism for the origin of chiral asymmetry in biochemical evolution. This invoked the cooperation of two molecules in autocatalytic formation of the same molecular species, and I proposed a territorial separation of systems of opposite chirality as an intermediate stage.

In 1973, I acted as chairman for the final oral examination of T. C. Lacalli for his Ph.D. degree. His thesis, "Morphogenesis in *Micrasterias*" (Lacalli, 1973), and more particularly his verbal presentation of it, led me to realize that the kind of kinetic equations I had been using in relation to optical resolution might be relevant to this much more extensive biological field. This led me to acquaint myself with the reaction-diffusion theory of pattern formation, as originated by A. M. Turing in 1952. From a pedagogic viewpoint, it led me also to recognize that the basic concept of kinetic generation of pattern can be more clearly understood in relation to optical resolution than in any other way. Therefore, I invite the beginner in kinetic theory to take my discussions of optical resolution not merely as a matter of personal biography, but as one of the best paths for anyone wishing to approach these concepts. It is the essence of symmetry-breaking.

My debt is enormous to both Pincock and Lacalli. If, on an afternoon in 1971, I had decided that I had something to do other than attend Pincock's seminar, or if in the summer of 1973 I had told the Faculty of Graduate Studies of the University of British Columbia that I had something to do other than to take the chair at an oral in a field of science quite unknown to me, then it is almost certain that I would never have worked in this field, nor written this book. Lacalli and I went on to develop a close collaboration which continues today.

Each of the three great divisions of physical chemistry – equilibrium, kinetics, and structure – is founded on a rock which was cemented firmly in place before I was born (1929): the universally accepted thermodynamics of Kelvin, Clausius, and Gibbs; the kinetics and statistics of Arrhenius and Boltzmann; and the quantum mechanics of Schrödinger and Heisenberg. In developmental biology I found something different, and immensely exciting: a field with a Great Unknown, and no firmly established conceptual basis. To pursue it is like trying to account for the rainbow in the fourteenth century, to do celestial mechanics before Newton, or to pursue quantum theory in the 1890s. There are many ideas around. Some of them are elaborately developed, and some will eventually be recognized as the correct concepts, but none has reached that status yet.

More specifically, the unresolved strategic question is whether the formation of pattern can be adequately described in the language of molecular biology and biochemistry (both of which deal with a spatial scale much tinier than that of pattern) or whether it requires description, in mathematical language, of the dynamics of interactions on a much larger scale. This book is concerned with the latter aspect. To me, it is a "scientific belief" (Polanyi, 1946, 1949) or "preconception" (Crombie, 1959) (see also the epigraph to Part III herein) or "paradigm" (Kuhn, 1962) that pattern formation on the scale of the organism cannot be accounted for without consideration of long-range dynamics. The kinetic preconception is that living pattern is generated by movement away from thermodynamic equilibrium, and therefore is explicable only in terms of rates of processes. It has been my experience that most physical scientists, given the problem of living self-organization, and with no prior knowledge of its theories, will instantly adopt the kinetic preconception, because they see nothing else in all our philosophies that seems suitable to this task. It is difficult to convey to them that most biologists are not envisaging explanations along kinetic lines: "But what else could do it?"

My primary purpose in writing this book is thus of an evangelical sort: to encourage adoption of the kinetic preconception, as a very promising working hypothesis on the strategic scale, among experimental developmental biologists. Why should I bother? Often a paradigm is, for a fairly long time, the preserve of a particular group among scientists, and this one has a good circle of adherents in the physical sciences. I was, however, brought up to accept as a credo that statement of the scientific method in which continuous interaction between theory and experiment is of the essence. That process of science cannot occur if the theoreticians and experimentalists are living in different worlds. The theories then become parts of pure mathematics, unrelated to the science of the physical universe. The experimental data become the dead body of science. A list of facts is as devoid of the intellectual life which constitutes science as a pile of assorted molecules, all in the right places, may be devoid of the life which makes a human being. The essence of both biological life and true science lies in processes and interactions.

This metaphor is not intended to deny the validity of the microstructural preconception nor the validity of the impressive living body of modern molecular biology which has grown out of it. The proper evolution of biological science needs more than one kind of body, with more than one conceptual basis. Essentially, the microstructural preconception or paradigm is that deterministic behaviour can stem from a single DNA molecule and extend in continuous deterministic sequence to larger scales of organization. The kinetic (and also thermodynamic) paradigm is that the single molecule behaves randomly (stochastically) and that deterministic behaviour on the macroscopic scale arises only as a statistical property of very many molecules. Work within this paradigm almost always requires the use of mathematical language in discussing and interpreting experimental results.

Both kinds of behaviour, deterministic-to-deterministic and stochastic-to-deterministic, as one goes from the molecular to the macroscopic scale, are well known from their widespread examples in nature. Surely both must be essential components of the complex sequences of events in biological development. But for the latter type, it is the physical scientist, rather than the biologist or biochemist, who is generally more accustomed to the type of discussion needed. In particular, the *experimental* physical chemist is accustomed to using fairly extensive mathematics in the discussion sections of most experimental publications. To my mind, advances in some large areas of developmental biology will be accelerated when many more experimentalists are using this conventional style of the physical chemist. That is why I presume to enter the field. I am not a theoretician – my theoretician friends keep reminding me of it.

Therefore, although I began moving into this field of pattern formation by doing only theoretical work, I was soon led into doing biological experiments myself. The advantages of using very large single-celled algae as systems for study were first brought to my attention by Lacalli's thesis on *Micrasterias*. This organism had been very carefully chosen in discussions between Lacalli and his supervisor, Dr. A. Acton, and I am grateful to both of them for orienting me toward the Chlorophyta. But in the event, my own experimental work has been on *Acetabularia*. This organism was in culture in the 1970s in the laboratory of Dr. B. R. Green, a plant biochemist whose laboratory in the Botany Department was just across the road from mine. For some years we collaborated on culture maintenance, and my first observations on the morphogenesis of whorls were made in her laboratory. Over an extended period the entire operation was gradually transferred to my own laboratory. I am greatly indebted to Dr. Green for enabling me to become, step by step, at least some sort of approximation to an experimental biologist.

An unrelated scientist with the same remarkably appropriate surname for a botanist, Dr. P. B. Green of Stanford University, has for ten years given me very substantial encouragement in two enterprises: pursuing experiment and theory together, and trying to express the theory in language suitable for the experimental biologist. He is entirely to blame for the existence of this book, which was his suggestion.

What manner of book is this, and what is its intended readership? When people ask me whether I am writing a textbook or a monograph, I am unsure how to answer. The intended readership can be found chiefly among research workers in developmental biology. But the purpose of the book, as discussed earlier, is to lend my weight to that of other practitioners of kinetic theory (reaction-diffusion, or mechanochemical, or other types) in seeking to bring about a "paradigm shift." The expositions of the mathematical material needed for those ready to accept the paradigm and go on from there will necessarily have something of the flavour of a textbook. But I have not sought to repeat the mathematical expositions of the books by H. Meinhardt (1982), L. Edelstein-Keshet (1988), and J. D. Murray (1989). A large part of the content of this book is essentially philosophical commentary on the various approaches to explanation of large-scale phenomena and their relationships to molecular phenomena.

Such considerations lie not within the realm of mathematics, but rather that of physical chemistry. This discipline has for more than a century been concerned primarily with the world immediately around us, a world that is at ordinary temperatures and therefore consists of large numbers of molecules organized into solid, liquid, and gaseous phases. This world includes all living material, and that presents a challenge to the phase approximation because of its content of a multitude of structures intermediate in size between the molecule and the macroscopic phase. Certainly the physicochemical concepts appropriate to this material are still at an embryonic stage, and their full development should give rise to one of the exciting fields of science in the twenty-first century.

There should be a place in this development for physical chemists and physicists, and people in those disciplines are also among my intended readership. Many are indeed interested in the theoretical concepts and already accept the kinetic preconception. But few are actually pursuing the interaction between theory and the experimental phenomena of biological development. The reason for this has to do with a different paradigm. For three hundred years, since Robert Boyle defined elements as "the ultimate limits of chemical analysis," a principal driving force in chemistry has been the simplification of systems by separations and purifications. The chemist likes to handle a system with very few substances present in it at any one time. Notwithstanding the well-known witticism that a physical chemist is someone who makes accurate measurements on impure substances, in fact the physical chemist is just as uncomfortable with an impure system as is any other kind of chemist.

The common denigration of living material as being irreproducible with respect to physical measurements is quite at variance with the definition of life in terms of the property of reproduction. Anyone who, like myself, has done experiments on the catalytic activity of inorganic solid surfaces must be well aware that they can be irreproducible beyond the limits of any trouble that one sees in the catalytic processes that support life. The modern surface chemist usually keeps everything except the one desired reactant away from the surface under investigation by the use of ultra-high vacuum. Anything approaching the multitude of substances present in a living being would quite destroy not only such ideally clean experimental systems but also the operation of industrial catalysts, which have to cope with much dirtier systems.

In living systems, unlike inanimate ones, chemical complexity is not synonymous with dirt. This is because the multitudinous substances of life are not a set of ignorant armies clashing by night, but participants in the most highly organized type of system in our universe. We may seek to study the processes of life one at a time because all the other processes interacting with the one of interest have been self-designed not to interfere, but even to provide assistance. That is the basic meaning of self-organization.

Physical chemistry from the 1920s to the 1950s was largely concerned with the properties of condensed phases, or gases at fairly high molecular concentrations. It has subsequently developed in a number of new directions, but many of them have involved either very clean conditions or sophisticated experimental techniques in which the physics of the technique itself occupies much of the time of the physical chemist. By comparison, remarkably little has been done to extend the conceptual basis of macroscopic organization from simple systems to ones which are complex but have acquired the knack of remaining organized, that is, living material.

Faced with the problem of chemical generation of time-order and space-order, physical chemists may tend to study, instead of life, the few simple inanimate systems which produce periodicities and patterns, such as the Belousov–Zhabotinski reaction (cerium-catalyzed oxidation of malonate by bromate). That particular system generates travelling waves which may have some correspondence to neural and cardiac electrochemical phenomena, but it does not generate the stationary wave patterns which are the essence of morphogenesis. Such patterns were, however, first produced in vitro in 1990 from two chemical reaction systems, and they appear to be in the category of Turing structures. Perhaps this new evidence will help to point physical chemists in the direction I would like to see them go: Study life itself! (For further information on all these chemical systems, see Section 10.3.4.)

The tripartite division (structure, equilibrium, kinetics) which I advocate as the first stage in classifying developmental mechanisms is also a classification of preconceptions or paradigms. Different basic attitudes of mind are needed to envisage the formation of shape as, first, small pieces fitting together to make larger ones; second, pieces aggregating so as to minimize the total free energy; or, third, processes acting kinetically to form shape as it is in a waveform of the surface of flowing water. This last is among a number of nonliving analogues of biological pattern formation shown by Meinhardt (1982). Very few biologists have entered into this third way of thinking. Among those who have encouraged me by doing so and who are not mentioned elsewhere in this Preface I must mention F. M. Harold (1990), J. Frankel (1990a,b), and especially Jay E. Mittenthal, with whom I have had substantial interactions on this topic over a number of years, including his comments to me on an early draft of this book.

I do not want this to be known principally as a "reaction diffusion book." Although that kind of kinetic theory is quite a broad field – the one which has been most extensively elaborated in regard to pattern formation, and the one which I describe in most detail here – nevertheless reaction-diffusion is still but one example of the triumvirate *activation–inhibition–communication* which collectively has the power to generate pattern ab initio. Other examples are mechanochemical theory, self-electrophoresis, and complex intercellular interactions such as mutual reinforcement of synapses in a self-assembling nervous system. It is the fundamental unity of all these kinds of theories which I most want to convey to the reader under the heading "kinetic theory."

In all my advocacy for this I remain essentially a physical chemist. I must express my gratitude to the Chemistry Department of the University of British Columbia and all my colleagues there, especially the Department Head for many years, Charles A. McDowell, for their tolerance toward (and often definite interest in and material assistance with) my chemically unconventional activities. Also, I am grateful to the Natural Sciences and Engineering Research Council (NSERC), Canada, for continuing financial support of my research through all stages of my switch of fields, in the face of an astounding diversity of referees' opinions. (One set of eleven referees assessing a certain grant application gave numerical box scores on my "originality" and "methodology" ranging all the way from 2/10 to 10/10; such is the range of views on the value of pursuing the kinetic paradigm and on how one should go about it.)

The "field" of a scientist may be classified in three different ways: (1) by experimental method, especially for those devoted to a complex time- and money-consuming machine (e.g., "an NMR imaging group"); (2) by the natural materials and phenomena studied ("a solid-state physicist," "a natural products chemist," "an invertebrate embryologist"); and (3) by the conceptual basis of one's thinking ("a physical chemist"). I have changed principally in the second of these three respects, from studies of inorganic gas–solid reactions to studies of living organisms. My biological knowledge, such as it is, has been picked up piecemeal, especially at diverse meetings and seminars, and I am indebted to more people than I can list for bits of my information. In addition to the people already acknowledged in this Preface, I would like to thank Dr. N. Auersperg for organizing many graduate seminars at this university and various symposia at scientific meetings elsewhere, from which I have received much factual knowledge and considerable intellectual stimulation.

Of the three bases for classifying one's "field" listed in the preceding paragraph, the most difficult to change is the conceptual basis of one's thinking. I remain a physical chemist, but, paradoxically, I am trying to influence biologists toward changing to that kind of thinking. To be steadfast in one's own mind-set while expecting other people to change theirs is the characteristic affliction of the evangelist, but some of them succeed!

At one point in a course on plant development which I attended, one of my botanical colleagues enquired, "Shouldn't we be talking about rates?" That query was inadequate to give the rest of the course a kinetic bias. In a multiple-author book on positional controls in plant development edited by Barlow and Carr (1984) the first chapter is by Meinhardt. Presumably the editors asked him to write it because they believed his approach to be important. But a conspicuous feature in comparison with the rest of the book is that although Meinhardt is referenced in several chapters, no one is actually *using* his approach. Likewise, in a recent set of papers on mechanisms of segmentation (French et al., 1988), also containing a Meinhardt contribution, both Meinhardt and Turing are referenced only very sparsely, and again the references do not represent extensive *use* of kinetic concepts by other authors. The most striking contrast remains that between Turing's account of "The Chemical Basis of Morphogenesis," which is entirely kinetic, and Lehninger's (1975) chapter "The Molecular Basis of Morphogenesis," which offers a completely structural account.

Let us now discuss the rates of processes and how they can work to form patterns.

PART I

Macroscopics without mathematics

Pour demeurer symétrique et beau, un corps doit se modifier tout entier à la fois. . . .
[To remain symmetrical and elegant, a body must modify itself all together at the same time. . . .]

—Pierre Teilhard de Chardin, *Le Phénomène Humain* (1955)

For Part I, this quotation is the first axiom of macroscopics. But it could also have been used as epigraph to Chapter 5, which seeks to suggest that the statement may not be an axiom. The existence of overall control may be observable if the smoothed symmetry of the whole is superior to statistical expectation based on the disorder of the parts.

1

Introduction

1.1   Philosophy: lumping, splitting, abstraction, and reality

D'Arcy Thompson (1917) wrote that "the *things* which we see in the cell are less important than the *actions* which we recognize in the cell." He expected that in the following few decades biology would advance in the direction of mathematical description of actions or processes. He, and most others at the time, believed that microscopy had reached the limits of its capacity to reveal microstructure, and few people believed that determination of the structure of genes was foreseeable. In the event, as everyone knows, developments that were not foreseen have made up a great part of the most spectacular advances in science over the past forty years. Meanwhile, those advances which Thompson anticipated have not occurred – to such an extent that Bonner (1961) omitted from his abridgement of Thompson's *On Growth and Form* the entire chapter containing the foregoing quotation. Was Thompson wrong?

My thesis is that Thompson erred only in regard to his expectation of the timing of an advance which would unite mathematical-physical science to biology in the same way that physics and chemistry had become united in the late nineteenth and early twentieth centuries. That unfulfilled union must take place, to my mind, in the twenty-first century if many of the problems of developmental biology, which today remain as mysterious to us as they were a hundred years ago, are ever to be solved. I cannot conceive of solutions excluding the extensive use of mathematical-physical science, and I see such approaches as being entirely complementary to the existing molecular biology and in no way antagonistic to it.

This thesis has two parts: first, that developmental biology is now at a different stage of the scientific method than are most other branches of science and needs different attitudes; second, that to establish the nature of a process can be as solid a scientific objective as to discover the nature of a concrete object, such as a molecule. Sections 1.1.1 and 1.1.2 respectively address these topics.

Neither of these statements finds ready acceptance among the generality of experimental developmental biologists today. Quite often, physical scientists attribute this to a reluctance among biologists to adopt mathematical language; and, over many decades, some biologists have from time to time vehemently rejected mathematical explanations. But this is not, I believe, the main problem. Scientists (like most other people) will do what they find necessary to reach a highly desired objective, even if this involves activities which are difficult, time-consuming, and not what they had expected to be doing. They must be convinced, however, that an unexpected line of approach is necessary.

Mathematics is not essentially different from verbal explanation. Mathematical reasoning is simply the continuation of verbal logic by other means, when the complexity of the logic makes its expression in words cumbersome and obscure. (For instance, puzzles of the following kind are designed to exploit the equivalence between trivially simple algebra and quite obscure verbiage: "Bill is twice as old as Joe was when Bill was ten years older than Joe is now; and Bill was thirteen years old when Joe was born. How old is Bill?") Occasionally, a view is put forward that mathematical reasoning is qualitatively different from verbal explanation and probably irrelevant to biology. I cannot argue against that viewpoint, because I have never even begun to understand it. The essential equivalence of mathematical logic and verbal logic is to me an axiom, a credo. My book can cater to readers who are not fluent in mathematical languages, but it can do nothing for definite unbelievers in this credo. All the words lead toward the equations, and the words are really useful only to people who are going to follow them that far.

If, then, the philosophical chasm between experimentalist and theoretician is not just a question of words versus equations as explanations, what is it?

1.1.1 Lumping and splitting

As a physical scientist, I was brought up to believe that the ultimate objective of science is unifying – that science is a climb toward some minute and distant shining summit which might turn out to be the one equation that describes everything. That objective may be far off, but those who have that basic attitude tend to feel that they have made a step upward whenever they find some common principle definitely or possibly present in two sets of phenomena which up to that point had appeared quite different. Thus, on my first encounter with the problem of morphogenesis, I became quite excited at the idea that something fundamental to it might be the same as something equally fundamental to the problem of optical resolution. In my writing on the latter problem, I pointed out that at the very earliest stages of biochemical evolution, an autocatalysis requiring the assistance of two product molecules was hardly distinguishable from sexual reproduction, and I indicated a number of other correspondences between processes which become quite different at later, more complex stages of organization.

Closer interaction with biologists over the next few years impressed upon me that many of them do not recite this unifying credo of the physicist's beliefs. Indeed, a referee commenting on one of my manuscripts wrote that several mechanisms might be mathematically similar, but that biologists would consider them different. The clear implication was that I had better do the latter if I wanted to publish in biological journals. Much more recently, a referee of another manuscript wrote that "I don't think the weaknesses of the paper lie in the mathematical aspects, but rather in the failure of the authors to appreciate how divorced their abstract entities are from the real entities which govern early body patterning." (This comment at least makes it clear that the use of mathematics is not the matter at issue.)

There are, of course, excellent reasons for this tendency of biologists to concentrate upon the analytical aspect of science as against the synthetic. One of the most striking features of life is its diversity, and the precise description of that is a sine qua non of biological science, a task great enough to command the total attention of large numbers of people. Yet it would be an insult to biology to suggest that it does not get beyond the first step of the scientific methods, the gathering of facts. Science has as its essential attribute the continual reworking of facts through all the steps of the scientific methods. And in that method, fact and theory are by no means as clearly separable as they are represented in elementary statements of the scientific method. Today's experimental results are commonly set down in the language of yesterday's theories, which would have been unintelligible a few years or decades earlier.

Within all the rigour of a definition of science as intellectual process, modern molecularly based biology is fully a science, and a rapidly advancing one. But the bias of its practitioners is analytical, and strongly so. They are splitters, not lumpers. (These succinct equivalents for analytically and synthetically minded scientists are well known in the conversation of physicists.) Surely, however, any complete scientist needs to give some attention to both lumping and splitting? The balance here is a matter of historical timing. In a field which seems just now to have all the unifying concepts it needs, there is room for the work of a multitude of splitters who want to give very little of their attention to lumping. For molecularly based biology, the general concepts of the structures of nucleic acids and proteins, as well as the nature of the genetic code and its transcription and translation, are thoroughly established and universally accepted. Secure in these generalities, the modern biochemical biologist can devote a lifetime to ferreting out what particular proteins or genes are responsible for a few particular phenomena within a single species of organism.

By contrast, developmental biology, concerned with the macroscopic organization of the organism, stands across a gulf from molecular biology, a gulf which structural concepts have been unable to bridge. Apparently the unifying concepts have not yet been found; if they are already in the literature, they have not yet achieved that consensus of recognition which is the foundation on which many splitters base their work. The lumping is needed first.

That noted champion of unpopular theories, the astronomer F. Hoyle, has remarked that when a problem remains unsolved, general opinion must be wrong. This, to my mind, expresses an important philosophic truth very relevant to the present state of developmental biology. Yet the neatly epigrammatic form of Hoyle's statement runs some risk of showing that brevity is the soul of unmannerliness. It is not my objective to suggest that the philosophic approaches of molecular biology are wrong. They are obviously right for the field, and have led to great triumphs of scientific discovery. I wish to assert merely that development is a different field and needs some *additional* approaches, which are likely to be radically different from (but in the end complementary to) the current microstructural emphasis.

The only opinion which I would call *wrong* would be one which denies that there is, in biological development, a Great Unknown, and therefore a new concept or set of concepts to be established. Surely it is evident that in living pattern and form, nature has provocatively concealed some essential underlying simplicity in an excess of ornament. There is no lack of diverse and fascinating experimental data. Yet phenomena which were meticulously described by embryologists of the 1880s remain, a century later, without generally accepted theoretical explanation – which cannot be said of atomic spectroscopy, Mendelian genetics, or a host of other century-old experimental topics.

1.1.2 Abstraction is *reality*

A few years ago, on giving a seminar in a series entitled "Simulation and Modelling in Science," I remarked to the organizer that the title was wrong. It should have been "simulation and modelling *is* science." The comment delighted him, but many scientists would not, I think, be equally pleased. The words "simulation" and "modelling" can be interpreted as having something of a pejorative cast, throwing doubt on their relation to reality. But when I first heard of the scientific method, neither of these words was being used at all in relation to it. Surely both are to be seen as parts of theory, the means by which science seeks to express a vision of the truth or of reality, whichever word one happens to prefer?

By the same token, the scientific enterprise could also be defined as an effort to take experimental facts and extract from them increasingly close approximations to truth by a process of abstraction. This implies, of course, that I deny totally the distinction between "abstract entities" and "real entities" in the quotation in Section 1.1.1.

The problem here is, I think, that sometimes a group of scientists will become preoccupied with the virtues of one particular kind of model, which they have found to be powerful, to such an extent that they come to regard this model as more "real" than the models used in branches of science less familiar to them. In more than one branch of science today the molecule has acquired such an exalted status. To be sure, the molecule is one of our best models, but it is in no sense an ultimate truth. For the purposes of the biologist and biochemist, a molecule often may be viewed in very concrete terms, as a geometrical object of definite size and shape. But to the chemical physicist, the reality of a molecule is that it is a solution of the Schrödinger equation. Many of its properties cannot be understood at all in classical terms, and the quantum-mechanical equations describing the molecule compose an altogether higher and more powerful model of reality than the concrete geometry.

One of my colleagues in chemical physics gave a seminar in which he was most adamant that he was a pure experimentalist; he sounded as though he might not even like theoreticians. He then proceeded to present some data in the form of drawings of molecular orbitals, not in ordinary space but in momentum space. To most experimental biologists, such a presentation would seem to belong to one of the most distant reaches of theoretical abstraction. What one sees as an abstract model, and what one selects as the concise language for immediate presentation and discussion of one's experimental data, will depend upon one's preconceptions or paradigms.

What, then, is the kind of "abstract explanation" which I am advocating that we adopt and regard as a down-to-earth description of the "real entities" of developmental pattern formation? When I described to a biochemist some results showing quantitative control of a morphogenetic feature of *Acetabularia* by some unknown bound state of calcium, he remarked that "it's going to be difficult to find out what is doing it." This is a problem which confronts most biologists most of the time: to find out what is doing it. I invite my readers to take a moment to formulate a conception of the kind of solution each is usually expecting; I anticipate that most modern biologists, for a very wide variety of problems, will most commonly be expecting that the nature and structure of a particular protein will turn out to be "what is doing it." For my part, I would usually be looking for a *process*, and I would be happy if I could measure kinetic rate constants, or diffusivities, or elastic constants, or conductivities, and show that they were attributes of that process, without necessarily knowing what molecules might be involved. I would hope that these things would eventually give some pointers to the biochemists as to where some kinds of molecules may be found, but only over a long time scale.

More recently I have found spatial patterns of bound calcium which correspond to my concept of a two-stage hierarchical process. I consider that to be a scientific step forward toward establishing the nature of the process. But I am not one step nearer to identifying the molecular species to which the calcium is bound, and the molecularly devoted scientist might say that I have made no advance at all.

1.1.3   Macroscopics and the structure of processes

A hundred years ago, the disputes between the positivist philosophers led by
Ernst Mach and the advocates of molecular reality led by Ludwig Boltzmann
were approaching a climax. Basically, the positivists believed that the su-
premacy in science of observation of macroscopic phenomena was incompati-
ble with a concept of molecules as real entities – they had to be abstractions.
The supporters of Boltzmann believed that macroscopic objects and molecular
objects were both real and that the connection between them could be readily
made provided that one recognized that connection as being statistical and
therefore requiring primarily mathematical discussion. The ultimate victory of
this viewpoint was achieved early in the twentieth century, and that provided
the philosophical foundation on which twentieth-century science has been
built.

Today in biology there is something of a dichotomy between the molecular
biologists and those who believe that the whole organism (or large parts of it)
should be the principal object of study. This division has something of the
flavour of a repetition of that debate of a century ago. What is it going to lead
to as a foundation for the biology of the twenty-first century? The hope, of
course, must be that macroscopic and microscopic studies will advance so as
to complement each other. But, in my opinion, that is very unlikely to happen
by simple extension of the structural approaches which have proved so powerful on the molecular scale. In the tripartite division of physicochemical concepts into structure, equilibrium, and kinetics, it is the second and third which
I tend to think of as comprising "macroscopics." I define this term (suggested
to me by my colleague Dr. R. F. Snider) as "the nature of change, and the
organization of matter in states above the molecular." I envisage that the study
of macroscopics for the present purpose is going to need, just as it did in
Boltzmann's day, statistical and mathematical treatment.

In a preliminary outline for this book I mentioned the word "macroscopics"
to Paul Green, the instigator for the writing of the book. His immediate
reaction, in slightly edited version, was as follows:

The issue of "macroscopics" is one I deal with a lot. The puzzle of development can
be likened to a multi-span bridge. One terminus is DNA and the other is a developmental progression. The first few islands joined are clear enough: RNA, protein, etc.; and
the spans are conversion processes, like transcription and translation. The problem we
both address is, "What's between the last island reached (self-assembly) and developmental progression? Most biologists subconsciously think that some "silver bullet" or
single protein will clear up everything in one stroke. A hard look at developmental
processes, however, shows that one has to account for a myriad of changes over large
distances (many cells). The idea of coupling one developmental change to one section
of the genome is inadequate because there are far too many developmental events. The
solution to this information paradox is that an organism inherits rules that spell out the
progression. The rules are, or are like, time-based differential equations which have

the ability to encode complex sequences with high efficiency. Thus one has to regard development as an integration through space and time, the genome providing the equivalent of the differential equations. Thus there is no escaping the calculus when studying development.

The difference here is that the molecular scientist, in seeking the explanation of some large-scale phenomenon, goes ever downward in spatial scale, and usually in time scale also, ending up in nanometres and picoseconds. The macroscopist, if I may so designate a devotee of macroscopics, goes through different levels of explanation, including many different concepts, such as differential equations, force fields, and so forth, but never changes the spatial scale or time scale, considering always the whole extent of the development. An analogy by Lacalli (1973), following J. Needham, concerns the study of a Swiss watch to discover how it functions. One may take the watch apart and examine, list, and diagram the springs, gears, shafts, and so forth, and how they fit together. Yet one does not have a full explanation without the application of equations of motion to the whole. These involve concepts of momentum, moments of inertia, and simple harmonic motion arising from a restoring force proportional to displacement.

If in the light of such a study of one oscillating system one were to set up a team to examine some other oscillating system of unknown contents, one might designate some people to take it apart and describe its parts in ever greater detail, and others to tackle other questions: What is the displacement that produces a restoring force, and what is the origin of that force? To be sure, these two parts of the team should exchange information, and the whole team is needed to produce the whole story. Also, a question of applied science versus pure science arises here. If one wishes to know *how to make* a Swiss watch, the information from the first part of the team will suffice; but if one wants to know *how it works,* the second part of the team is vital, with only a limited amount of the structural information being necessary. This analogy would seem to give the edge to the molecular biologist for practical utility, in conformity with current developments and expected advances in so-called genetic engineering. Yet consider: Suppose that one wishes to *design* an oscillatory system other than those already studied. To which part of the team should one have paid attention?

The relation of the Swiss-watch analogy to Green's comments on biological development cited earlier is that in a biological system we know that the genome decrees the manufacture of a number of enzymes, and thereby specifies the kinetic rate constants for a number of chemical reactions; but one cannot therefrom predict how much of each reaction product is going to be produced at various places in the system until one has written down, and solved, the differential equations containing those rate constants. In the Swiss-watch analogy, the "genome" would specify the force constant of the balance spring and the moment of inertia of the balance wheel. From these, we could get the frequency of oscillation simply by solving equations of motion.

In Part I of this book I have tried to use as few equations as possible. Therefore, the full comparison between the activities of molecular biologists and those of kinetic-theory practitioners will not emerge until Part II, which will get closer to the daily work of the latter group. There it should become apparent that these people can be as much splitters as are the molecule hunters. Equations also have complex details. The theorist will often refer to the "structure" of a dynamical mechanism and will think of the terms in the equations as components of that structure, just as the molecular scientist will think of a carboxyl group as a component of the structure of a molecule. This, of course, obscures my distinction among structure, equilibrium, and kinetics. But I hope it tends to make clear that two groups of scientists may be engaged in essentially parallel enterprises which appear different because the two groups have different perceptions of the "ultimate realities" they are seeking: molecule versus process, matter versus motion.

The structure of processes is taken up again in Section 6.4.3, particularly with regard to the concept of the "structural stability" of equations of motion. For instance, an equation for oscillations which will continue undiminished forever may be converted into one for oscillations which, more realistically for most observed processes, will die away as time goes on. This requires one additional term in the equation, and the immortal oscillations are destroyed by that term; the equation is structurally unstable with respect to that addition, which effectively is a poison for the oscillations. It is, of course, a velocity-dependent term representing viscous resistance (or friction) in the case of mechanical oscillations; in more general terms it would be called a relaxation process, a term well known in magnetic resonance.

Molecularly minded scientists at once want to know the nature of the viscous substance, or the promoter of relaxation. The kineticist would like to know that, but not necessarily now. For complex systems, such as biological material, a magnetic-resonance experiment can establish thoroughly, scientifically, that there are, say, five different relaxation processes for water protons, with a quantitatively measurable relaxation time for each, and clearly characterized changes if one takes the system to pieces or makes other disturbances. All this can constitute years of good publishable science, throughout which the chemical nature of the relaxing species remains a puzzle, probably to be solved much later by someone else.

1.2   Strategies of research

1.2.1   On starting to build a bridge from both ends

In a referee's report on a recent reaction-diffusion paper by T. C. Lacalli, the following comment appeared: ". . . the paper could be published as a demonstration of a strategy for modelling these phenomena. However, at the present pace of inquiry, such models will soon be supplanted by models that relate in detail to molecular processes." This was a favourable report, but nonetheless this quotation provides another illustration of the conceptual gap in the field. If I may continue Paul Green's metaphor, one cannot supplant one end of a bridge by building the other. They are planted in different ground, and neither will ever occupy the place of the other. But ultimately, one has a bridge when they meet in the middle. The phenomena referred to in this example were those of *Drosophila* segmentation. To be sure, molecular information is being accumulated at a quite astonishing rate. But how the molecules participate in *processes* is an inquiry which is hardly advancing at all. The discoverers of molecules often postulate some processes, but usually with insufficiently rigorous descriptions of their dynamics to permit the essential tests, in the computer, of whether or not the models actually work. Meanwhile, as the referee quoted earlier pointed out, the modellers who are studying dynamics with some scientific rigour usually do not have precise and detailed ideas of which molecules are participating in the processes. There is still a big gap in the middle of the bridge.

Figure 1.1 is my strategic overview, a surveyor's map of the islands to be joined by the bridge. The arrowed lines at top and bottom are the engineering supervisor's notes regarding who is currently doing what in the construction work. They illustrate the contrasting approaches of physicists and biologists. They show also a limitation in the analogy. The two groups building from opposite ends are using quite different materials and ways of joining them together. There is no resolution for this which can completely save the analogy. But it is partly resolved by my discussion in Chapter 4 of developmental control of the shapes of crystals. There I indicate that one should enquire in regard to any shape-generating phenomenon what *aspects* of it fall into each of the three divisions – structure, equilibrium, and kinetics – rather than trying to classify the whole complex phenomenon into one of these categories exclusively. This means that at island 2 (second from left) in Figure 1.1 the builders may discover that though their materials are different, there are ways of fitting them together which make sound engineering to complete the bridge.

1.2.2 Making the join: bridges versus brains

In the language of the bridge-building analogy, there are two problems to be addressed: First, how shall the workers from opposite sides go about making the join in the middle? Second, and not so obvious from the analogy, will they recognize the join when it has been made? On the latter, my experience has been (especially in relation to *Drosophila* segmentation stripes) that when I have become excited about what has seemed to me to be the first girder in place linking the two sides, the value, relevance, and methodology of what I have seen as the linking step have been vehemently denounced by some prominent experimentalists in the field.

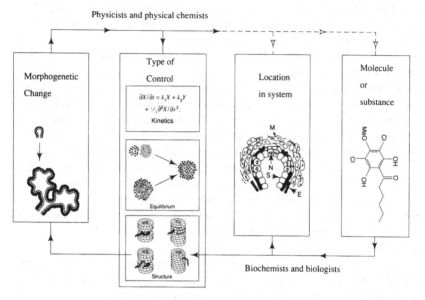

Figure 1.1. Contrasts in strategies of research. All scientists have the ultimate objective of linking together the contents of all four boxes (1 to 4 from left to right), but different kinds of scientists differ on, first, the chronological order in which they expect to build the bridges and, second, which subdivision of box 2 they expect to become most significant. The illustrations used in boxes 1 and 3 are of the morphogenesis of the mouse submandibular salivary gland, modified from Bernfield et al. (1984). For this example, see also Figure 4.14 and Table 4.1. In box 3, the possible alternative locations for crucial pattern-forming events indicated are as follows: E, extracellular matrix; S, cell surfaces; N, nuclei; M, mobile mesenchyme cells as the "particles of the system." Chapter 4 explains the meaning of the threefold subdivision of box 2. Molecules of many different kinds and sizes, from ammonia to glycoproteins, have been mentioned in the literature as putative morphogens. One of the most clearly established is illustrated in box 4: the substance DIF-1, one of a morphogen pair in the patterning of *Dictyostelium discoideum;* see Section 10.3.1.

The bridge-building work, of course, involves the use, by both physical and biological scientists, of the scientific method, with its characteristic alternation of experiment and theory. Both problems raised in the preceding paragraph require for their resolution an examination of the different ways in which that alternation may proceed. The scientific process goes most smoothly under the aegis of a universally accepted preconception or paradigm. Two things are then possible. First, an individual practitioner of the field can be classified as theoretician or experimentalist, and for most of the time can pursue activities entirely of the one kind or the other, knowing that the experiment–theory interaction exists on a social scale. In other words, the bridge has been built, and it is possible to drive over and talk to the people on the other side whenever one wants to. This is the dynamical structure which obtains in most of modern chemistry, in relation to the preconceptions of atomic and molecular quantum mechanics established in the 1920s.

Second, the general acceptance of a theoretical paradigm makes possible for the experimentalists a dynamic in which most of the hard work is done at the bench, and much less effort needs to be spent on interpreting the data. One set of experiments can be expected to lead smoothly to the next, with minimal theoretical linkage. By "smoothly" I mean that the scientist is rarely faced with the obstacle: "I don't see what the next step is." Rather, a multitude of next steps can be seen, and the problem becomes to find sufficient grant funds to assemble the people and equipment to pursue them all at once. This is the dynamic of modern molecular biology and biochemistry.

When, as in the matter of pattern formation, there is no generally accepted preconception, the dynamic of the field is altogether different. There may still be groups of people who may be described as theoreticians or experimentalists. But there is no bridge between them. This has, for the present discussion, two consequences. First, the field needs some people who cannot be neatly classified as belonging to the one group or the other, to try to bring them together. It is because I believe that the classical physical chemist is this kind of person that I presume to consider myself potentially useful in this enterprise, and persist in trying to pursue it. E. A. Moelwyn-Hughes wrote in the first chapter of his textbook of physical chemistry that "the complete physical chemist [glassblows] his own apparatus and solves his own equations." That balance, in which one is as likely to spend a few days of concentrated effort at the bench, or with paper and pencil immersed in algebra, is to my mind the principal feature missing from this field of biology, the absence of which is limiting its dynamic.

Second, the bridge analogy, though useful to a certain point, has a serious inadequacy for this philosophical analysis. No one is in any doubt about when the girder is placed and rivetted which at last joins the two sides of a bridge together. Everyone knows when it is time to throw a party celebrating the junction. That is not so for the experiment–theory junction in biology. Different people have quite different perceptions of whether or not a join has been made. This arises from their differing conceptions of the scales of time, number of workers, and complexity of data and theories which should properly be the nature of the scientific experiment–theory alternation.

Should one, for instance, expect to think mainly in terms of minor hypotheses relevant in the first instance to a particular month's data on a particular developmental phenomenon in one organism, and capable of being confirmed or rejected by the next month's crucial experiment? Or should one expect that if a crucial experiment is somewhere performed, it will be seen as such only in a historical perspective from at least fifty years later? For instance, in the complex and difficult subject of thermodynamics, one thing which nowadays seems comparatively simple and straightforward is the first law, the conservation of energy. Historical accounts commonly give the credit for the crucial experiment to Benjamin Thompson, Count Rumford, in 1799; but, as W. J. Moore (1972) has written, "the times were not scientifically ready for a mechanical theory of heat." The concept was still being aggressively put forward to a generally unreceptive scientific community in the 1840s by such as J. R. Mayer. Its general acceptance is dated from J. P. Joule's 1849 paper, which, among other things, showed that Rumford had performed a remarkably good first determination of the mechanical equivalent of heat.

Another famous half-century gap was that between the statement of Avogadro's law in 1811 and the publicity given to it by Cannizzaro at the Karlsruhe Conference in 1860; here again, the times were not ready at the earlier date. Scientists simply were not prepared to distinguish between an atom and a molecule of an element and to recognize that two atoms of the same kind could combine. The instance best known to biologists probably is the thirty-five-year gap between Mendel's work in 1865 and the publicity given to it independently by three people in 1900. Here, it is not clear to me to what extent the delay was caused by lack of the appropriate *Zeitgeist* and to what extent it was due simply to poor publicity at the earlier date. The generalization from all of this is that major conceptual advances commonly take thirty, forty, fifty years or more to be assimilated by the scientific community; but once they are assimilated, the writers of brief historical paragraphs in textbooks are careful to give credit to the originator, thereby promulgating an impression that the whole thing was clearly established and generally accepted decades before it really was.

What perspective should one seek to have in the midst of such a period of conceptual uncertainty? To my mind, at such a time it is quite premature to expect that the first contacts between experiment and theory can be put to the test of fine-tuning the fit between the theory and all the known data. Rather, one should see a theory as being promising enough to deserve extensive follow-up efforts if it does a fair job of accounting for some rather general features of the phenomena. Perhaps this is the best test at this stage: With how many different specific examples of pattern formation is the theory making a promising initial contact? My answer here to the matter of perspective is that it should be an overview of very much more than just one organism. I make this statement for both of the well-known sequential steps in scientific theorizing: the establishment of generalizations from data, and the devising of models to explain the existence of these generalizations.

For the former, a good example is the evidence that various living structures of more or less cylindrical shape have a system of "positional information" which corresponds to a set of polar coordinates. It is impressive that the work of French, Bryant, and Bryant (1976) embraced both insects and amphibians. It is more impressive that work on ciliates (Frankel 1989) points to a similar system in these single-celled organisms. For the latter, kinetic theory of the reaction-diffusion type has shown definite promise in relation to the following: formation of complex shape in single-celled algae (*Acetabularia* whorls, repeated dichotomous branching in *Micrasterias*); morphogenesis of cellular slime moulds (patterns of differentiation in *Dictyostelium*, whorl formation in *Polysphondylium*); grafting and regeneration in *Hydra;* mammalian coat patterns and lepidopteran wing markings; patterns of activation of pair-rule segmentation genes in *Drosophila;* and many others. See Chapter 10 for references, and see Meinhardt (1982) for some of these and other examples.

In a few instances, more than one kind of kinetic theory has been suggested for the same phenomenon. In his original formulation of reaction-diffusion theory, Turing (1952) envisaged it as accounting for gastrulation; but Odell et al. (1981) proposed a mechanochemical theory for this. Likewise, for *Acetabularia* whorls, I have used reaction-diffusion theory (Harrison and Hillier, 1985; Harrison, Graham, and Lakowski, 1988), whereas Goodwin and Trainor (1985) have used mechanochemical theory. Both types of theories are within the general scope of kinetic theory (Harrison, 1987); see Section 8.1 and especially Table 8.1. A few controversies between advocates of different models within the kinetic paradigm could be quite good for the paradigm itself. It might escape people's attention that in following the controversies they have tacitly accepted the paradigm.

Controversy is much more serious when it concerns the applicability of the paradigm in any form. For instance, *Drosophila* segmentation has been treated by reaction-diffusion modelling by several workers. Such theories involve communication by diffusion between nuclei in the syncytial blastoderm. But many drosophilologists believe that the phenomenon can be explained by the "reading" by each nucleus separately of positional information in preexisting gradients. Such a concept of "no crosstalk" between nuclei of course denies totally the applicability of the kinetic preconception to this phenomenon.

Among the numerous phenomena of pattern formation for which attempts are made to apply various types of kinetic theory, surely some will ultimately be found to be properly described by one or another of the available models, others perhaps by none of them. For the dynamic of a field in which numerous contacts are made, some to be broken, some to be strengthened and made permanent, the bridge-building metaphor is quite inappropriate. Rather, the scientific process has its analogues in biological development itself, as, for instance, in the establishment of connection between the optic nerve and the brain (at the optic tectum, in the lower vertebrates). Here, the brain and the eye are two separate structures. I envisage the brain as the analogue of the body of experimental biologists, and the eye as the theoreticians, because of the greater number and diversity of the former, and also because it is the theoreticians who are trying to make contact with the experimentalists, just as the optic nerve grows out to make contact with the brain, whereas the experimentalists are largely a self-interacting group.

In this metaphor, I see each current attempt to apply kinetic theory to biological pattern formation represented as the formation of a synapse between a branch of a retinal axon and a tectal cell. The synapse is at first weak. Some of them go on to strengthen and become permanent. Others are in the wrong place, and the contact breaks. For the properly patterned eye–brain junction to be established, there must be very many tries at synapse formation, many of them eventually successful, many unsuccessful. The attempt to unite biological experiment and theory needs to be seen in the same way. There needs to be a lot more going on than there is today for the proper union to be formed in the twenty-first century.

A body of theoreticians and a body of experimentalists may each be seen as a remarkable structure, in some ways complete within itself, as is an eye or a brain. But, as for everything in the history of science, until the two are firmly and correctly joined together, something essential is totally lacking: vision.

1.2.3 Visions of the invisible

Scientific explanation involves a continual tension between the concrete and the abstract in which, on a long historical perspective, the abstract is usually the deepest and most precise level. Thinking in concrete terms, however, is often simpler, very useful because it can be done quickly, and quite legitimate provided that the ultimate justification of the concrete in terms of the abstract is always appreciated, so that the abstract can be appealed to whenever the concrete may mislead. To most modern scientists, the most obvious example of this is the quantum-mechanical description of electrons in atoms and molecules. The Schrödinger equation (or its mathematical equivalent, Heisenberg matrix mechanics) is quite necessary to show, for instance, that an O_2 molecule is held together with the strength of a double bond, and yet contains two unpaired electrons. Many chemists of the molecule-making majority, however, do not spend their time maintaining fluency in the language of the Schrödinger equation, with its full panoply of spherical harmonics, Legendre polynomials, orthogonal sets, Hermitian operators, and so on. Rather, they would "prove" the foregoing statement about O_2 pictorially, by drawing an energy-level diagram and putting electrons into the levels. Such diagrams are powerful, permitting one to conduct many correct derivations without having to resort to equations. But in the end, the equations have led to the diagrams, and are the only justification for them. Much the same applies to the drawing of the so-called shapes of, for instance, p_x, p_y, and p_z orbitals and the use of such pictures to show how they can combine into molecular orbitals. The pictorial representation is again a powerful way of approximating some essentials of the mathematical treatment so that it can be done much more easily. But such a picture, though it looks like the shape of an object, is in fact a graph of a "wave function," which is *by definition not an observable property*. They are visions of the intrinsically invisible.

The hope that drives me in writing this book is that the same kind of relationship which has been established between 1920s quantum theory and experimental chemistry will eventually be set up between the equations for the dynamics of biological molecules and experimental developmental biology. We are not there yet, but in Chapter 3 I attempt a preliminary discussion on the use of pictorial reasoning in the kinetic theory of pattern.

The dynamics of molecules, comprising the movements and changes of individual molecules envisaged as accounting for rates of reaction and transport processes, may seem a much more concrete matter than the mathematics of quantum theory. After all, individual molecules are potentially (and sometimes actually) observable, unlike wave functions. This apparently different kind of reality for the single molecule, as compared with the mathematical "abstraction," does not, however, imply that one no longer needs visions of the invisible. First, the macroscopic properties of matter, which are essentially Boltzmann-statistical, are dependent for their explanation on the properties of an assembly of molecules all of the same kind in which, in fact, individual molecules are indistinguishable from each other. This is very important, philosophically, in regard to the preconceptions or paradigms of biochemists versus those of classical physical chemists. It is therefore taken up again in Section 1.3.1.

Second, even macroscopic motions, for which there is no doubt that direct observation would be possible with sufficiently sophisticated machinery, have sometimes been inferred from indirect evidence. The classic example is the concept of circulation of the blood, established by William Harvey (1578–1657) [Crombie, 1959, vol. II, sect. II(3)]. The way this concept was established, in relation to the ideas of preceding centuries, is a type for the scientific enterprise. I call attention here to two aspects only. First, Harvey was a strong advocate of the comparative approach: "Had anatomists only been as conversant with the dissection of lower animals as they are with that of the human body, the matters that have hitherto kept them in a perplexity of doubt would, in my opinion, have met them freed from every kind of difficulty." Second, the dynamics remained for centuries unknown, although the anatomical information on structure was profuse, down to a spatial scale quite adequate for correlation with the dynamics: For example, the valves in the veins were known, but it had not been appreciated that their structure betrays their function and hence the direction of flow of the blood. How many unappreciated structure–function relationships may there be in the wealth of structural information in modern molecular genetics?

1.2.4 Preconceptions and experimental programs

Though I call myself a lumper, I do not believe in a unique path toward the ultimate lump. At the levels of partial truth at which we must continually work, there is no uniquely correct way of describing some piece of the universe in which we have a scientific interest. Many apparently different descriptions and explanations might have equal validity. Science is in human minds, individually and collectively. A different concept established at some time in the past might have led us to be describing things very differently today; science is a product of its history.

(In my own high-school education in England, I was led toward this view especially by the accounts given of the phlogiston theory. As it was told, English scientists in the seventeenth century had been well on the way toward discovering oxygen and its role in combining with metals. Then the concept arose that a metal, when heated in air and converted to what was called a calx, had not gained anything, but had lost, to the air, something called phlogiston. That concept dominated chemistry through the eighteenth century. Priestley discovered the substance that Lavoisier would later name "oxygen," but Priestley called it dephlogisticated air. This was taught as being a delay in the progress of science because of the introduction of a misleading concept. It was a comfortable idea in England because it made the English look better than the Continentals who devised the phlogiston concept. What nobody pointed out in the 1940s was that chemists were increasingly adopting a more general description of oxidation as a loss of electrons. Was phlogiston the electron concept ahead of its time?)

Different concepts will lead to different interpretations of data and thence to different designs for future experimental programs. This can lead to serious difficulty when two groups of scientists try to interact but start from different preconceptions or paradigms. Neither group will like the experimental programs which the other devises. If one group is working theoretically with a new paradigm, it helps if they can do their own experiments. An experimentalist with a different paradigm is unlikely to want to put a lot of effort into the theoreticians' suggestions, because the experimentalist most probably will not like the shape of the experimental program suggested.

What does this imply, specifically, for the interaction between empirical developmental biology and kinetic theory? Here I shall describe a few things which have happened in my own attempts to do both experiment and theory, and how my ideas on experiments seem to differ from those of many biologists, without claiming that I am giving a very general answer to the question.

First, as a physical chemist thinking about kinetics, I want to do experiments on the kinetics of processes as they happen. That, for the living system, means working on it while it is still alive – mostly in vivo experiments. To be sure, many experimental developmental biologists like to do that. But the field is becoming increasingly molecular, and that implies increasing fractions of experimental time spent processing dead material.

Second, it is a commonplace that the classical physical chemist wants to turn everything in the world into a straight-line plot. This is not merely jocular. It means that the physical chemist, studying macroscopic properties, wants to make changes in variables which will produce quantitative responses as continuous functions of the change. The contrast here is that biologists commonly are much more concerned with discontinuous responses, on–off switchings. In my work, for instance, I inflict much experimental drudgery on people in my group by getting them to measure, quantitatively, the spacings between adjacent hairs in an *Acetabularia* whorl, in numbers of cells large enough to give reliable averages. This property shows continuous variations with temperature, with the calcium concentration in the culture medium, and with the presence of the calcium chelator EGTA. Beyond certain limits, further changes in these variables cause cessation or gross abnormalities of the morphogenesis. I am interested in those threshold values only to keep well away from them in most of my experiments.

This work has recently been attracting increasing numbers of citations and quotations, which is gratifying. But I do not yet find that biologists interested in this work express their interest in a form such as the following: "Harrison and Hillier (1985) found that the spacing between hairs in an *Acetabularia* whorl change linearly with $1/[Ca^{2+}]$. This linear variation presents a challenge which *any* theory of the morphogenesis must seek to explain." I believe that that thought would be uppermost in the minds of most physical chemists reading the same material.

Third, and perhaps most important, is the question of what kinds of entities the experimental program is designed to characterize. There are many kinds of entities around, and different programs will disclose the nature of different ones. (The phlogistonists were postulating an entity which could not be tracked by weighing, and they were therefore at the beginning of a road which has at the end of it the electron, but not the nucleus. Lavoisier changed the emphasis to tracking by weight, and thereby discovered the chemical elements. The modern theory of atoms and molecules embodies both kinds of entities. In the oxidation of a metal, it *gains* the weight of oxygen nuclei, but the metal atoms *lose* electrons. Theories of both loss and gain were in some sense correct. But this could not have been anticipated at the end of the eighteenth century, and there could have been different paths to where we are today – or to somewhere else.)

My approach to the study of pattern formation is that each pattern is an entity in its own right, with certain properties to be discovered and characterized. To this end, the most interesting patterns to me are those which are most clearly formed all at once, that is, in a single simultaneous event over the whole spatial region occupied by the pattern. In *Acetabularia,* the hair initials (or ray initials, for the reproductive cap) are all first seen simultaneously. There is no hint of the sequential formation of parts that one sees in the segmentation of vertebrates and some insects and in phyllotaxis, and less markedly in *Drosophila* segmentation and in flower formation. Therefore I find *Acetabularia* an ideal system in which to study a pattern as a thing, an entity.

Does the kind of fit between experiment and theory which I find in *Acetabularia* have any relevance to *Drosophila* segmentation? Several people (including me) have applied reaction-diffusion theory to the latter phenomenon, and such work is currently receiving a very mixed reception. (Meinhardt, 1977, 1982, 1988; Harrison and Tan, 1988; Lacalli, Wilkinson, and Harrison, 1988; Nagorcka, 1988; Lacalli, 1990; Lyons et al., 1990). In brief, the controversy involves the following two attitudes. Normal *Drosophila* development features, as the first clear indication of the segmented body plan, seven stripes of expression of certain genes – see the cover picture on the book by Scott F. Gilbert (1988). This pattern can be disturbed by mutations in numerous other genes. On the one hand, the theorists generally start by envisaging the seven-stripe pattern as an entity, and try to account in general terms for its formation, using a minimal number of chemical substances and interactions between them to do the job, on the assumption that the pattern is an interacting whole, with communication by transport of substances. Extension to account for the details of the normal pattern formation and its disturbance in mutants should be done by gradual minimal additions, in which the number of substances used in the theory will at first be many fewer than the known genes and their products. The full detailed match is a long way down the road. On the other hand, many of the experimentalists, however, now maintain that there is increasing evidence that each stripe is autonomous from a very early stage and should be thought of as a separate entity, that any theory must show a good match to a large fraction of the current wealth of detailed information, and that interaction across the pattern-forming region is not necessary. (They envisage the pattern as having its origin in known monotonic gradients of certain gene products along the egg. Nuclei in different positions could independently read the levels of concentration in such gradients and respond by on–off switching of the segmentation genes in a manner very critically governed by the primary gradients.)

I cannot here resolve a controversy which is likely to continue for some years and in which I am definitely committed to one side. Two rather general considerations strike me as worthy of reflection and further work:

1. In gradient-reading models, two adjacent nuclei on opposite sides of a stripe boundary and only about 10 μm apart must respond quite differently to the level of a primary gradient. Over that distance, the change in concentration in one known gradient in *Drosophila* (*bicoid* protein) (Driever and Nüsslein-Volhard, 1988) corresponds, in free-energy terms, to an energy thousands of times smaller than the energy of most chemical bonds – in fact, less than thermal energy, which is commonly capable of randomizing anything that does not have much greater binding energy. My conclusion from this is that the local-gradient-reading models, with no interaction between nuclei, need testing in carefully designed computations (with random input added to mimic thermal "noise"), just as much as reaction-diffusion models need testing in a computer. To do the job properly and show that the qualitative suggestions about what is going on actually work needs the approach of the theorist. This is a neglected field at the moment, because most theorists are attracted to the other kind of model.

2. Reaction-diffusion mechanisms sometimes are ruled out on grounds which are spurious – because they fail to take into account the possible complexity of control of such a pattern mechanism by the inputs to it. It is well known experimentally that *Drosophila* segmentation patterns are hierarchically controlled by patterned inputs from genes which have been activated earlier. If I make a Brusselator reaction-diffusion model produce seven stripes, I can make any of them appear or disappear independently of the others, and so make each appear autonomous, simply by suitably patterning the inputs. In this regard, often it is also not realized that a given input to such a mechanism may fail to produce pattern when its concentration is either too high or too low. Thus, a given substance can produce positive or negative regulation. The advocate of kinetic theory would tend to avoid the use of these terms as *qualitative* attributes of a chemical substance: The concentration range must be specified. Perhaps a good test for whether or not a biologist has been converted to the kinetic paradigm (upon reading the rest of this book, or in any other way) might be whether or not the person has stopped using the terms "positive regulator" and "negative regulator."

1.3   Mixing and unmixing: molecules versus large objects

To this point this book has been a philosophical commentary on the kinetic paradigm, and hence on why (or whether) the reader may see a sufficient degree of promise in the paradigm to justify reading on. Here I begin the elaboration of the paradigm. In what follows, we must often consider assemblies of two kinds of objects, perhaps represented simply as A and B, and the processes which we might characterize as "chaos-out-of-order" (mixing) and "order-out-of-chaos" (unmixing, sorting out, pattern formation). Sometimes one may legitimately consider such processes without first deciding whether A and B represent molecules or large objects, but not always. Molecules are "real," but their quantum-mechanical nature still sets them apart from large objects in ways which are sometimes significant. To what extent is it legitimate sometimes to think of a biological cell as if it were a huge molecule, and the sorting out of an assembly of two kinds of cells as being like the separation of two liquid layers when a binary solution has been taken past an immiscibility threshold? If one takes an interest in the computer models called "cellular automata," does the adjective have its biological meaning, or could each "cell" represent a molecule?

1.3.1 Molecules and individual identity: classical physical chemistry versus modern molecular biology

In the education of the chemist and biochemist, an impression is often given of the solid establishment of the atomic theory in the early years of the nineteenth century, and its continuous progress ever since. In this view, the structural organic chemists of the 1860s already saw the molecule and its constituent atoms as objective realities, and have never changed their outlook since.

But if one looks toward the physics of the very late nineteenth century, one sees a quite different view. The reality of molecules and atoms was being seriously questioned by an influential group of physicists, foremost among them Ernst Mach. Their philosophy, known as "positivism," saw objective reality only in the macroscopic behaviour of matter. Explanation in terms of atoms and molecules was seen as explanation in terms of abstractions which were in no sense themselves concrete objective realities. By the 1890s, some prominent scientists who had earlier believed in the capacity of the kinetic molecular theory to account for thermodynamic properties no longer did so (Fenby, 1981).

What was seen in the 1890s as the imminent demise of the atomic theory appears today as a brief illness in the adolescence of a concept which went on to healthy adulthood after the 1910s. Its recovery was brought about in part by union of experiment and theory in relation to Brownian motion, especially by Einstein and Perrin during the years 1905–13 (Fenby, 1981; Lavenda, 1985). But earlier, the conceptual foundation had been laid by the statistical mechanics of Boltzmann and his persistent advocacy of it, culminating in the Lübeck Conference of 1895 (Moore, 1972). Ninety years later, most biologists are fully convinced of molecular reality and of its overwhelming importance to their field, but they are not yet sure that mathematics has a significant role in biology.

To such, it may be interesting that Boltzmann's supporters in the cause of molecular reality were the young mathematicians of the day. Mathematics, and plenty of it, is needed to handle the concept that the transition from chaos to order, and from random to deterministic behaviour, is statistical and coincides spatially with the transition from molecular to macroscopic scale. This concept of "molecular randomness, macroscopic determinism" provided the basis for acceptance of both molecules and large objects as concrete realities. It became an essential element of the philosophy of physical chemists, and remains so today.

If the physical scientist and the biologist are to build bridges between their territories, it behooves them both to contemplate the different kinds of bedrock at the two ends. The modern molecular biologist recites a credo in which it is asserted that a single DNA molecule can produce deterministic macroscopic consequences through a continuous chain of deterministic steps. This contrasts strikingly with the Boltzmann-statistical view described earlier. Yet, without Boltzmann and his supporters, molecular science might have remained longer in disarray and today be less advanced than it is.

How may the concept of randomly acting molecules producing order out of chaos only through action in great numbers be reconciled with that of the deterministic single molecule? First, one must recognize that in a certain restricted sense the positivists were right: A water molecule in the body of a dog is not an individual in the same sense that the dog is an individual. If we have two dogs named Rover and Fido, there are always differences between them which allow us to distinguish them. If they are separated and one is later selected at random and shown to us, we can give it the correct name. In a study of population dynamics, they might be regarded as identical items in a count of the population. But they are never *quite* identical. The same applies to objects which are smaller, but still macroscopic. The developmental biologist knows that the fate-mapping of cells in an early embryonic stage to show to what tissue or organ each will give rise may be *difficult,* but the biologist always believes this to be *possible,* and sometimes succeeds in doing it.

A part of the problem for Boltzmann and others in securing acceptance for classical statistical mechanics was that if one similarly treats molecules as distinguishable individuals, calling one H_2O molecule Rover and another Fido, one gets a wrong count of the number of their arrangements and cannot calculate a statistical entropy which is an extensive property of the system; that is, if one doubles the size of the system, the statistical entropy does not double. To resolve this, the classical statisticians had to assert the individual *indistinguishability* of molecules and divide their calculated numbers of states by the rearrangements of all molecules of the same species. The entropy and other thermodynamic properties then played the game according to the rules. But particles moving in the manner of Newtonian mechanics are not fundamentally indistinguishable. Given the present location and motion of each molecule in a gas or liquid, one can in principle fate-map back through the equations of motion to where each of them was at any earlier time.

The methods of classical statistical mechanics, therefore, were not put on a proper foundation until the quantum mechanics of the twentieth century arose, including the Heisenberg uncertainty principle. This established the approximate nature of position and motion. When a water molecule comes back to us, the information as to whether it is Rover or Fido may be forever lost. The question of what its name is has in fact no meaning. Molecules of H_2O have no individual identities and thus lack one essential ingredient of the nature of large objects.

If, however, two molecules are not of exactly the same species, but have some difference, albeit a minor one, in atomic composition, then they have two separate individual identities. Two DNA molecules, even of the same molecular weight and with the same overall base composition, but with a minor difference in sequence, can be securely identified as Rover or Fido, no matter how often they are removed, randomly interchanged, and returned to the observer. Thus the worlds of Boltzmann and of Watson and Crick are different places, far apart; but the former was at least an essential staging point on the way to the latter.

I note in passing that the foregoing account obviously raises the question whether or not one can use analogues of thermodynamic variables in discussing assemblies of cells, or populations of organisms. The concept is attractive a priori, especially in relation to evolution, but the route to its development is full of extremely dangerous pitfalls, especially in relation to the distinction between molecular identity and macroscopic identity. For a rigorously treated concept of entropy in relation to populations and evolution, see the writings of Demetrius and co-workers: Demetrius (1985), Demetrius and Ziehe (1984), and earlier work there cited.

My topic remains, however, morphogenesis of the organism and its parts. It is a commonplace that every living organism consists principally of liquid water. A marine alga, such as the *Acetabularia* on which I have done experimental work, may be contemplated philosophically as a picture sketched out in three dimensions by a variety of organic substances, but with the water of the ocean as its three-dimensional canvas, passing right through it and very little impeded in molecular motion. Water exchanges between the inside and outside of most cells much more rapidly than does anything else. To the other molecules, constituting most of what we perceive as the living organism, that water is a gusty hurricane of overwhelming force and erraticism. The great question of developmental theory is this: Do the self-organizing structures of living organisms paint themselves into three-dimensional pictures by joining together structurally, bit by bit in building-block or jigsaw-puzzle fashion, in sufficient strength to overcome the hurricane, or does development at some stage make use of the hurricane and produce order out of chaos in the Boltzmann-statistical manner? This book is for anyone who feels that the latter is sufficiently probable to be worth thinking about.

1.3.2   *Arrangements, aggregations, amplifiers, automata*

$$BAABABBBABAABBABAA \rightleftharpoons BBBBBBBBBBAAAAAAAAAA \qquad (1.1)$$

Here we have two kinds of hypothetical objects in a disorderly arrangement and in an ordered one, with transitions between the two in either direction: left to right, order out of chaos; right to left, chaos out of order. Both kinds of changes occur in natural systems. What is it, in the initial state of a system, which determines its destiny: being ordered, to stay ordered; being chaotic, to stay chaotic; to go from order to chaos; to go from chaos to order? In the first two of these four possibilities, a major feature of the initial arrangement is stable; in the other two, some minor feature has been *amplified*. When either of those drastic changes in structure occurs, the thing one must search for in the initial state is the *amplifier*. This sort of amplification has the essential attributes of catalysis. It is the acceleration of certain processes, brought about by something which is not usually an obvious major feature of the gross structure of the starting configuration. Like catalysis, it never violates the laws of thermodynamics by producing thermodynamically impossible changes in equilibrium states. But it can sometimes act to take a system away from equilibrium, and to hold it out of equilibrium, provided that there is sufficient linkage between what is going on in the system and the continuous changes in its surroundings to satisfy the laws of thermodynamics.

A well-known example in which the dynamics of a system make an extremely effective amplifier of chaos is the idealized set of perfectly elastic billiard balls on a perfectly frictionless table with perfectly elastic ends and sides with which the balls can collide. If a number of these balls are moving in parallel paths exactly perpendicular to the ends of the table, they may continue to bounce back and forth in unchanging order forever. But clearly, introduction of the slightest error into the path of one of the balls will lead very soon to collisions, and a total breakup of the parallel paths into a system of motions in random directions at a distribution of assorted speeds. In a currently active branch of mathematical physics, the word "chaos" is used in a precise sense to mean the kind of disorder produced by such an amplifier. For an account at the level of the nonexpert, see Crutchfield et al. (1986). Of the idealized billiard game, they write as follows: "For how long could a player with perfect control over his or her stroke predict the cue ball's trajectory? If the player ignored an effect even as minuscule as the gravitational attraction of an electron at the edge of the galaxy, the prediction would become wrong after one minute!"

Thus, in this system, the presence of the amplifier of chaos in the dynamic properties of the system is overwhelming. It can take as its starting disturbance the minutest irregularity in the initial state and deliver the system to the same ultimate chaos. This implies that the ultimate state is an equilibrium one; it will be the same, regardless of initial state and disturbances thereto, given the same set of balls and the same total kinetic energy. This model, of course, is intended, in its commonest usage, as an analogue to the behaviour of molecules in a gas, and the final state is the Maxwell distribution.

A question thereby arises: Are the billiard balls in this model, or the A's and B's at the beginning of this section, analogous to molecules only, or macroscopic objects only, or both? Are the uncertainties which lead to final chaos quantum-mechanical properties, related to the Heisenberg principle, and therefore relevant to molecular and smaller scales only, or are they properties of large objects? These philosophical questions are raised here not in an attempt at final resolution of them, but to warn readers from different branches of science that they may look at things like the sets of A's and B's and have quite different concepts in mind, according to their backgrounds.

Crutchfield and associates devote some space to this matter and conclude that "some large-scale phenomena are predictable and others are not. The distinction has nothing to do with quantum mechanics." That is, in a system of large-scale objects, "chaos," or "exponential amplification of errors due to chaotic dynamics," "ensures that the uncertainties will quickly overwhelm the ability to make predictions."

In this context, the modern biologist who would like to look at my arrangements of A's and B's and think of them in much the same way, whether each
letter represents a molecule, or a cell in some particular differentiation state,
or an organism in a population, may be closer to being in step with some of
the thinking of modern physicists than is the classical physical chemist. The
cell-as-molecule concept is indeed an important one in morphogenesis, and is
addressed in several places in this book (see Sections 4.4.1, 4.4.4, 4.5, and
5.4). The essential message to the biologist is one of reassurance. Order-out-
of-chaos unmixing processes, such as the forward direction in equation (1.1),
may be thought of in much the same terms for molecular units or for large
units provided that one avoids a certain pitfall. Assemblies of large units have
essentially no entropy of mixing (otherwise called "configurational entropy"). For them, it is not useful to try to distinguish the left- and right-hand
sides of equation (1.1) in terms of an entropy. But for molecules, that entropy
is very important. Generally, a big interaction between system and environment must provide a driving force beyond some threshold value for the unmixing to occur, if A and B are molecules. This kind of threshold is discussed
in Section 6.4 and is there distinguished from various other thresholds which
arise in morphogenetic theory but are not concerned with satisfying the laws
of thermodynamics.

In the preceding paragraph I have switched from discussing chaos-out-of-
order to the reverse, that is, from the backward to the forward sense of
equation (1.1). (The backward sense is just the idealized billiard-ball dynamics with a starting configuration of two kinds of balls, differently coloured and
at first spatially separated; it is diffusion.) The forward, order-creating, process also involves amplifiers. An amplifier is something within a system
which selectively exaggerates certain of its initial features, so that the system
develops to express the properties of the amplifier much more than the main
properties of either external interactions or the initial state of the system,
except that the initial state contains the amplifier.

This definition is sufficiently broad that any morphogenetic mechanism can
be called an amplifier, and that is legitimate, and perhaps useful. But if one
seeks to classify mechanisms, another question arises: Does the amplifier act
essentially statically or dynamically? The answer "statically" leads toward a
picture of geometric fitting, for which one is more likely to adopt the word
"assembly," as in the most restricted sense of the term "self-assembly." The
alternative "amplifier" can reasonably be given a restricted sense, for mechanisms in which the dynamics govern the outcome. Such amplifiers lie within
the scope of classical physical chemistry and the perspective of "molecular
chaos, macroscopic order." They are made of randomly moving molecules,
and their deterministic properties are such statistical properties as concentrations in solution and rates of reaction as functions of concentration. In relation
to a seemingly chaotic initial state, what a kinetic amplifier of order perceives
is not, as in the case of the chaotic amplifier, some tiny detail of the system,

but some long-range order, often deeply hidden. What it is, for instance, in the left-hand side of equation (1.1) is discussed in Section 5.3.2, especially Figure 5.6. But again, as in the case of the chaotic amplifier, what is finally expressed is much more a property of the amplifier than a property of the initial state.

Does the observed microstructure of organisms suggest, a priori, static or dynamic amplifiers? To many biologists it has seemed to be the former; to me, the latter. I have long and consistently been impressed by the fact that both the organism and the protein molecule have precise geometries, but that as one goes down from the organism to microstructures such as the cytoskeleton one sees structures with some general order but a lot of detailed disorder. Good geometric fitting of small parts, sufficient to explain the larger-scale regularity, seems to be absent. This topic, especially in relation to symmetry, is taken up in Chapter 5. The account focuses attention on the approximations which one makes in describing the shape of an object. It is related to the currently popular topic of fractal geometry, as discussed and publicized by Mandelbrot (1977, 1982). Early in his later book he raises the question of the true shape, and more particularly the length, of a coastline as one takes one's degree of approximation downward from that of a small-scale map to that of individual sand grains. The same perspective is much more important in the shapes of living things and how they are formed, without excessive accumulation of error in form, in the reverse order of those degrees of approximation.

That reverse order is aggregation. In the inanimate world, much is known of the geometry of crystal growth, in both simple forms and complex, variable forms such as the stellate snowflake, as well as the much more irregular dendritic growth in such things as colloidal aggregates and deposits on electrode surfaces. What do these tell us of order, chaos, and amplifiers in the mechanisms of their formation? In brief, what one might call "sticks-where-it-touches" aggregation, with no subsequent movement of the particles, produces irregular, irreproducible dendritic growth matching the randomness of the diffusion which produces it. Therefore, it is usually known as diffusion-limited aggregation (Ramsay, 1986; Sander, 1987). Everything else requires movement of particles within or upon the surfaces of the aggregates. In Section 4.2 I discuss crystal growth and show that some simple shapes may be equilibrium shapes, needing motion only so that the shape can find its way to equilibrium. But complex shapes are carved out by molecular kinetics and must be explained kinetically. Thus, the bridge between the molecular and the macroscopic must be built with the classical physical chemist's resources of a Boltzmann-statistical world.

If the statement that ends the preceding paragraph is accepted as at least a promising possibility, the need for mathematics in pursuing that possibility should be at once evident. This topic is taken up at greater length in Section 6.1, in which some attempt is made to set the kind of mathematics required into the framework of the general philosophical strategy of the approach. This kind of mathematics is of the seventeenth and eighteenth centuries, and Boltzmann's advocacy of microscopic-to-macroscopic-via-mathematics is of the nineteenth. But in recent years, a new means of expression for formal logic has become accessible to everybody: the computer program. In particular, a kind of model known as a "cellular automaton" has been indicated as relevant to problems of biological development and evolution. There is clearly some kind of relationship between automata and the forward direction of equation (1.1). The automata commonly produce patterned distributions of two symbols. Three general questions arise, and they are addressed at various points in this book (e.g., Sections 3.2.1, 5.2, and 5.3). First, what is the correspondence between such computer programs and the equations earlier published for kinetic models of morphogenesis, especially reaction-diffusion theory? Second, can such computer modelling be used effectively without full mathematical analysis as a component of the scientific method, leading to prediction and further experiment, and the linking together of a priori apparently different concepts? Third, when the word "cellular" is used to describe these automata, should the biologist take it to mean cellular in the biological sense, or can the models represent patterning on the molecular scale? This is just a new slant on the matter of whether the A's and B's in equation (1.1) represent molecules or large objects, or possibly either. In brief, my attitude is that most automaton programs do not include a random choice at any place where it could represent a diffusion process. Without such a step, the automaton does not represent reaction-diffusion at the *molecular* level, but its cells could be biological cells.

1.3.3   The kinetic preconception

Aristotle's universe was, as the world immediately around us appears, full of matter. The stars and planets moved by being carried on rotating spheres, and each sphere derived its motion from that of the one outside it. An object moved up or down to find its right place, which was a function of its material environment.

In the realm of physics, especially in relation to astronomy, that kind of thinking was being replaced by less concrete, more abstract ideas from the fourteenth century onward, culminating in the Scientific Revolution of the seventeenth century. Planetary motion became thoroughly understood as kinetic process, although its explanation involved forces between widely separated objects seemingly unconnected to each other, a concept previously seen as having an air of total unreality. Likewise, light was understood for centuries as a vibration, in a fully scientific interplay of theory and experiment, without any notion of what the vibrating medium might be. At one point, those who yearned for a more concrete Aristotelian concept tried to postulate a "luminiferous ether," but that idea has long been dead.

Physicists (with whom I include physical chemists) are socially conditioned by this historical perspective to accept strategies of research in the sense of the upper arrow in Figure 1.1. They are prepared to study process and motion as a scientific objective, with the question of what material is undergoing the process as a secondary consideration, perhaps far enough away that it may occupy the efforts of different people even in later decades.

Such thinking has been slower to enter biology because the biological world appears to be Aristotle's. Every part of space in a living organism is full of something. When speaking, for instance, of a pore in a desmid cell wall, I often have to point out that a pore is not an empty hole. It is just a region full of something different from that in the surrounding region. For a proper description, one should know what the filling substance is. By the same token, Aristotle's law of motion, that velocity is proportional to force, is quite correct for motion in a viscous medium. It is used today, as Stoke's law, often without the recognition that it is Aristotle's and that he remains correct for a material-filled world. Micron-sized objects such as bacteria, in a liquid medium, have in fact no recognizable acceleratory motion at all when acted on by a force; Newton is irrelevant.

Why, therefore, can one not deal with all phenomena within living organisms in a concrete, Aristotelian way? Indeed, one can get much further with this kind of thinking in biology than in physics, and that is why so much of biological theorizing is concrete. The main theme of this book is that some parts of the explanation of development require the more abstract, and thus usually more mathematical, thinking associated with motion as the ultimate reality. As the first theme of classical thermodynamics was that heat is a mode of motion, so the present theme is that biological form is a mode of motion. D'Arcy Thompson (1917) gave instances of a great variety of transient shapes assumed by rapidly flowing water. Meinhardt (1982) drew attention to a stable periodic pattern produced by a continual flow of water, and the seemingly stable forms assumed by water issuing from a sluice-gate, flowing over rocks, or rounding a bend in a river are well enough known. For whitewater rafting, particular water structures can be marked on charts and given names as if they were objects. As if? The whole point is that they are objects, and living organisms must sometimes be appreciated as objects in the same sense. This is not easy to visualize; nor, applied to our own bodies, is it congenial, for it uncomfortably stresses our transience.

A major obstacle to recognition of the great question which kinetic theory seeks to answer is the very static appearance of the microstructural information in electron micrographs. To be sure, many structures in a cell are firmly secured. But much material is not; and for untethered protein molecules, the sea within a cell rages beyond anything experienced by a small boat trying to round Cape Horn. For undergraduate classes, I like to point out that a 40-foot boat driven 75 miles off course in a storm is the analogue of a 30-kd globular protein, diameter 40 Å, randomly moving 40 μm (i.e., the length of a large cell). At normal diffusivity for such a molecule, this would take an average of

3 seconds. How can any spatial control be maintained in such a storm? The general answer is that molecules, in large numbers, have statistical properties which are deterministic. The randomness disappears when the numbers are large enough.

(One might ask, Why does the molecular approach work at all? Surely a large fraction of these boats must get smashed up in the storm. The point here is that, in relation to the intensity of the storm, the internal structure of a molecule is held together about 10 million times more strongly than the boat in the analogy. Without that strength, there would be no structural chemistry.)

In biological evidence, one does not easily see either the raging storm or the deterministic flows and transformations of material which overcome it. The electron micrograph is like a mariner's chart or a map in an atlas. A famous story among opera lovers is that the conductor Franz Lachner, while rehearsing Wagner's *The Flying Dutchman* in 1864, complained about "the wind that blows out at you wherever you open the score." Anyone can feel this wind on hearing the remarkably evocative music played. Only a fully trained musician can feel it when looking silently at the printed score. When one contemplates an electron micrograph, and then a diagram of chemical changes in biochemical cycles, one should feel first the storm which raises the question and then the great determined flows through it which shape the organism.

2

Morphogen: one word for at least two concepts

William of Ockham's name is known to most scientists only for the principle of economy in theoretical explanations – Ockham's razor – most succinctly stated as follows: A plurality must not be asserted without necessity. It is less well known that three centuries before Newton, Ockham asserted that there was no objection to the concept of action at a distance (Crombie, 1959).

Kinetic theory of living pattern and form is essentially concerned with both of these preoccupations of Ockham. It stems from a recognition that a kind of action at a distance must be operating whenever, in biological development, a shape appears in the absence of any antecedent of the same shape or of a precise geometrical fitting together of parts. The action must be some form of communication over the entire scale of the new pattern at, or slightly before, the time of its first morphological appearance. The scale of the morphogenetic field may be the size of a single cell, or even part of a large cell, or the span of many cells in a tissue. In very diverse examples, pattern often becomes visible at a size of tens of micrometres. Kinetic theory can readily accommodate at least the range from 1 μm to 1 mm.

Of the various possible kinds of communication, molecular diffusion is given the most prominence in this book because reaction-diffusion theory is the most extensively developed branch of kinetic theory. This part of the general theory well exemplifies the whole, because other communication systems (such as mechanical stress) usually will have to be combined with feedback loops (negative and positive, inhibition and activation) of similar nature to those in reaction-diffusion theory to give a complete account of pattern generation.

It should be recognized, then, that diffusibility may not be a sine qua non for the intimate involvement of a substance in a morphogenetic mechanism. Diffusion may sometimes be the actual means of communication, and sometimes, at the present early stage of attempts to correlate theory and experiment, a metaphor for other means of long-range communication.

Many biologists seem to regard the use of differential equations as a complex, difficult, and obscure manner of setting up a theoretical explanation. I maintain that they usually represent, rather, the vigorous use of a well-honed Ockham's razor which has shorn away all the complexity *not demanded by*

31

necessity. The equations themselves usually express, in concise form, fairly direct and simple postulates. Difficulties arise in *solving* the equations. Surely this is to be expected if the equations indeed express an intrinsic simplicity which is capable of explaining a number of the complexities of biological development.

In devising theories for the production of pattern and form, one must be careful to put in at least the minimum that will have the capacity to do the job. Ockham's razor must be used to shave the theory's whiskers, not to chop off its head. Many well-known theoretical concepts, such as "a diffusible morphogen," fall short of pattern-forming ability. They may be significant parts of a complete mechanism, but gain that significance only within the context of the whole. In this chapter, I try to take some of these pieces of the body of kinetic theory and show what part of the explanation each represents and what is missing. This discussion is intended to lead to a view of the countenance of a clean-shaven pattern-forming theory and to take the reader as far into it as one can go without using the calculus.

The word "morphogen" appears rather frequently in the following account. It has, just now, at least two usages which appear, superficially, to be consonant, but which actually are conflicting. The account develops this point and also indicates a change from my earlier viewpoint (Harrison, 1987), in which I argued for restriction of the term "morphogen" to one of these meanings. The use of the word by experimental biologists for the other meaning is now so prevalent that I suggest drawing a distinction between two types of morphogens, which may be called I and II.

A type I morphogen might otherwise be called a Wolpert positional signaller, and it is the kind of substance that experimental biologists usually are postulating when they use the phrase "a diffusible morphogen." The substance is envisaged as being distributed in a monotonic gradient from local sources which it does not itself control by any kind of feedback. Because its concentration is a single-valued function of position, such a substance is appropriate for conveying positional information. But the essential topology or symmetry of the pattern is already present in the arrangement of sources of a positional signaller. Therefore, it does not have the *form-generating proper-ty* which, etymologically, is surely the most precise significance of the term "morphogen." By the same token, the type I morphogen is not the right kind of thing to be a central feature of any answer to the great question of how pattern is formed ab initio.

A type II morphogen, or Turing morphogen, is one of *at least two* diffusible substances (X and Y, or Gierer and Meinhardt's A and H) taking part in interactions of "activator–inhibitor" character (Turing, 1952; Prigogine, 1967; Gierer and Meinhardt, 1972). These interactions confer on the pair of substances the ability to set up, for instance, patterns of repeated concentration peaks which become the effective sources of X and Y. Thus they control the positions of their own sources and are prime movers in the generation of pattern. Anyone thinking in terms of this kind of theory, and having found some evidence for the presence of one substance which may play a role in such a mechanism, will always be thinking next about the other substance. The pair interaction is of the essence.

2.1 Type I morphogens: Wolpert positional signallers

The concept that a cell somehow knows its position within a developing assembly of cells, and differentiates or regulates its functions in response to that knowledge, is more than a century old and has been discussed for both plants and animals. The botanist Hermann Vöchting (1877, 1878) seems to have priority for the concept; see a number of references to his work in the volume edited by Barlow and Carr (1984). Lewis Wolpert (1970, 1981) has promulgated the concept strongly, with particular reference to phenomena of the animal kingdom.

How does a cell measure its position in a developing system? In other words, what specifies the coordinate system? It is most usual nowadays to postulate a gradient of some diffusible substance. One must further suppose that a cell located somewhere in this gradient is able to respond differently to different concentrations of the diffusing messenger. Interaction between cell and messenger might take a variety of forms, but could readily be of the known kind between a hormone and a cell, involving binding of the messenger to a receptor on the cell surface. The postulated messenger in fact falls within the definition of a hormone as a substance produced in one place and transported to another where it initiates a change, provided only that one is prepared to envisage a much shorter scale of distance for the transport than one normally does in relation to hormones. Consequently, the transport may be by the slow means of diffusion, rather than by fast bulk transport, as, for instance, through a bloodstream.

This kind of messenger may be envisaged as operating in morphogenetic fields of one, two, or three spatial dimensions. Usually, some more or less localized source is envisaged for the messenger, together with either a localized sink or destruction of the messenger everywhere at a rate proportional to its concentration. Such models lead to steady-state concentration profiles of simple form for the messenger, such as those sketched in Figure 2.1. The curve of Figure 2.1b may, according to the spatial dimensionality of the system and the distribution of sinks for the messenger, assume a variety of detailed shapes, not all exactly the same as the exponential falloff with distance shown here. But all these shapes must, if the model has only the features just specified, have one feature in common: The concentration of the messenger, in the steady state, must fall off monotonically with distance from the source.

This is the essential requirement for a substance which must specify a spatial coordinate without ambiguity. A distribution with peaks or troughs in it

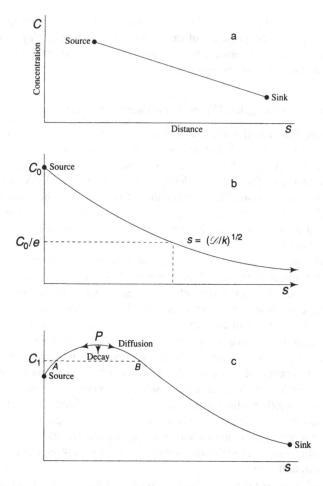

Figure 2.1. (*a*) For a one-dimensional system, the steady-state distribution of a diffusing material between a localized source and a localized sink. (*b*) For a diffusing material with a localized source, no localized sink, but decay everywhere at a rate proportional to its concentration (first-order decay constant k), steady-state distribution; $\mathcal{D}$ is diffusivity. (*c*) A distribution which could not occur in the steady state for diffusion alone or diffusion combined with first-order decay. The peak at P must be dropping because material is being removed by both diffusion (in both directions) and decay, and is not being supplied by any means. This distribution is unsuitable to specify a positional coordinate in the morphogenetic field extending from source to sink, because concentration is not a single-valued function of position (e.g., the concentration C_1 specifies both position A and position B). For the proof that gradient (*b*) is exponential in distance, see Problem 6.5.6. For a gradient of this form in a known substance (*bicoid* protein in *Drosophila*), see Section 10.2. For a similar (but not yet so precisely quantitated) gradient of retinoic acid in the chick limb bud, see Eichele and Thaller (1987), following the discovery of retinoic acid as a Wolpert positional signaller or type I morphogen (Tickle et al., 1982). For a review contrasting monotonic gradients

fails to do this (Figure 2.1c); but this cannot arise in the steady state by diffusion and first-order decay alone, except for the single peak corresponding to the source position. This is normally envisaged as being at one extremity of the morphogenetic field, and therefore causes no ambiguity.

By the same token, however, the diffusible signaller of this kind is a passive messenger which conveys unaltered the information of how far away the source is. It does not play an active role in converting one kind of information (genetic) into another (morphological), because it cannot *create* sources and sinks. Consider, for analogy, an artist who has designed a large mural and is employing a number of art students to paint in some of the areas he has mapped. If one observes the work in progress, seeing only these assistants and what they are doing, one sees a number of people who clearly have some of the attributes of an artist (e.g., they have acquired the technique of painting). But there is no evidence that any of them yet have the ability to create a good design; and it might not be easy, from observation only of the work in progress, to find out who is the designer.

Although I would much prefer to see the word "morphogen" reserved for substances which have an active role in generating form, I now bow to the increasing usage of the word for Wolpert signallers by suggesting the classification into two types. What did Turing mean when he introduced the word? He wrote that the word was "intended to convey the idea of a form producer. It is not intended to have any very exact meaning, *but is simply the kind of substance concerned in this theory*" (italics added). This is read in different ways by different people. I take "very exact" to refer to chemical structure, and the italicized words to tie the definition to his mathematics (i.e., to my "type II").

Living organisms have a remarkable capacity to generate quantitative measures of distance within themselves as they develop. They can, so to speak, not only draw coordinate systems but also put numbers on the axes. Any respectable mechanism proposed to describe morphogenesis should address explicitly the problem of how a physicochemical system can do this. A great strength of kinetic theory is that it is the only extensive part of the literature of developmental biology which has so far done this.

The passive diffusible messenger which sets up a monotonic gradient between a localized source and a localized sink can partition the distance between them, but does not determine its magnitude. If, however, there is a source only, and removal of the messenger is by its decay everywhere, then the messenger itself can set up a quantitative scale of distance from the source. To see this most clearly, it would be nice (though not essential to the argument) to see this scale as a linear one, that is, somehow to convert the curve in Figure 2.1b into a straight line. This is easily done, simply by plotting ln C

Caption to Figure 2.1 *(cont.)* and patterns of repeating parts, with much the same attitude as mine regarding the problems of their formation, see Nagorcka (1989).

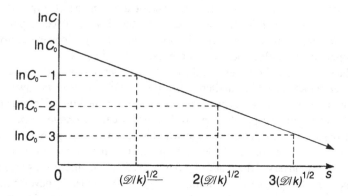

Figure 2.2. Steady-state distribution of a diffusible messenger with first-order decay (Figure 2.1b) replotted with a logarithmic scale of concentration, which may also be taken as a linear scale of free energy of the messenger, in the approximation of ideal behaviour.

instead of C versus distance s (Figure 2.2). Every change of one unit in the natural logarithm of the concentration corresponds to a movement of a distance $(\mathscr{D}/k)^{1/2}$ along the system; this quantity can be regarded as the effective unit of distance.

Physicists who believe that everything significant can be described on the back of an envelope are commonly addicted to dimensional analysis. In those terms, if one notes that a diffusivity $\mathscr{D}$ has dimensions of (distance)^2 (time)$^{-1}$ (e.g., cm^2 sec^{-1}) and that a first-order decay constant has dimensions (time)$^{-1}$ (e.g., sec^{-1}), then the obvious combination giving a measure of distance is $(\mathscr{D}/k)^{1/2}$. All quantitative measures of distance in reaction-diffusion theory are more or less complicated variations on that theme.

If the messenger is a substance in solution which behaves as an ideal solute in the thermodynamic sense, then its free energy per mole (chemical potential) changes linearly with ln$C$. The distance scale is then marked out as a free-energy scale; equal intervals of free energy correspond to equal intervals of distance.

This is the simplest example of a reaction-diffusion model. In lacking the capacity to generate sources, it is very far from a complete morphogenetic model, for, as set out in this and later chapters, what remains to be added is rather a lot. The model, however, besides representing fully the properties of substances which may exist as positional messengers (type I morphogens), exemplifies two matters of strategic importance: (1) An active role for the messenger, quantitation of spatial scale without external reference, appeared only when reaction (as decay or self-destruction) was added to diffusion; both properties are essential for an active role. (2) The reader is being asked to take my word for the exponential shape of the curve in Figure 2.1b and the expression $(\mathscr{D}/k)^{1/2}$ for the distance unit. Even this simplest case cannot be discussed fully without calculus (see Chapter 6 for details).

2.2 Activators versus morphogens

The missing component in the foregoing model is an origin for the source, or for the difference between source and sink. The relative positions of these regions are manifestations of polarities in the developing organism. Some of these may be traceable back to the beginning of development (e.g., to the animal–vegetal polarity of an oocyte). This cannot, however, be the case for every polarity which is seen during development, for if all of these could be found at the outset, in proper spatial relationship, then, first, all development would be attributable to the inheritance of macroscopic templates, and there would be no role left for the genome, and, second, the initial state would contain a miniature of the developed form. This would be equivalent to the long-discarded concept of the homunculus in the sperm.

Morphogenetic mechanisms, therefore, though they may in some instances build upon existing polarities which are macroscopic templates, must in many instances be capable of generating polarities ab initio with the aid of substances specified in their chemical nature, but not directly assigned to spatial positions, by the genome. Even where a rudimentary polarity exists at the outset, it usually needs great amplification. The mechanisms which can act as sufficiently good amplifiers usually turn out to be the same ones which have the capacity to produce a polarity ab initio. These concepts are further discussed in relation to the making and breaking of symmetry in Chapter 5.

The obvious way for a localized region to become markedly different from surrounding regions is by what has variously been termed positive feedback, self-enhancement, or autocatalysis. Of these three terms, the last would commonly be taken to imply action on the molecular scale, that is, a feedback loop indicated by a single step in a chemical reaction mechanism. But the other two terms have no specific implication of a uniquely defined spatial scale. They could include action above the molecular scale, by which, for instance, (1) differentiated cells could cause adjacent ones to differentiate in the same way (homeogenetic induction) and (2) in a developing nervous system, the conduction of impulses across a newly formed synapse could cause neighbouring synapses to become stronger – the theory of Willshaw and von der Malsburg (1976) for retinotectal specificity.

Self-enhancement is an essential feature of most kinetic theories of morphogenesis which are complete in the sense I defined earlier. A substance possessing this property on the molecular scale of its formation reactions can grow in concentration at differential rates which reflect and amplify initial irregularities in its distribution. Thus, concentration peaks, such as P in Figure 2.1c, can build up. The property of self-enhancement, however, by itself confers no control on the positions of these concentration maxima. If the input to the mechanism consists of those random fluctuations of concentration which are everywhere in chemical systems, this "concentration noise" will simply be amplified without increase in order. The result will not look much like Figure 2.1c. That diagram, taken in the context of the discussion in this section and the preceding one, suggests that a combination of diffusion, decay or destruction, and self-enhancement may be necessary for full morphogenetic potential.

This is indeed the drift of my argument. But it will turn out that something more usually will be needed for proper control of pattern formation: more than one substance with this combination of properties. It is by no means obvious a priori, that such an increase in the complexity of the theory is a necessary head which Ockham's razor is not permitted to chop off. To see this, it is necessary to go into some detail regarding the behaviour of a single substance with this combination of three properties.

To this end, as indicated in Chapter 1, my favourite example, for both autobiographical and pedagogical reasons, is spontaneous optical resolution by stereospecific autocatalysis. Even this instance could be argued to concern more than one substance, if the two enantiomers are counted separately. But, as will emerge later in this chapter and in the more extensive account in Chapter 5, their mutual interaction is insufficient for full control of a pattern, and they essentially exemplify the capacities of a single substance.

Any elementary account of chemical kinetics starts with the examples of first- and second-order reactions, that is, those in which the rate of reaction is proportional to the concentration of a reactant or to its square. The discussion of feedback, or autocatalysis, involves the recognition that reaction rates may depend also upon the concentrations of the products. An elementary discussion must start in like manner by assuming that two probable dependences of the reaction rate are upon the concentration of the product or upon its square. The latter, second-order autocatalysis, turns out to be the one which confers substantial morphogenetic power upon the reaction product (the "activator"). It has been used in almost all extensive elaborations of the kinetic theory of pattern formation (e.g., Prigogine, 1967; Gierer and Meinhardt, 1972). In terms of mechanism, second-order autocatalysis is likely to arise if a potential catalytic site requires the attachment of two product molecules as ligands to activate it.

A reader who already has some acquaintance with Turing's (1952) suggestion which is the original source of reaction-diffusion theory may at this point be suspicious, because Turing's equations do not contain squared concentrations. This is a false appearance in relation to the mechanistic implications of the equations. They involve departure from equilibrium *in both directions,* and for this to be symmetrical and first-order in the displacement is most easily achieved by a mechanism which contains second-order autocatalyses. This statement may be quite mysterious to the mathematically uninitiated reader. Its exposition is a major theme of the introduction to mathematical formulations in Chapters 6 and 7. But we may make a start here, with the instance of optical resolution, which well exemplifies the point.

W. H. Mills (1932) seems to have been the first to indicate clearly, in relation to this example, how chemical kinetics may act to take a system away from equilibrium; in this case, the movement is from the racemic equilibrium of equal concentrations of D- and L-enantiomers. Mills pointed out that if a system is already displaced from that ratio, say, for example, such that there is twice as much D as L, then a second-order stereospecific autocatalysis would lead to D being produced four times as fast as L. Thus the system is moving away from racemic composition toward resolution as D. One need only add a removal process, such as a flow system in which a solution passes through a catalyst bed bringing fresh reactant and removing product not attached to the catalyst, to see that progress to complete resolution, with total loss of one enantiomer, is possible by this mechanism. It is not required that the enantiomers be diffusible; the system can, in fact, be taken as well stirred. Also, the starting imbalance could be much less than the 2-to-1 ratio that Mills assumed for the sake of simple numerical illustration. He discussed statistical fluctuations as a starting point and concluded that they would be sufficient to start a system at least as large as a modern eukaryotic cell on the road to resolution.

This model illustrates well the basic kinetic behaviour needed for a substance to be an autocatalytic activator. Among the important aspects of this behaviour is the relation of squared terms in the catalysis to linear departure from equilibrium, which the model very easily shows. The rate of formation of D is proportional to D^2; the rate of formation of L is proportional to L^2, and with the same constant of proportionality, because of the symmetrical relationship between enantiomers. Hence the optical asymmetry $(D - L)$ is growing at a rate proportional to $(D^2 - L^2)$; and by the familiar formula for the difference of two squares, this is $(D - L)(D + L)$. The washout of all excess production ensures that $(D + L)$ is constant on the catalyst surface. Hence the optical asymmetry $(D - L)$ is growing at a rate proportional to $(D - L)$. Suppose we call this asymmetry U. Then U, in the rate equations, has dimensions of concentration and looks as if it were the concentration of a single chemical substance designated by the same symbol. But, unlike an ordinary concentration, U can be either positive or negative, according as there is an excess of D or of L in the system. If U is positive, its growth is proceeding in the positive direction; if U is negative, its growth is proceeding in the negative direction. (Throughout this book, the symbol for a hypothetical substance is used, without square brackets, for its concentration.)

The preceding paragraph has a concealed philosophical purpose. It conforms, at least approximately, to my promise not to use mathematics in Part I. But several of the sentences are the exact logical equivalents of equations, and could be written as such (Chapter 6). I maintain that mathematical theories

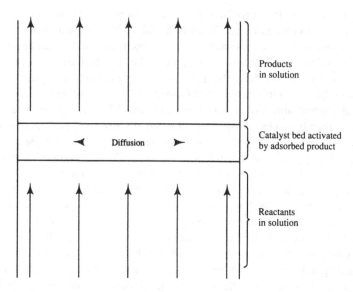

Figure 2.3. A simple geometry for theoretical discussion of kinetic models for pattern formation. Reactants flow upward in a tube which is envisaged as narrow in the direction perpendicular to the plane of the page. Thus the catalyst bed has one dimension much longer than the other two, and if pattern is formed at all it probably will be only in that direction – coordinate s. The pattern-forming reactions and diffusion take place within the region of the catalyst bed. Product concentrations in solution in the region of the catalyst bed are always sufficient to keep all sites on the catalyst saturated with product. Thus the total amount of adsorbed product is constant. Product made in excess over this amount is continually being washed away in the upward flow, an equivalent to first-order decay. Rapid exchange between product in solution and adsorbate is assumed, so that the product attached to the catalyst has a composition reflecting changes in production. For the optical-resolution model discussed in this chapter, (D + L) is constant over the catalyst bed; the possible pattern is of variable (D − L).

and logical arguments expressed in words are not different in kind. One resorts to equations when the logic has become too cumbersome for convenient verbal expression. The preceding paragraph approaches this threshold. Because I tend to think of an equation as a sentence, I am always puzzled when some biologists argue that mathematical explanations may not be appropriate in their fields. They all use words, and logic.

A kinetic model has no pattern-forming ability across a morphogenetic field in the absence of a communication system extending throughout the field. If the optical-resolution model is to generate a pattern or regions resolved as D contrasted with others resolved as L, those two species must be diffusible. Figure 2.3 shows a suitable geometry for a simple discussion of this possibility. The problem is reduced to one spatial dimension by making the catalyst bed much longer in one direction than in the other two, and the inflow of reactants and washout of products are at right angles to that direction. The catalyst is always saturated with D and L at a total concentration C_s.

Suppose that the system is initially racemic, with adsorbed D and L both $C_s/2$, and that a small disturbance is somehow introduced in the form of a sine wave, of small amplitude, with one full cycle across the width of the system (Figure 2.4a). The total amounts of D and L in the system are unaltered; only the spatial distribution has changed. If one has a Mills mechanism at work in the catalyst bed, then, in the absence of diffusion, every displacement above or below the equilibrium line at $C_s/2$ will grow at a rate proportional to its current value, that is, it will grow exponentially in time, a familiar growth law in biological systems. This converts Figure 2.4a into Figure 2.4b. On the latter diagram, the diffusion of D is suggested by arrows along the curves. Clearly, if this is occurring, it must slow down the rate of upward movement of the peak *P* and the rate of downward movement of the trough *T*. As always, diffusion tends to ensure that every valley shall be exalted and every mountain and hill made low.

I cannot, short of the mathematical discussion, conveniently show that for a pattern which is a sine function of the distance, the rate of this smoothing by diffusion is also proportional to the displacement from equilibrium at every point. But in fact it is, and the sine wave remains a sine wave, but with changing amplitude.

At this point the reader may well wonder why I have chosen to discuss a sine wave as the initial disturbance. Its presence imposes a defined pattern on the system. As it grows in amplitude, its form (e.g., the symmetry of the concentration distribution) does not change. Thus, in this growth of amplitude the pattern may have become more obvious, but there has been no pattern formation, no symmetry-breaking. How does this example help us toward the objective of showing how living things extract orderly form from chaos?

First, this discussion of a diffusible activator is intended only as a second step along the road toward that distant objective. A diffusible activator, even in the form of two enantiomers, does not have all the required attributes for getting order out of chaos that a complete pattern-forming mechanism must have, and we must start with something orderly to see in any simple way what it can do. Second, although pattern is here amplified from an existing rudiment rather than formed ab initio, what is done by the mechanism is in fact a very significant generation of order. As the amplitude of the sine wave increases, the enantiomers are unmixing. One is increasingly becoming stacked up in the left-hand half of the system, and the other in the right-hand half. Anyone with an elementary knowledge of thermodynamics will recognize this as the type species of a nonspontaneous process, which cannot occur without some interaction with the rest of the universe outside the system.

Third, and most important for the exposition of how reaction-diffusion works, the sine wave is a very special pattern with which to measure pattern-forming ability. When we have gone the next few steps along the road and arrived at the Turing equations, we shall find that the dynamics specified by them are still incapable of distorting a sine wave into anything else. That property is unique to the sinusoidal disturbance. *Anything else* will be changed

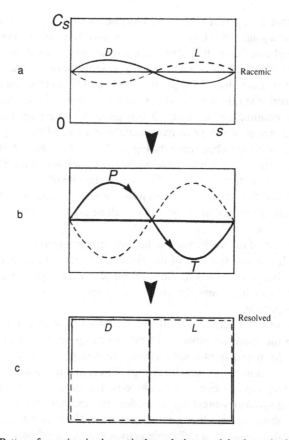

Figure 2.4. Pattern formation in the optical-resolution model, along the long dimension of the catalyst bed in Figure 2.3. (*a*) A small initial disturbance in the form of a sine-wave pattern of partial resolution. (*b*) The same disturbance some time later, as it would grow in the Mills autocatalytic mechanism without diffusion or with insufficient diffusion to stop the growth. The diffusive flow of D from peak *P* and toward trough *T* is shown by arrows on the curve. (*c*) "Nonlinear" distortion of the sine wave at the limit of complete resolution. For clarity, (*b*) has been drawn as sine waves, although the amplitude is quite close to $C_s/2$. At so large an amplitude, distortion from sinusoidal form toward the final square-wave shape of (*c*) would in fact already be evident. For clarity of presentation, the boundary condition used in these diagrams is that equilibrium is maintained at both ends of the system. This means that the ends, or regions beyond them, are working actively to equilibrate the system at those points, since D and L are leaving or arriving by diffusion at unequal rates, but the boundaries are morphogenetically passive in the sense of not defining any polarity for the system. The same end conditions would remain appropriate if the diagrams were all turned upside down.

in shape as it is amplified. The essence of the matter is that more complex, more chaotic, and therefore more realistic inputs can be resolved into a sum of sine waves of a variety of wavelengths. At least in the early stages of pattern growth the amplifying mechanism treats each of these separately, as if the others were not there. And because components of different wavelengths are amplified or suppressed at different rates, the shape of their sum changes qualitatively. Examples are given later in this chapter (Figures 2.5, 2.6, and 2.7).

Meanwhile, however, if the reader stays with one sine wave and considers the *later* stages of its increase in amplitude, it can be seen that there is some trouble ahead. The concentration of D or L can nowhere rise above C_s nor fall below zero. That is, the exponential growth of amplitude must somehow stop. It was not evident from my verbal argument for the occurrence of the exponential growth that the argument contained an approximation which must break down at large amplitudes. If one casts the argument in equations, it is evident enough, and it involves what the mathematician calls nonlinearity. It is, however, pictorially quite clear that the symmetry of the initial disturbance in Figure 2.4a must lead to two oppositely resolved patches, as in Figure 2.4c.

The Mills mechanism is thus capable of amplifying a rudimentary pattern to a pattern of fully resolved patches. Because it does this most quickly in the absence of diffusion, it may not be obvious why I have mentioned diffusion at all. The problem lies in the total lack of discrimination of the autocatalytic amplifier per se. Without diffusion, the wavelength of the initial disturbance in Figure 2.4a would be quite irrelevant to the exponential growth rate of the pattern. Every disturbance would be amplified at the same rate. Therefore, concentration noise, containing a mixture of many wavelengths, would be amplified as noise. The nondiffusing activator is not a pattern generator; it is not a morphogen.

By contrast, consider the three waveforms in Figure 2.5 (left-hand side) if the substance diffuses. The peak *P* has the same height in all cases and is therefore growing by autocatalysis at the same rate in all. But diffusion rate depends upon concentration gradient, and therefore, for the same amplitude, rates are different at different wavelengths. Specifically, the diffusion rate becomes faster as the wavelength is shortened. Hence, as the resultant of catalysis and diffusion, short-wave patterns grow more slowly than long-wave patterns. Indeed, at a sufficiently short wavelength, diffusion will win, and the pattern will decay exponentially instead of growing. Figure 2.5 (right) shows what might have happened, at the same later time, to each of the three waveforms; the middle one is taken to be at the threshold at which diffusion just counterbalances catalysis, and the pattern neither grows nor decays.

Figure 2.6a shows some irregular-looking "concentration noise" and its analysis into the sum of three sinusoidal waves. An autocatalytic activator which did not diffuse would simply convert this chaos into a larger-amplitude

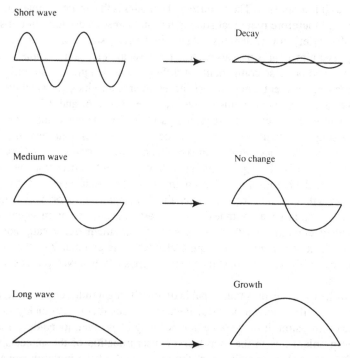

Figure 2.5. Illustration of the wavelength dependence of amplification of a spatial-concentration waveform by a diffusible activator. The three changes shown all refer to the same time interval, which is called $t = 2$ in Figure 2.6. The initial concentration noise in Figure 2.6 is the superposition of the three initial waveforms shown here.

version of the same. But a diffusible activator, having the same wavelength discrimination as that in Figure 2.5, would produce the sequence of changes shown in Figure 2.6b–d. Evidently, as time goes on, different components of the initial noise arise to transient dominance of the pattern; but the final destiny of the system is dominance of the longest-wavelength pattern in the input. In fact, as discussed in more detail in Chapter 6, the system must eventually become uniform, as exemplified in nature by the universality of L-amino acids and D-sugars and the absence of the opposite enantiomers.

The fundamental properties of a diffusible activator are, then, that it grows and spreads; and because the spreading diminishes concentration gradients and hence diminishes the adverse effects of diffusion on the growth rate, the widest patches of activator grow most rapidly, for comparable peak concentrations. Pattern arises transiently and then vanishes. The activator which gave early promise of being a morphogen shows that it is not quite that; something else is still needed for pattern generation. These simple conclusions could have been reached by considering a single activator without the complications of the relation between a symmetrical pair that arise in the optical-resolution

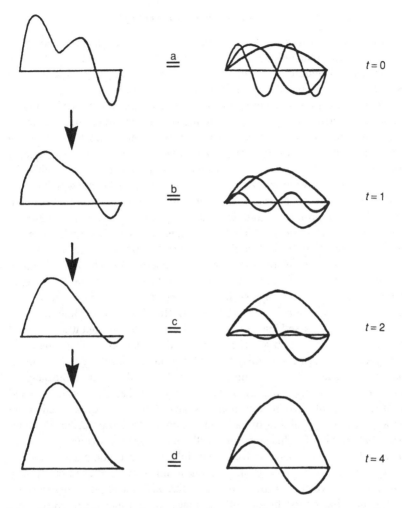

Figure 2.6. Amplification and smoothing of concentration noise across a one-dimensional system by a kinetic model of the optical-resolution type which gives greater exponential growth rates for longer waves, without limit. The left-hand column shows overall concentration, the horizontal line being the equilibrium value, and the right-hand column shows its sine-wave components. The rate constant for exponential growth of each component is $k[1 - (\lambda_0/\lambda)^2]$, where λ_0 is the wavelength of the component which neither grows nor decays. The time unit is $k/3$. See Chapter 6, especially equation (6.45) and Figure 6.5, for mathematical details. Compare *Polysphondylium* data (Section 10.3.2, Figure 10.9a,b).

model. But the latter has an advantage in that it exemplifies very clearly the symmetrical two-way departure from equilibrium which, as we shall see later, remains a feature of kinetic systems lacking the symmetrical relationship in molecular geometry.

2.3 Inhibitors, such as chalones, versus morphogens

The rather obvious lack in the model of the preceding section is something to
put the brakes on the indefinitely wide spread of the activator peak. One may
expect any such additional feature to bear some relationship to the discussion of
establishment of a spatial scale, in Section 2.1. There, we see that addition of
decay, or destruction, to the properties of a diffusible messenger confers upon it
the ability to quantitate spatial scale. To the specific end of limiting the size of
an activated region, however, there is a better model set of properties, involving
inhibition. The common feature between this and the previous example of
Section 2.1 is that delimitation of spatial scale by kinetic action within the
system seems always to require a negative aspect to the kinetic activity.

The point may be exemplified by the simplest version of the essential
property of the postulated mitotic inhibitors known as chalones. These are
tissue-specific inhibitors of mitosis in the same tissue which produces them.
The two words "hormone" (Starling, 1905) and "chalone" were coined early
in this century to designate, respectively, stimulatory and inhibitory chemical
messengers. [E. A. Schäfer coined the word "chalone" at the Seventeenth
International Medical Congress in 1913; see the extract from the *British
Medical Journal* of that year reproduced by Houck (1976).] While the word
"hormone" soon passed into general use, the other did not, but it was revived
by Bullough (1962) with a more restricted definition as a mitotic inhibitor
"produced by a tissue with the primary function of controlling the growth of
that same tissue." There remain doubts in the minds of some biochemists
about the existence of chalones. They have resisted chemical purification and
identification, although biological assays have shown the requisite activity in
extracts and the possibility of carrying out concentration procedures to aug-
ment that activity. Of this situation, Bullough (1983) has written that "from
time to time biochemists and cell physiologists have complained that no
reality should be accorded to the chalones until they have been chemically
characterized, which is a point of view that will not impress anyone who
knows anything of the history of hormone research and usage. Biological
evidence can be fully as impressive as biochemical evidence."

The following is a much oversimplified view of the action of a chalone in
regulating homeostatically the size of a tissue. But it contains the gist of the
matter, both for that function and for the relationship of chalones to morphogens. The cells in a tissue are initially in an activated state in that they all
undergo mitosis at regular intervals. In that state, as well as in the later
inhibited state in which they have ceased to divide, they all produce a chalone
at a constant rate per cell. The chalone diffuses throughout the tissue at a rate
that is rapid compared with the rate of mitosis, and so it can be regarded as
uniformly distributed. It can also escape across the surface of the tissue at a
rate proportional to its concentration inside the tissue and to the surface area.
(Every one of these assumptions can be substantially modified in a variety of
ways without removing the essential property to which they lead.)

If we suppose the tissue to be three-dimensional (e.g., a sphere in the simplest visualization), the rate of inhibitor production will be proportional to the volume, that is, to r^3 (r = radius), while the rate of escape will be proportional to Cr^2 (concentration times surface area). Hence, if the rate of adjustment of the inhibitor concentration is fast compared with the rate of change of r due to mitosis and cell growth, the inhibitor will always be at a steady-state concentration, increasing in direct proportion to the radius. The final assumption needed is that a particular threshold concentration of the inhibitor will switch off mitosis in any cell exposed to it. It is then clear that the tissue must stop growing when it reaches a definite radius, which we could express in terms of kinetic constants if we turned all the foregoing statements into equations. Further, if cell loss later diminishes the size of the tissue, the steady-state inhibitor concentration will drop, and mitosis will begin again, until the former size is restored. Thus the size of the tissue is stable against disturbance, or is homeostatically controlled.

Nothing in this argument is specific to three-dimensional geometry. For a single-layer epithelial sheet in the form of a circular disc, the chalone production rate would vary with r^2, and the escape rate with r, and again the balance of these terms would leave steady-state C proportional to r – and likewise if the system were a one-dimensional filament, the rates then being proportional to length r and to C.

There also is nothing in the foregoing argument which is specific to a spherical or circular shape. The proportionalities to r, r^2, and r^3 remain correct if any tissue, as it grows, develops into a shape which is, in the strict geometric sense, similar to the original shape. If, however, a tissue grew by cell division and subsequent growth, with indiscriminate close packing of the cells, its shape could not be other than a sphere. The chalone could determine its radius, but could not govern its shape or form. A chalone is not a morphogen.

2.4   Type II morphogens: Turing morphogen pairs

In each of the three preceding sections, hypothetical substances have been discussed which would have properties appearing, a priori, likely to be useful in the matter of pattern formation or morphogenesis: first, the communication necessary to organize the activity of a morphogenetic field; second, activation of some kind of chemical behaviour; third, inhibition to limit the size of the activated regions. But in each case the postulated set of properties fell short of the requirements for the substance to *generate form or pattern*.

The essential postulate of the kinetic theory of living pattern and form is that for pattern-generating activity, the substances possessing the sets of properties which I have discussed separately must interact *with each other*. It then becomes rather obvious why it is so difficult to identify any one substance as a morphogen. The word is akin to the words "twin," and "husband" and "wife." To identify a person as such, one must know of the existence of another person. To identify a chemical substance as a type II morphogen, one must know its relationship to at least one other chemical substance.

Suppose that a substance has the activator property discussed in Section 2.2 and illustrated in Figure 2.4. A peak such as that shown on the left-hand half of that diagram is a system in which growth is occurring. But this growth, being in the total amount of a chemical substance, is somewhat different spatially from the case of growth in the number of cells in a tissue discussed in Section 2.3. Because a chemical substance can exist over a range of concentrations in solution, the total amount can increase without the need to occupy extra territory, at least until the saturation condition of Figure 2.4c is reached. The diffusibility of the substance, however, gives it the power to spill over into adjacent regions, and as explained and illustrated in Figures 2.5 and 2.6, extra spatial size leads to faster growth, just as it does for the tissue of close-packed cells all undergoing mitosis at regular intervals. In this sense, the optical-resolution mechanism of Section 2.2, in which the largest and therefore fastest-growing region finally takes over the whole system, is akin to Bullough's (1983) concept of cancerous tumours as being tissues lacking the homeostatic control of a mitosis-inhibiting chalone.

Suppose now that the activator assists catalytically in the production of a substance which acts back on the activator to inhibit its production. This is not much different from the chalone mechanism for homeostasis, except that the activator and inhibitor are now both envisaged on the molecular scale, rather than the former being cells and the latter molecules. The essence of the action of the inhibitor remains the same: Instead of the growth rate of the activated region increasing indefinitely with its size, there will come a point beyond which increasing size will give slower growth.

In the mechanism for tissue homeostasis, this point is reached because the cells move slowly, defining a slowly increasing region of space through which the inhibitor can move rapidly, and from which it can escape rapidly by continuation of its diffusion across the boundary of the tissue. The same feature is necessary for the two molecular species now envisaged as a morphogenetic control mechanism, and for essentially the same reason. The adjustments of inhibitor concentration which permit it to control the growth of the activator can be properly made only if the inhibitor moves much faster than the activator.

Theories of pattern, form, and morphogenesis must, however, cope with phenomena which have no precise parallel in the homeostatic control of a single tissue, particularly the formation of complex patterns with a number of repeats of the same kind of unit. In some instances such development occurs in a time sequence (e.g., vertebrate somites, spiral phyllotaxis), but in others the repeated parts appear all to be specified in a single pattern-forming event (e.g., whorls, and segmentation in insects at the cellular blastoderm stage). For the latter type of event, to apply kinetic theory one must envisage a number of complete cycles of the concentration waves occupying the whole system. The actions of activator and inhibitor must extend throughout this region. This is a contrast to the chalone theory, in which the inhibitor is tissue-specific and has no effects beyond the limits of the single activated region.

To secure this proper organization of a complex pattern, it is also necessary that both activation and inhibition be continuous functions of the concentrations of the two substances causing them. In the tissue homeostasis model, the growth rate of a cell mass in which all cells are dividing is proportional to the number of cells, and hence exponential; but the postulated inhibitory effect of the chalone has a simple switching character (i.e., it is a discontinuous function of the concentration). This is good enough when the inhibitor has nothing to do after escaping from the activated region. But if the situation is like that of Figure 2.4, an inhibitor which has escaped from the left-hand half of the system into the right is there going to have to play a role which is in some sense a mirror image of what it did in the left-hand half. This requires the inhibitor to produce effects when it is at low concentrations, as well as when it is at high concentrations.

In the preceding four paragraphs I have presented a model having five features: a self-enhancing activator; an inhibitor produced at a rate controlled catalytically by the activator; the inhibitory effect of this back upon the activator; slow diffusion of the activator; and rapid diffusion of the inhibitor. These are five of the six terms in the pair of differential equations published by Turing (1952) under the title "The Chemical Basis of Morphogenesis." The sixth term is a self-decay of the inhibitor; the model will work without it, and the advantages which it confers are not easily seen without the mathematical analysis of the equations.

Provided that all biosynthetic pathways are in action to provide the activator and inhibitor with the substrates they need for their catalytic activities, their mutual interactions are capable of generating pattern and form. Together, as a pair, they are morphogens. If one of them is lacking, the other also ceases to be a morphogen, though it may remain an important substance in one of the categories of positional messenger, activator or inhibitor, hormone or chalone. In kinetic theory, morphogenetic character is conferred upon substances by a kind of espousal between them.

2.5 The two-morphogen interaction

The essence of any kinetic theory of pattern formation is that initial input is always to be found everywhere in the form of a chaotic jumble of small disturbances containing rudiments for pattern of any spatial dimension, or wavelength. The job of a pattern-forming mechanism is to amplify or suppress these rudiments selectively. Both the shortest and the longest must decay, while those in a narrow intermediate range are amplified. In this sense, a pattern-forming mechanism is a kind of band-pass filter for the random input.

In two-morphogen reaction-diffusion models, the following occur: the activator amplifies indiscriminately; diffusive spreading is fastest at short wavelengths, and converts the amplification into attenuation of these; and the inhibitor does the same for long activated regions, because its diffusive escape is then slower and it builds up within the activated regions.

As qualitative generalizations, these aspects of the working of a reaction-diffusion mechanism are, I hope, obvious enough in the context of the earlier parts of this chapter. Quantitatively, there are many different reaction-diffusion models showing substantially different behaviours in detail. All the well-known two-morphogen models, however, reduce to the same one when the amplification of pattern is just beginning, out of small-amplitude rudiments. That one is the linear Turing model. To my mind, the novice in this field is well advised to understand the working of the linear model as a basis for approaching the nonlinear models, such as the Brusselator of Prigogine (1967) and the model of Gierer and Meinhardt (1972).

The meanings of "linear" and "nonlinear" are illustrated in Section 2.2, especially in relation to Figure 2.4. There, the growth rate of a quantity U, which happened to be a measure of displacement from equilibrium in that case, is indicated as being proportional to U. This is a linear term in the growth equations; any other function of U, such as U^2, or U^3, or $1/U$, or $\ln/U$, would be a nonlinear term. For reasons which require the calculus for their exposition, and which are therefore deferred to Chapters 6 and 7, reaction-diffusion models in which all terms in the growth equations are linear have the general property that they cannot distort the shape of a pattern which is a pure sine wave in the distance variable. The dynamics of the model can lead only to an increase or decrease in the amplitude of the sine wave, as between Figures 2.4a and 2.4b, and the three changes shown in Figure 2.5.

Those diagrams are for an activator only. The inhibitor in the linear Turing model gives similar kinetic terms, linear in its *displacement from equilibrium both ways*, often called V. We may now alter Figures 2.4, 2.5, and 2.6 to the equivalent representations of behaviour when the inhibitor is also present. The assumed starting disturbance, Figure 2.7a, has activator and inhibitor dis-

Figure 2.7. (*opposite*) Some morphogen distributions and their evolution in time. Vertical distance is concentration; horizontal distance is position along the system. (*a*) Putative initial state, a sine-wave distribution with X (solid line) and Y (broken line) in phase with each other. (*b*) Evolution of such patterns (X only shown) according to wavelength; both short and long waves decay, but a wave of intermediate length grows. (*c*) A somewhat chaotic initial state evolves to a more orderly pattern, because the initial state is a sum of three patterns of different wavelengths like those in (*b*), and only the one of intermediate wavelength grows. (*d*) Typical steady state of a Gierer-Meinhardt pattern; the inhibited region is very long and flat compared with the activated region. The wavelength selection shown in (*c*) is a closer match than is Figure 2.6 to the *Polysphondylium* data (Section 10.3.2, Figure 10.9a,b).

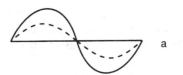 a

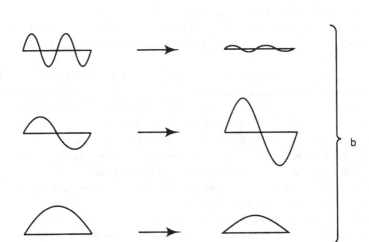 b

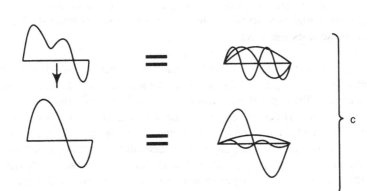 c

 d

tributed in sine-wave patterns of the same wavelength and in phase with each other. Let us now consider the model hypothetically as a machine in which we can switch on each of the important processes separately, in sequence:

1. The activator starts to depart from equilibrium in proportion to the value of U at each point. The U wave grows in amplitude exponentially, without distortion from the sine-wave shape.
2. The activator starts to diffuse. The rate of growth of its amplitude is slowed down, but it remains exponential growth, without distortion of the sine-wave shape.
3. The activator, in proportion to U at each point, starts to catalyze the growth of V (upward and downward, two-way departure from equilibrium, just as it catalyzes its own growth). The V wave now grows exponentially, but at a different rate from the U wave. Thus both will stay sinusoidal, but the ratio of their amplitudes will change as time goes on.
4. The inhibitor starts to diffuse, faster than the activator. The rate of growth of the V amplitude is slowed, perhaps quite drastically.
5. The inhibitor starts to inhibit the growth of the activator, in proportion to the value of V at each point. The growth of U is slowed down, and both U and V remain distributed in undistorted sine-wave form. Without mathematical discussion, however, it is not at all obvious what is now going to happen to the relative amplitudes of U and V waves. A clearly plausible statement, which turns out to be correct within certain quantitative limits upon relative rates, is that U and V are now so locked together by mutual interaction that their amplitudes must settle down to a constant ratio. Both then grow together, so that if in a certain time the U wave doubles in height, so does the V wave.

Consider now how the system will behave at a variety of wavelengths: the analogue of Figures 2.5 and 2.6, but with the inhibitor added. At very short wavelengths, diffusion will be sufficient to swamp out all possible growth, and the system will decay, just as did the activator alone. At longer wavelengths, the effect of diffusion is smaller, and the system *may* be able to grow. (Whether it *will* grow or not depends on quantitative considerations which need the mathematical analysis; short of that, one can see qualitatively that at least the rate of decay is going to become smaller as the wavelength increases, and the reader will have to take my word for it that the decay can change sign and become growth.)

But because the inhibitor diffuses faster, an increase in wavelength gives more advantage to the inhibitor than to the activator. Thus, as wavelength increases, the steady-state ratio of inhibitor to activator increases. This must bring about an effect of wavelength on growth rate in the direction opposite to that of diffusion. As wavelength increases, the inhibitor becomes more and more capable of putting the brakes on activator growth because there is more and more of the inhibitor in relation to activator. Here, we are looking at a more complicated version, with continuous variables, of the way in which a chalone can switch off tissue growth at some critical size. Thus, if the increase in wavelength is pursued far enough, growth will turn back into decay. This is the band-pass aspect which gives the model the power to select a medium wavelength out of initial chaos containing both longer and shorter ones (Figure 2.7b,c).

The reader may at this point have some suspicions about the assumptions that have been made regarding the chaotic input. It is plausible enough that the total of U disturbances can be Fourier-analyzed into a sum of sinusoidal components of various wavelengths, as shown in Figure 2.6a, and, likewise, that the V disturbance can be similarly analyzed. But my discussion has depended upon an assumption that the components of the same wavelength in the U and V disturbances are exactly in phase with each other. This seems, to say the least of it, improbable for a chaotic input in which there is no reason for U and V disturbances to be correlated at all, because no correlating mechanism has yet started to operate.

For linear waves, the solution to this problem is suggested by an illustration due to Maynard Smith (1968) of how the Turing model works. I defer the detailed presentation of this illustration to the reintroduction of this model in Chapter 7 (Figure 7.1). But the gist of it is that if one takes a system at equilibrium and introduces a local U disturbance only, with no initial V disturbance, the U disturbance will spread and itself generate a V disturbance which soon will adjust into phase with the U waveform. The in-phase character of these two disturbances is, in short, something which is soon commanded by the strong kinetic locking-together of U and V whether it exists in the input or not.

Investigation by computation of the properties of some of the nonlinear models (Chapter 9) gives even more striking results. For the Brusselator model, if a single peak is generated and the system is then doubled in length to see if the model will turn one peak into two and thus give some account of dichotomous branching, indeed it occurs. But the time sequence is that the U pattern splits into two peaks, while the V pattern looks essentially unchanged. Then, slowly, the V pattern also begins to readjust into two peaks. It is not at all easy to understand, from study of the algebra of the models, how the U and V waves seem somehow to "know" their ultimate joint destiny while they are so far out of kilter with each other at intermediate stages. But it is fortunate for ready comprehension of what is going to happen that they do.

2.6 Activation and inhibition versus activated and inhibited regions

Figure 2.7a may be described as showing an activated region as the left-hand half and an inhibited region as the right-hand half. Commonly, when reaction-diffusion theory is applied to generate a chemical concentration pattern of this kind, it is then assumed that some kind of developmental activity is associated with a high concentration of activator, such as a high growth rate of tissue or of cell surface, or differentiation of cells in a particular way, or formation of a specialized structure such as the preprophase band which seems often to determine where a plant cell is going to divide.

The part of a morphogenetic field which is activated for a particular specialized activity often is much smaller in extent than the half of its length which is suggested by the positive half-wave in Figure 2.7a. But the sine wave does not give a very clear boundary to any smaller activated region, because it is rather flat at the peak. Some nonlinear models easily distort the waveform to give much more "spiky" peaks; to this end, the Gierer-Meinhardt model is outstandingly good (Figure 2.7d). Because of this special property of producing, in effect, isolated mountains in the midst of a vast plain, the Gierer-Meinhardt model has given rise to the slogan "short-range activation, long-range inhibition," which is sometimes transferred to the entire body of reaction-diffusion theory.

I am very much in two minds about the value of the slogan. On the one hand, it describes succinctly a property of one particular reaction-diffusion model, and it is an important property. On the other hand, it can mislead in regard to the general workings of reaction-diffusion. The problem is that the word "inhibition" is attached to what is going on in the vast plain. But in any reaction-diffusion model of the two-morphogen type, the primary function of inhibition is to limit the size of the activated regions, and it is *within* those regions that the inhibitor is most concentrated and most actively at work. The precipitous slopes of the mountains are not buttressed and shored up from outside; they are glued and cemented from within, by the inhibitor–activator interaction. A more precisely correct phrase, which is unfortunately a little cumbersome for a slogan, is "short high-interaction regions, long low-interaction regions." The interaction has to be understood as involving both activator and inhibitor.

Demographic analogies spring to mind, in which, instead of mountains and plains, cities and rural areas are used for the high- and low-interaction regions. One may, for example, think of the police force as an inhibitor of lawless activity and recognize that its strong inhibitory effect is within the cities. The rural areas are more law-abiding, not because of stronger inhibition but because of a generally lower level of the interactions that tend to lead to lawbreaking. Analogies of this kind, however, generally fail if pursued too far, as, for instance, into the precise details of the activator–inhibitor ratio, an important feature of the Gierer-Meinhardt model, as discussed in Chapters 3 and 9.

If we return to Figure 2.7a, we see that the inhibited region looks just like the activated region upside down. This diagram was presented as a putative initial state, before the reaction-diffusion dynamics had had a chance to work on it. But there are certain mechanisms in which, as the pattern develops through reaction-diffusion dynamics, the upper and lower parts of the pattern always remain the same shape. It can then be seen that the processes going on in an inhibited region and an activated region are not necessarily different from each other in any qualitative sense, but are in fact two versions of exactly the same thing, symmetry-related through a change of an arbitrary algebraic sign. The obvious instance is optical resolution, as in Figure 2.4. A region inhibited for formation of D is activated for formation of L, and vice versa.

Some types of reaction-diffusion models have a particular kind of non-linearity which maintains this property – that the inhibited regions are just like the activated regions upside down – even when they contain everything needed to stabilize pattern in a Turing-like manner (which the simple optical-resolution mechanism discussed earlier does not). One of these is the "hyper-chirality" model we have devised (Harrison and Lacalli, 1978). Models with this property have a particularly interesting feature in regard to two-dimensional pattern: They tend to produce stripes rather than spots. This may be quite significant in relation to such morphogenetic processes as *Drosophila* segmentation (see Section 9.2 for further discussion).

3

Pictorial reasoning in kinetic theory of pattern and form

Practitioners of reaction-diffusion theory are frequently presented – off the cuff, in some informal discussion – with some example of morphogenesis and asked whether that type of theory can explain it. On an epithelial sheet, can it account for the arrangement of feather follicles, or bristles, or whatever, in a regular hexagonal array? Can it account for periodically repeated structures only 1–2 μm apart, such as xylem rings? Can it account for instances of unequal division of plant cells? If so, what about the more numerous instances of unequal division? Can it account for morphallactic regulation, in which a chopped-off piece reproduces in miniature the former pattern of the whole, as in the differentiation of a slime-mould slug? If so, what about epimorphic regulation, in which a structure with a piece removed grows out to restore it, as in many phenomena of insect development.

(This last question is very important and can be expressed much less polysyllabically: Can reaction-diffusion both measure and count? Experimentally, both clearly happen, even in normal development. The whorls of hairs produced by the alga *Acetabularia* have variable numbers of hairs, but constant spacing between adjacent hairs: Morphogenesis measures. The cellular blastoderm of *Drosophila* always has 14 parasegments in normal development, though the length of the egg can vary by at least 30%: Morphogenesis counts. The answer needs mathematics; see Section 10.4.)

The theorist, so questioned, tends to feel somewhat like the Delphic sibyl; and the questioner often goes away as unsatisfied with the answer as many pilgrims were with the signal-to-noise ratio of the supposed transmissions from Apollo. To the theorist, the answer must be equivocal because the question is too broad. It contains the name of a vast field, reaction-diffusion theory. Various specific versions of this theory are capable, taken together, of accounting for most of the phenomena of development. But any one version is much more restricted in its scope.

Perhaps a biochemical analogy may help. A cell biologist may suspect that some behaviour of a cell is caused by actin, and thus may ask the question: Could actin do it? Upon seeking an answer by immunoflourescence and finding that there is no actin present in the significant regions, the biologist must abandon that idea, but is quite unlikely to abandon the question: Could a protein do it? That question has hardly even been formulated in the biologist's mind. It is too vast. The answer is almost always in the affirmative, but that by itself doesn't tell anyone very much.

Much the same is true of reaction-diffusion theory. It is a generic term covering a broad range of possible explanations. The *detailed* form-generating properties of two different reaction-diffusion models can be as different as the properties of two proteins. The *general* features which most reaction-diffusion models have in common, in the scheme of mutual interactions and transport of two substances, are analogous to the *general* concepts of protein structure, such as the amino acid residue, the peptide link, and the various concepts of secondary, tertiary, and quaternary structure. And reaction-diffusion per se is only part of the more general field of kinetic theory. Thus, I am asking the reader to accept Chapters 2 and 3 of this book in the same spirit as one accepts the early chapters of many biochemical texts, in which the general structures of proteins, nucleic acids,and suchlike are set forth – not, that is, in the spirit of reading a specialized monograph about actin and myosin and their relatives.

For example, in the matter of formation of regular hexagonal arrays on a surface, Meinhardt (1982) indicated some considerable difficulty in persuading his model to do this. But this doesn't mean that there is trouble persuading reaction-diffusion to do it. Lacalli (1981) found the same trouble with the Gierer-Meinhardt model, but showed that the Brusselator (Prigogine, 1967), in a modification by Tyson and co-workers (Tyson and Light, 1973; Tyson and Kauffman, 1975), produced hexagonal arrays out of random input with no difficulty.

The various questions posed in the first paragraph of this chapter are a very mixed bag in regard to the levels of theoretical treatment needed to answer them. Several of them require the mathematical treatment which is approached gradually in the next few chapters and finally given in Chapter 7. Some of these questions are discussed in that context in Chapters 7–10. For instance:

1. The question of the reasonable limits of spacings which can be produced by a reaction-diffusion mechanism requires the algebraic solutions of the Turing equations (see Section 7.3.2).
2. The measuring-versus-counting question, which is quite crucial to a grasp of the broad scope of the theory, requires not only the algebra but also some consideration of what kind of chemical mechanism may give rise to the rate equations (Section 10.4).
3. The matter of the hexagonal arrays has to be gone after with a computer programmed to solve the kinetic equations (Section 9.1.5).

But there are many features of the applicability of reaction-diffusion which are not like this. In regard to the effects of chopping a developing system into pieces, whether it is done artificially with the scalpel or naturally by the appearance of the cell plate in a dividing plant cell, much can be done pictorially. One needs some appreciation that different mechanisms respond differently, but only in a very general way. This chapter discusses this feature of development in some detail. The intent is to show the experimentalist how to think usefully about whether reaction-diffusion might account for the phenomenon of immediate interest. Thus, the oracle circumvents the question by putting the pilgrim in the oracle's seat, for in the matter of equivocal prophecy, it is much more satisfying to give than to receive.

In the preceding chapter the basic workings of reaction-diffusion models were discussed. A few things were lacking from that account, however, things one must know about to start upon detailed application of the theory. The most important of these is the fit of pattern to the size, shape, and boundary conditions of the morphogenetic region. But also, the dynamics of pattern development were treated in Chapter 2 largely as a matter of unrestricted exponential growth of amplitude. There are only two indications, in relation to Figures 2.4c and 2.7d, that, like most exponential growth, this must in practice always come to an end at some point, and that the system then enters a regime of substantially different behaviour. It is there, in the nonlinear regime, to use the mathematicians' adjective, that the diversity of different reaction-diffusion models arises.

The remainder of this chapter is divided into three parts. The first gives the rest of the theory needed before one can start about application. The other two parts are designed for a double contrast: between the plant and animal kingdoms, and between linear and nonlinear regimes of the mechanisms. The account of cell division in plants is of my own devising, and not previously published. The account of a few features of insect development gives what I consider to be the gist of the important specific behaviour of the Gierer-Meinhardt model. The contrast between the two corresponds to my present view that a number of significant features of plant development can be accounted for without going outside the linear regime, whereas Meinhardt (1982) is correct, at least for such examples as insects and *Hydra*, in laying stress on the strong nonlinearity of their dynamics and hence almost ignoring the linear regime. Pedagogically, this leads to a problem which I have made some attempt to solve by the balance of material in this chapter and this book. One cannot understand the theory of nonlinear dynamics without first spending a lot of time thinking about linear dynamics.

3.1   The fit of pattern to boundaries and the dynamics of pattern growth

We continue to be concerned with the mutual interaction of two substances, often referred to as activator and inhibitor, or X and Y, or A and H in the writings of Gierer and Meinhardt. For the reasons set forth in Chapter 2, these substances must have positive and negative catalytic influences, enhancements and inhibitions, upon themselves and upon each other, of the kinds there specified. All these catalytic influences share a feature which may look peculiar to the beginner in this type of theory, in relation to anything that one commonly learns in elementary chemical kinetics. This is, as explained in Chapter 2, the two-way departure from equilibrium. If X_0 represents an equilibrium value of X, or a spatially uniform steady state which is not true equilibrium because it is maintained by supply, production, destruction, and removal of X, we may write $U = X - X_0$; and similarly for the other morphogen, $V = Y - Y_0$. Linear self-enhancement of X is growth of U at a rate everywhere proportional to its current value, which may be positive or negative, so that peaks in a concentration waveform move upward, and troughs move downward. Diagrams of these waveforms are sometimes labelled with the variables U and V, and sometimes with the true concentrations X and Y, or A and H, which can only be positive. When reading any account, one must be careful to check that one has understood which kind of variable is being used. There is no generally accepted convention on terminology. In my own earlier writings I used X and Y for what I am now calling U and V.

The apparent peculiarity of the catalytic interactions, thus baldly stated, in which the catalytic effect appears to be proportional to only a part of the concentration of the catalyst, is in fact quite a normal feature of relatively simple chemical reaction mechanisms. It is illustrated in Chapter 2 by the example of stereospecific catalysis in a matched pair of enantiomers, because this is the easiest example to present without plunging into equations. Mathematical analysis later in this book shows that the property is very far from being uniquely attached to that example. X can quite well be a single molecular species, and still display the same dynamics for its self-enhancement. This term, or autocatalysis, or positive feedback loop, should also be recognized as not far from synonymity with the two most fundamental properties of life: reproduction and assimilation, which imply feedback on a grand scale, involving thousands of chemical reactions. The equations of reaction-diffusion should be understood in that spirit. A self-enhancement written simply as

$$2X + Y \rightarrow 3X$$

may represent a single step in a reaction mechanism, or a feedback going right round one of those multistep cycles which are so commonly written in biochemical schemes.

3.1.1   The fit of pattern to boundaries

Once again, it is assumed for simplicity that the system is elongated and is producing pattern only along its long axis, an essentially one-dimensional

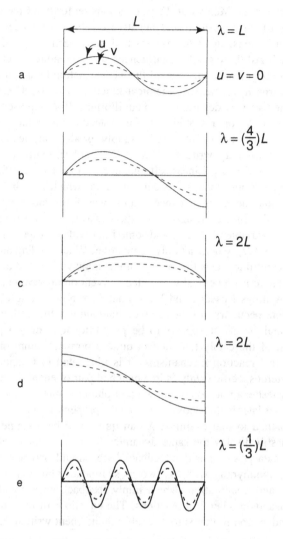

Figure 3.1. Examples of linear (i.e., sine-wave) two-morphogen patterns along an elongated system, with various ratios of wavelength to system length, and two kinds of boundary conditions; λ, wavelength.

situation. Figure 3.1 shows a few patterns which might occupy the length of the system. These are all sine waves (i.e., they would arise in the linear region of behaviour). They differ in the number of wavelengths occupying the complete system and in the conditions satisfied at the ends: nodes ($U = V = 0$), or crests or troughs. In no case has a waveform been depicted which has, at the ends of the system, anything other than those three special points on the waveform. Therefore, a full cycle of the wave may be fitted to the length of the system, as in (a), or three-quarters of a cycle, as in (b), or half a cycle, as in (c); but nothing intermediate between these three special cases is shown.

This limited discrete set of choices has arisen because it has been assumed that each boundary of the system demands one or other of only two boundary conditions. The first is that diffusion of the morphogens ceases at the ends of the system; that is, the ends are impermeable barriers. This is called a no-flux boundary condition (or, in the jargon of the trade, a Neumann boundary condition). It is mechanistically simple, and very plausible in many systems, and hence is widely used in theoretical studies. When there is such a diffusion barrier at the ends, the concentration profile always adjusts so that there is no delivery of material to the boundary by diffusion, and no removal of material from the boundary by diffusion. Because the rate of diffusive flow is proportional to the concentration gradient, which is the slope at any point of any of the curves in Figure 3.1, that slope must be zero at a no-flux boundary (i.e., there is a crest or a trough at the boundary). This is shown only at the right-hand end of (b) and both ends of (d).

Figure 3.1d, variously modified in detail by nonlinearities which distort it from the pure sine-wave form, is a widely discussed type of pattern. It shows the establishment of a simple head-to-tail gradient in a elongated organism. It has been applied, for instance, to the hypostome-to-basal-disc gradient in *Hydra* (Gierer and Meinhardt, 1972), to the differentiation of a slime-mould (*Dictyostelium*) slug into pre-spore cells at one end and pre-stalk cells at the other (Lacalli and Harrison, 1978), and to the head-to-abdomen gradient as set up at the cellular blastoderm stage in insect embryogenesis (Meinhardt, 1977, 1982). For this kind of pattern, which end is "up" and which is "down" might be specified by some preexisting polarity. But if L represents the length of an entire free-living embryo in which we are looking at the first significant gradient-forming event and it is immaterial which end happens to become head, then the pattern can arise with no antecedent of specific polarity.

Figure 3.1b, on the other hand, shows a polarity arising from preexisting asymmetry between the two ends of the system. The right-hand end is a no-flux boundary, but the left-hand end is something else. It is a point at which departure from equilibrium is not permitted (i.e., $U = V = 0$), or, in other words, X and Y are held at their equilibrium concentrations X_0 and Y_0 at all times. Although in most accounts this is acknowledged as the most likely alternative to no-flux boundaries, it has not been as widely used for specific biological examples. The reader who approaches this topic from a background of classical elementary courses in chemical kinetics and thermodynamics is entitled to be surprised. In all that has been described of these models in this and the preceding chapter, this boundary seems to be the only point which is behaving in a chemically "normal" fashion, that is, getting to equilibrium and then staying there. But there are snags. First, the so-called point at the end of the system cannot be a point. The slopes of the U and V curves just within the system are such as to indicate either rapid delivery of material to the end of the system from within [left-hand ends of (a), (b), and (e); both ends of (c)] or withdrawal of material into the system from the end [right-hand ends of (a) and (e)]. Thus the end of the system must be an effective supplier or remover of rapidly transferred material, and this activity must extend into a finite region beyond the marked end of the system. Second, what is going on is mechanistically more complicated than the no-flux barrier. Why should out-of-equilibrium behaviour be permitted up to some point and forbidden beyond it?

I defer the discussion of this mechanism to Part III. For the nonce, my principal reason for paying attention to this boundary condition myself is that as an experimental scientist working on plant cell surfaces, I find it forced upon me by evidence. Especially in the phenomenon of tip growth, in which a cylindrical cell elongates by action principally at a dome on one end of it, there can be a complex pattern of high growth rates over the dome, sometimes leading to branching morphogenesis (in the single-celled algae *Micrasterias* and *Acetabularia,* for instance). But at the equator of the dome, where it joins the cylinder, the growth rate usually sinks to a minimum which is not zero, and which is continued down the cylinder. The concept that new cell surface supports out-of-equilibrium activity, while old cell surface displays growth corresponding to equilibrium concentrations of catalysts, is clearly an attractive a priori postulate in such situations.

The constant-concentration boundary condition goes under the name of Dirichlet, to mathematicians. The constant value need not actually be the equilibrium value; but other values become mechanistically more complicated to justify, and I do not use them in this book.

The next question in regard to Figure 3.1 is, What determines the number of cycles of the waveform fitted between the boundaries? Diagrams (a), (c), and (e) all have the same Dirichlet condition at both ends, but the numbers of cycles of the pattern between the ends are, respectively, $1, \frac{1}{2}$, and 3. How does the pattern-forming mechanism decide which one to produce? The answer lies in the property of the mechanism as a band-pass filter for concentration waves, as discussed in Chapter 2 and illustrated in Figure 2.7. If it happens, for instance, that a wavelength equal to the length of the system, as in Figure 3.1a, is close to the condition of maximum growth rate, then it may readily occur that both the long wave of (c) and the short wave of (e) will be unable to grow and in fact will decay away.

A proper understanding of how pattern is fitted to system size needs a somewhat more detailed account of how the Turing mechanism works as a band-pass filter. Waddington (1956) scornfully characterized Turing models as "inherently chancy" and likely to "play a part only in the quasi-periodic dapplings and mottlings which often fill up relatively unimportant spaces." He thought that the model required an exact fit of a fixed "chemical wavelength" to the size of the system. If that were so, it would indeed severely limit the applicability of the model; but the Turing dynamics are in fact the much more versatile band-pass effect discussed in the next section.

3.1.2 The dynamics of pattern growth

In the linear regime, the amplitude of a morphogen waveform grows as $\exp(k_g t)$. The growth rate constant is k_g; if it is negative, the pattern is decaying rather than growing. Now k_g is a continuous function of wavelength. Figure 3.2 shows one form that it can take. The shape of this curve, showing decay of both short and long waves, and positive growth, or amplification, only of the intermediate range of wavelengths between λ_0 and λ_f, was anticipated qualitatively in Chapter 2. The so-called chemical wavelength, λ_m, is that of the pattern which will grow most rapidly; but it is not the only pattern which can grow. Suppose that the length L of a system is somewhat offset from that maximum, as shown in the diagram. A Dirichlet boundary condition constrains the possible patterns to a discrete set with wavelengths $2L$, L, $(2/4)L$, $(2/5)L$, $(2/6)L$, . . . , $(2/n)L$. Several of these are marked in Figure 3.2, and it is evident that all except one lie outside the band-pass and have negative growth rates. That one, with a full wavelength filling the system as in Figure 3.1a, will easily become established as the pattern despite the offset from the "chemical wavelength."

As a pattern of chemical concentrations grows, it eventually approaches the limits of exponential growth. Mathematically, this means that rather complicated terms in the growth equations, of the kind called "nonlinear," become dominant. In relation to the biological relevance of the models, the significant generalization is that models which were essentially all the same through the linear region become individually different in the nonlinear region. A model which is very appropriate to explain one developmental phenomenon may be quite useless for another (again, my analogy to individually different proteins).

Nonlinearity always involves some distortion of the chemical pattern from the sine-wave shape, though again the distortion may be much more marked for some models than for others. Rather obviously, one may classify the distortions into two categories: ones which make the peaks or troughs of the pattern "spikier," like the Gierer–Meinhardt activator peak in Figure 2.7, 3.13, or 10.11, and ones which involve flattening, like the optical-resolution regions in Figure 2.4c or the flat, low-lying "long-range inhibition" region of the pattern in Figure 2.7. Evidently, the Gierer–Meinhardt model produces one of these kinds of distortion for the peaks and the other for the troughs.

By contrast, the Brusselator model manages to stop the growth of pattern amplitude and bring it to a steady state without nearly such striking distortions from the sine-wave shape. In consequence, for reasons discussed in the next

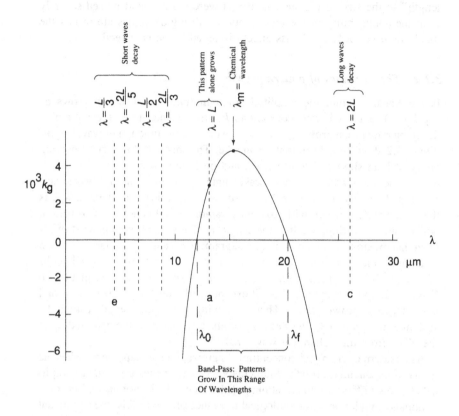

Figure 3.2. A possible variation of the exponential growth rate of pattern amplitude (k_g) with wavelength λ. Patterns with wavelengths between λ_0 and λ_f will grow; this is the band-pass character of the model. The wavelength marked λ_m has the maximum growth rate. Multiples and fractions of L represent patterns of various complexity, with nodes at the ends, fitting a system of length L. Three of these correspond to Figure 3.1 (*a*, *c*, and *e*), and are marked with those letters. The curve is adapted from Harrison, Snell, and Verdi (1984) and is for a particular numerical choice of constants in the linear Turing model.

section, that model retains most of the important dynamic features of the linear Turing model while getting rid of the unrealistic one, the unrestricted exponential growth.

The contrasts in dynamics between the linear and nonlinear regimes, and between different nonlinear models, become particularly important when a pattern is already established and some natural or unnatural disturbance interferes with it. The kinds of responses, known in experimental biology as "regulation" whenever they are substantially effective in overcoming the disturbance, are discussed in the next section.

3.1.3   The guests of Procrustes: responses to chopping and stretching

Procrustes, in the Greek myth, had a guest bed of standard length and was in the habit of adjusting his guests to it by chopping pieces off or stretching them out. His standard fit and procedure for dealing with misfits were as restrictive as Waddington believed the Turing wavelength to be. The Procrustean intervention is a common kind of experiment in developmental biology. But in quite normal development, a pattern may effectively suffer Procrustes' attention. In the division of plant cells, the new cell plate can give as effective a chop to the system as anything done with a scalpel. And because a pattern-forming region often grows in size during development, stretching of patterns is also very common.

If one is a prospective guest of Procrustes, then to stand any chance of surviving the night, it is important that one know one's height. Does a developing pattern in any sense "know" how long it is? Thus stated, the question is not very useful, for a pattern does not have a "height" but rather a discrete set of heights; it can attempt to fit the Procrustean bed with half a wavelength, or one wavelength, or $\frac{3}{2}$ wavelengths, and so on. The significant question is whether a pattern, in its passage through the linear and nonlinear regimes, retains "knowledge" of its original linear wavelength, acquires a new wavelength in the nonlinear region, or totally loses track of its original wavelength. Even more precisely, because different parts of a nonlinear pattern behave rather differently, if one examines some small local region of a pattern, can one from this information find the wavelength?

In this respect, the extreme contrast is between a square wave and a sine wave (Figure 3.3a,b). For the former, either of two local regions such as A and B will give exactly the same information (the same value of U and no change of U with position s), but nothing to indicate that one of them is closer to the middle of the wave than the other. This pattern does not "know" what its wavelength is. But for the sine wave, different values of U and different curvatures can be measured in two different regions, such as C and D. Here, I can no longer avoid one brief reference to the calculus. The property of interest is not actually the curvature but a closely related property, the second derivative of U with respect to s. It is a well-known formula of elementary differential calculus that the second derivative of a sine function is proportional to the function itself: $d^2U/ds^2 = -(4\pi^2/\lambda^2)U$. Thus, if one measures, for any one small region of the pattern, both U and its second derivative, one can calculate the wavelength. This information is carried everywhere in the pattern.

What is the significance of this contrast in relation to chopping and stretching? Essentially, if a developed pattern which still retains strong and variable curvatures somewhat like those of a sine wave is chopped or stretched, it readjusts to the pattern which would have arisen ab initio at the new length. A

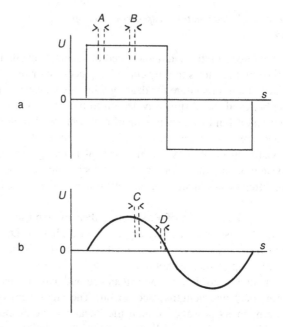

Figure 3.3. Parts of (top) a square-wave pattern and (bottom) a sine-wave pattern, to illustrate that local regions of the latter contain information about its wavelength, whereas local regions of the former do not.

linear Turing pattern is, in short, very labile. (I am here, as often in this book, writing of a macroscopic pattern as an entity, and applying to it a word for which the common chemical usage is in relation to a molecular entity.) But if the pattern has flat regions which represent loss of "knowledge" of its wavelength, various quite different things may happen. One of these is the establishment of an additional peak at the wrong end of the system for normal development, leading to teratogenesis.

This linear/nonlinear contrast is presented in the following two sections in such a way as to combine it with the contrasts plant/animal and single cell/multicellular organism. The nonlinear behaviour can by no means be fully understood merely from the shape of the distribution of one of the morphogens. As always in kinetic models, the interactions between the two must be considered. This is exemplified as simply as possible in Section 3.3. There it becomes evident that the activator/inhibitor ratio is an extremely important determinant of the local behaviour of parts of a Gierer–Meinhardt pattern, and also that this kind of pattern does show some local behaviour, and hence does not always behave as a single entity.

As a preliminary to understanding these aspects, one should understand two further attributes of a long, low-lying flat region in a morphogen distribution. First, it breaks communication across the pattern, because communication is by diffusion, and there is none where the concentration profile is flat. Second, because concentrations are low, they cannot be decreased very much at all. This means that nothing can "dig a pit" beside an existing "mountain." But short of digging a pit, for diffusion to do the work of shovelling the mountain into it, one can never get rid of a mountain in a reaction-diffusion pattern. This is at least one reason why Gierer–Meinhardt patterns cannot easily readjust by the destruction of existing peaks, to be replaced by new ones elsewhere. Thus it cannot find its way from an initial disorderly array of peaks on a plain to a regular hexagonal array. This is not the whole story. The rest of it has to do with the stabilization of the peaks from within. Much of the detailed behaviour still has to be found out by computation, and no model has yet been studied exhaustively in this way. Even the mathematicians cannot yet arrive analytically at accounts of all the biologically significant properties of nonlinear models.

3.2   Modes of cell division in plants

The mechanisms of development of multicellular systems are determined both by what each individual cell can do and by what it cannot do. In this respect the plant and animal kingdoms are sharply distinguished. Animal cells have flexible surfaces. Hence, on the one hand, they are capable of locomotion and can change their relative positions in an assembly, but on the other hand, their surfaces often cannot hold developmental information in the form of precise geometric shape or orientation of division. Plant cells do not change their relative positions, but the rigid cell wall which deprives them of locomotion also confers on them the ability to hold geometrical and directional information.

Thus the geometrical niceties of how individual cells divide are of great importance in plants. There are two general aspects to this topic, from the viewpoint of developmental controls. First, on the multicellular scale there must be long-range processes commanding some cells to divide in one direction and others in another; such control is fundamental to organogenesis (Green, 1980). Second, on the scale of the single cell there must be a control process determining the position of the plane of division. Both scales are quite possible ones for the operation of kinetic controls in general and reaction-diffusion in particular; the latter is quite appropriate to wavelengths from about 1 μm up to at least 1 mm or so (Section 7.3.2).

3.2.1   *One dimension: filaments*

In this section, attention will be given chiefly to events within the single cell. The first examples to be discussed concern the extension of filaments by equal or unequal division of cells, and especially the sequences observed in the latter case. These observations overlap the plant and bacterial kingdoms, via

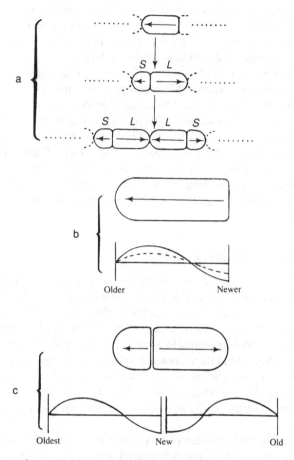

Figure 3.4. *Anabaena* division sequence and morphogen model.

those denizens of a border state, the blue-greens. In that group, *Anabaena catenula* was found by Mitchison and Wilcox (1972) to show an interesting sequence of unequal divisions (Figure 3.4). It is not a very complicated sequence, but the rule for its formation has been expressed in a formidable variety of ways (Lück and Lück, 1976; Lindenmayer, 1982). The algorithms of Lindenmayer are written in terms of polarities of the cells (arrows), changes of polarity (plus and minus signs), states of the walls (numerals 1, 2, and 3), and states of the cells (*a* and *b*). Such complexity becomes useful when one wishes to go to more complex examples, such as two-dimensional sheets or three-dimensional stacks of cells, and test whether or not a rule is operating consistently throughout what often ends up looking like a rather irregular set of successive divisions. These algorithms are closely related to the currently fashionable computing devices known as "cellular automata" (Lindenmayer and Rozenberg, 1976; Wolfram, 1984). Among these, the systems which have been applied to plant development are becoming known, after Lindenmayer, as L-systems (Prusinkiewicz and Hanan, 1989). They have a certain completeness in themselves in showing the presence of orderly behaviour where it was not at first apparent. But they do not seem to me to lead very readily toward the question of what chemical mechanisms may be operating to produce the rules. To this end, the very simple statements in Mitchison and Wilcox's (1972) account are more pertinent. First: "if a given cell has arisen as the left (right) daughter of a division, then, at its own division, its left (right) daughter will be the smaller daughter. . . . we can assign an arrow to each of the two daughters . . . pointing away from the newly formed septum. A cell retains this arrow until it divides." Second: ". . . a new septum site always appears further from the most recently formed septum and we might conjecture that this is because the latter is surrounded by an area of disorganization or inhibition."

For reaction-diffusion as a possible mechanism, this suggests a simple interpretation. Consider a morphogen distribution along the filament axis, and suppose that the peak of this distribution determines the plane of the next division of a cell. This division will be asymmetric if the morphogen distribution is like that shown in Figure 3.1b. The asymmetry of this distribution is the reality, in chemical terms, of the polarity-indicating arrow (Figure 3.4b). This asymmetry must arise from a difference in chemical interaction with the morphogens between new and old cell end-walls. Suppose that old end-wall acts to bring the morphogens to equilibrium (Dirichlet boundary condition), but that new end-wall is first formed without morphogen and, immediately after division, defines a new minimum where the morphogen maximum existed just before division. This pattern of boundary interactions requires the morphogen distribution in the two daughter cells to change from that of Figure 3.4b (divided at its peak) to those of Figure 3.4c. These correspond to polarity arrows both pointing away from the new plate, in accord with the experimentally known rule for division.

The effect of the new cell plate might be simply to act as a morphogen sink. Alternatively, one might follow the suggestion of Mitchison and Wilcox and take the plate to be an instantaneous source of a large amount of the inhibitor Y. Although X and Y are ultimately in phase with each other, sudden introduction of a high concentration of Y at any point would lead to a quick decrease in X, commanding the same polarity change as a morphogen sink. Thus the reaction-diffusion description puts some flesh on the bones of the inhibitor concept. But it must be relatively easy to destroy a peak. Reaction-diffusion models are not all alike in this regard. The Gierer-Meinhardt model, as discussed in Section 3.3.1, does not behave appropriately to explain this division sequence.

Mitchison and Wilcox indicate that the division pattern of *Anabaena* seems to be unique. For contrast, the green algae *Ulothrix* and *Chaetomorpha* (the

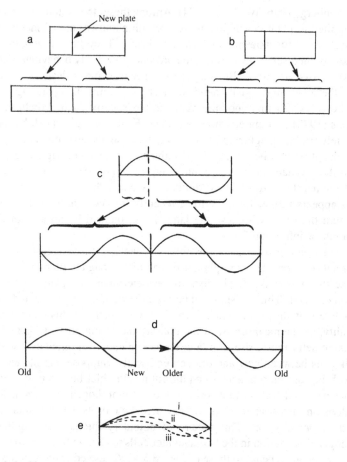

Figure 3.5. *Chaetomorpha* division sequences.

latter under slow-growing conditions) show the opposite pattern of polarity changes: The new cell plate is near to the younger end (Figure 3.5a) (Lück and Lück, 1976). In fast growth, *Chaetomorpha* divides as if the same polarity were maintained for all cells (Figure 3.5b). Of course, unequal division is in general something of a curiosity, equal division being quite common.

Such a range of behaviours imposes no strain upon an algorithmic or "cellular automaton" representation. An algorithm can be written for any division sequence which is meticulously following some rule, whatever that rule may be. But variety in behaviour severely tests any attempt at mechanistic explanation; some variants are going to be much more easily explicable than others. There is, for example, no difficulty in accommodating the sequence of Figure 3.5a within a reaction-diffusion explanation. In both daughter cells, the morphogen maximum remains close to where it was before division. This implies that the new cell plate is only a *weak* sink for morphogen, and doesn't bring its concentration down nearly so markedly as in the case of *Anabaena*, or, on the basis of inhibitor production, that the *Ulothrix* and *Chaetomorpha* cell plates don't produce one. They are morphogenetically neutral, simply chopping the morphogen distribution into two parts, which are unaffected by the cell plate until it becomes old enough to establish the equilibrium boundary condition (Figure 3.5c).

In this connection, for the examples in Figures 3.4 and 3.5 it is immaterial whether, by the time of the next division, the morphogen concentration at the end remote from the event has (1) remained low or (2) already relaxed back to equilibrium (Figure 3.5d). Of course, for an ideal sine wave, (1) implies that a cell will divide at one-third of its length from one end, and (2) implies that it will divide at one-quarter of cell length (Figure 3.5d, left and right, respectively). I am not at this point trying to quantitate the inequality of division. Deviations from the sinusoidal form caused by nonlinearities (in the mathematical sense) in the reaction-diffusion mechanism could easily make quite large changes in the position of the division plane.

Equal division is not at all difficult to account for in the same mechanistic context. If the relation of morphogen wavelength to cell length is somewhat different from that envisaged for the preceding two examples, a cell may not be able to accommodate more than half a wavelength. If one boundary, the older end-wall, already requires this, there is no way that a morphogen pattern can establish itself until the other end-wall permits the same boundary condition [Figure 3.5: (ii) just after division; (i) and (iii) both fit Dirichlet conditions at both ends, but (iii) has too short a wavelength to grow].

The division sequence which seems to be a misfit to this mechanism is that of the optimally cultured and fast-growing *Chaetomorpha* (Figure 3.5b). Retention of polarity in the same sense for all cells of a filament does not correspond either to the asymmetry of the sequence of new end-walls or to the asymmetry of the supposed morphogen distribution as it is chopped up by neutral end-walls. An additional feature seems to be needed in the form of a persistent gradient from which the morphogen distribution arises anew in each new cell. Postulation of such a gradient, for this and many other instances of morphogenesis, has two contrasting strategic aspects. On the one hand, it can appear something of a deus ex machina, which serves only to push the crucial question one stage back. What mechanism controls the gradient? Turing's (1952) objective in his original proposal of reaction-diffusion theory was clearly to show how an apparently uniform system can develop a precisely controlled nonuniformity for which the system has no antecedent other than random disturbances. On the other hand, especially in some multicellular animal systems such as *Hydra* and insects, there is overwhelming experimental evidence for the existence of such persistent gradients. A substantial part of the work of Gierer and Meinhardt (1972; Meinhardt, 1982) has been to show how particular types of reaction-diffusion models can work to build complex structure out of rather simple directional information contained in such preexisting gradients.

My aim in the foregoing account is not to try to establish with certainty that the location of the plane of division of a cell in all or most filamentous structures is determined by reaction-diffusion in particular or kinetic mechanism in general. Clearly, this can neither be proved nor disproved on the basis of what we now know. My aim is to show the reader, and most especially the reader unversed in differential equations, some ways of thinking, with the aid of a few scribblings of sine waves, which will permit the reader to make some assessment of the probability of such a mechanism and, when faced with a variety of experimental behaviours, to assess what part of this range of phenomena can easily be accommodated within a common mechanistic scheme, and what part represents complications or snags for the mechanism.

3.2.2   Location of determinative events within the cell

For filamentous structures, it was possible to beg the question of just where the event takes place which determines the plane of division by supposing that pattern is formed one-dimensionally in the sense of the filament axis, which is often (though not invariably) the long axis of each cell. The next example to be discussed, in Section 3.2.3, is the root of the water-fern *Azolla,* for which Gunning and co-workers (Gunning, Hughes, and Hardham, 1978; Gunning, Hardham, and Hughes, 1978b; Gunning, 1981, 1982) have traced a complete and complex series of cell divisions. This involves a number of types of division of a number of quite different shapes of cells. To discuss the phenomena in relation to theory, we need a more precise postulate in regard to the location of the pattern-forming event.

A plant cell is often a polyhedron, or a cylinder, or other form having the same general property that the three-dimensional region occupied by the cell is bounded by a finite number of two-dimensional faces (some plane, some curved) which join at well-defined one-dimensional edges. A priori, the bulk, the faces, and the edges must all be considered as possible locations of the crucial control processes in cell division. One may, however, assess relative probabilities for these various locations on the bases of both experiment and theory. The well-known identity, in most cases, of the plane of division with the equator of the mitotic spindle (including, for a few animal cells, the observation that rotating the latter rotates the former) focuses attention on the interior of the cell. But for plant cells it has been recognized, since Pickett-Heaps and Northcote (1966) discovered it in wheat, that the plane of division is often clearly indicated in interphase by the appearance of a "preprophase band" of microtubules, long before the mitotic spindle has started to assemble. For the *Azolla* root, Gunning (1982) has stated that the plane (or curved surface) of every cell division in the complex sequence is precisely marked in advance by a preprophase band at the cell surface. The cell plate ultimately joins the existing surface at the midline of this band, to submicrometre accuracy, the band being itself 1.5–3μm wide. When this happens, at the end of telophase, however, the band is no longer present. Its microtubules disappear at the onset of prophase, but some sensitivity to drugs which would be expected to affect microtubules remains in the intervening period. There are also instances of plant cells (other than *Azolla* root) in which no attempt to observe a preprophase band has succeeded. All this experimental evidence leads to the view current in cell biology texts: ". . . the significance of the band to the division mechanism, other than providing indications that *unknown events preceding mitosis* may fix both spindle position and cytoplasmic division in plants, is obscure" (Wolfe, 1981; my italics).

In the sense of the enquiry for locations of morphogenetic control, therefore, all this evidence leaves the cell looking as structureless as it appeared before the invention of the electron microscope or even before the optical microscopic observation of the details of mitosis. All one seems to know surely is something about timing: that the determinative event occurs during interphase, preceding the appearance of the preprophase band. It is, however, reasonable to suppose that the event may be rather closely related to the preprophase band and hence to its location. This takes our attention to the cell surface. Further, Gunning indicates that for all arrays of microtubules in the *Azolla* root, including the preprophase band, first appearance is at edges of the cell, followed by spreading over the faces; see Gunning (1982) and the earlier sources cited there.

From the viewpoint of general kinetic theory, comprising both reaction-diffusion theory and anything else which, mathematically, leads to a formulation as wave equations, the fewer spatial dimensions the game is played in, the better for precision of control. Elsewhere (Harrison 1982) I have called attention to the advantage which a developing system would acquire in reducing the morphogenetic field from two dimensions to one, by using the analogy of the sound of a violin string versus that of a cymbal. This case of acoustic oscillations has only very partial correspondence to that of morphogenetic waveforms as envisaged in reaction-diffusion theory. The sound of a three-dimensional acoustic oscillator is actually somewhat better organized than that of a two-dimensional one. But in reaction-diffusion, as the system goes up in dimensionality from one to two to three, one finds successively more different patterns which will be competing on more or less equal terms by having very similar values of the exponential growth constant k_g. Thus more dimensions implies less chance of achieving simple overall order. Hence the practitioner of this kind of theory is led to make a first attempt to understand the behaviour of a polyhedral cell by regarding its morphogenetic field as most probably a framework of edges, with the faces as a second possibility, and the whole bulk of the cell as an unlikely field for the playing of this game.

In a broader view of mechanistic possibilities for the positioning of the cell plate in each successive division, a question arises which is almost the standard cliché of physical chemistry in approaching any new phenomenon: Is it governed by equilibrium or kinetics? D'Arcy Thompson (1917) discussed the geometry of successive cell divisions in relation to equilibrium configurations of intersecting soap films. That discussion, largely omitted from Bonner's 1961 abridgement of Thompson's work, was revived by Green and Poethig (1982); Goodwin and Trainor's (1980) "field description" of the cleavage process likewise relies upon free-energy minimization as the controlling process. The contrast between the equilibrium and kinetic concepts will be taken up again in Chapter 4. The immediate purpose, in Section 3.2.3, is to show only how one may think about the capabilities of the kinetic explanation, in the particular form of reaction-diffusion.

3.2.3   *Division sequence of the* Azolla *root meristem*

Much of plant development is significantly dependent upon the control of direction in a sequence of cell divisions; but this sequence is more regular and reproducible in some plants than in others. Mosses, liverworts, and some ferns show greater regularity than the higher plants; and among the latter, roots show more regularity than other tissues. Regularity is greatest when a structure is generated from a single apical cell. An example in which the sequence has been determined in great detail is the root of the water-fern *Azolla* (Gunning, 1981, 1982; Barlow, 1984).

The single apical cell is roughly an octant of a sphere, with the curved surface facing outward. In the earliest stages of development, while the future root is still within the shoot, there are one or two divisions parallel to the curved surface (Figure 3.6, *BCD*), one of which produces the root cap initial. But most of the root development arises from a sequence of 50–55 successive cleavages parallel to the flat faces (e.g., *ABC*, producing a pie-slice-shaped cell *ABCA'B'C'* and regenerating the geometry of the apical cell in the smaller version *A'B'C'D*. Each flat face has a division parallel to it in rotary sequence, so that each face is cleaved off 17 or 18 times. This sequence defines a spiral, the chirality of which is not genetically determined. Both clockwise and anticlockwise sequences occur, depending on the orientation of the particular root within the shoot from which it emerges. The rule for which face cleaves next is simple: It is always the oldest of the three flat faces.

If we regard the cell as a framework of six edges which are the possible morphogenetic fields for specification of the preprophase band, the rule is that the three edges bounding the oldest face are inactive, while the other three obey the same rule as applies along the long axis of *Anabaena catenula* (Section 3.2.1). If indeed this identification can be made, *Anabaena* is far from unique, because *Azolla*'s division pattern is the norm for single apical cells. The rule suggests a reaction-diffusion model similar to that proposed for *A. catenula*, but there are complications. Clearly, new morphogen sinks (or inhibitor production) at A', B', and C' would tend to drive the maxima close

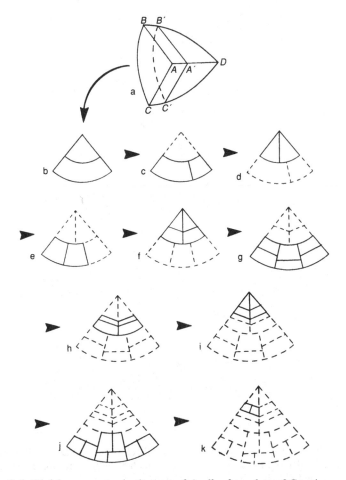

Figure 3.6. Division sequence in the root of *Azolla,* from data of Gunning.

to point *D* on three edges, leading to a chopping-off of a small corner at *D* as the next division. This never happens. To account for the observed sequence, we must assume, first, that the interior corner *A* is always held at equilibrium morphogen concentration by some interaction with the tissue surrounding it. The division plane will then never be remote from point *A* along any of the three straight edges.

Suppose that one pie-slice division has just occurred (Figure 3.7a). The three corners *B'*, *C'*, and *D* are all different in age characteristics, in terms of the two flat faces which meet at each of those corners. In Figure 3.7, these characters are represented by the letters y, m, and o for young, middle-aged, and old. If these dual characters govern the depth or persistence of a corner as a morphogen sink (or its activity as a source of inhibitor), then *C'* clearly has

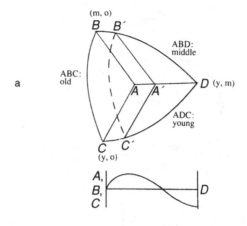

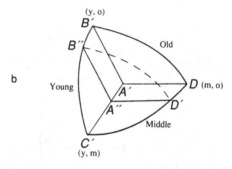

Figure 3.7. Two successive divisions of the apical cell in the *Azolla* root, illustrating the rationale for determination of the next division plane in the sequence from young, middle-aged, and old characters of cell surfaces.

that "young" character in greatest measure, specifying the next morphogen maxima as A'', B'', and D', and the continuing division sequence in like manner as clockwise viewed from within the root.

Gunning noted, and considered it significant, that the cell plate is not exactly planar as drawn in Figures 3.6 and 3.7. It is consistently slightly bent, so that $D'D$ is shorter than $B''B'$ (Figure 3.7b). In the most general terms, this is quite compatible with any theory which accounts for age-sensitive pattern, because $C'B'$ and $C'D$ do not have the same age profile. This means that any kinetic theory is more promising than any equilibrium theory (such as the positioning of cell plates like sheets of soap bubbles, Section 4.3), because kinetics deals with time-dependent phenomena, and equilibrium does not.

Figure 3.8 shows possible time sequences for the reestablishment of reac-

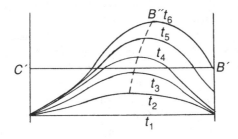

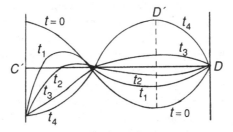

Figure 3.8. Hypothetical time sequences of development of the morphogen profiles along two edges (*C'B'* and *C'D*) of the apical cell of the *Azolla* root in the time interval between the two divisions shown in Figure 3.7. The final states are the same, but intermediate states are not. *B"* and *D'* might be differently positioned along their respective edges if division occurs before the final form has been reached.

tion-diffusion pattern along *C'B'* and *C'D*. In keeping with the spirit of this chapter (and all of Part I), these are freehand sketches, with no attempt at precise computation from any particular reaction-diffusion model. The final distribution, at t_6 in Figure 3.8a and t_4 in Figure 3.8b, is the same for both edges; but the initial distributions are different. *C'B'* starts with a morphogen minimum all the way along, and hence symmetrical end-conditions. The new peak starts to arise in the middle of the edge, but moves toward *B'* because the dual character (y, o) of *B'* is older than that of *C'* (y, m), and hence *B'* starts to relax gradually back toward equilibrium concentration. *D*, however, is already so old that it is keeping concentrations at equilibrium. The change from the premitotic distribution ($t = 0$) to the immediate postmitotic one (t_1) amounts to a decrease in pattern wavelength which should cause the pattern to decay and then recover its original wavelength with inverted form, and quite probably with very little shift in the position of the peak as it grows. Thus, if the new division occurs before the end of the sequences here shown, the peak *B"* may be caught farther away from *B'* than *D'* is from *D*.

Without computation, we have now reached the limit of pictorial conjecture and arrived perilously close to "arm-waving." But pictorial reasoning has pointed the way toward a possible kinetic-theory explanation of even the fine-tuning of the detailed shape of the cell plate.

The division sequence of a "pie slice" is complex. It consists of a series of "formative" divisions, which carve the pie slice into a number of cells of different sizes and shapes. These are followed by "proliferative" divisions, in the plane of the pie slice, converting each cell into a file of cells disposed longitudinally along the root. We shall not be concerned with the latter. The formative divisions, as given by Gunning (1982), are redrawn in temporal sequence in Figure 3.6. At each division, the inactive cells are shown by broken lines, and the dividing cells, with their new cell plates, by solid lines. Before trying to cope with irregularities, I note that division activity tends to alternate between inner and outer parts of the pie slice. This is what one would expect if one started with a system of two cells, one larger than the other, with all parts of the system growing at the same fractional rate and any cell dividing perpendicular to its longer dimension when that length reaches a critical value. Such a sequence is shown in Figure 3.9, with initial cell dimensions "cooked" so that the series continues perfectly regular indefi-nitely. Each cell has its long dimension $2^{1/2}$ times the shorter one. The smaller cell starts out $2^{1/4}$ times smaller (in linear dimensions) than the larger one, and in each time step shown the linear growth is by a factor $2^{1/4}$. Thus, at the start, the larger cell has just divided. At the next step, the smaller cell has reached the same length, and therefore has divided in the same direction. Because of the special shape chosen for each rectangle, when it divides, the two daughter cells have the same shape, but with 90°-rotated orientation. Hence, the next two divisions again alternate as, first, outer half and then inner half, but both are now with horizontal cell plates – and so forth, through the whole series.

A general tendency for a similar alternation of divisions of inner and outer parts, and of horizontal and vertical divisions, is evident in the experimental data of Figure 3.6. The overall size increase is, however, not as large as that shown in Figure 3.9. Evidently, in the later stages of the *Azolla* sequence, somewhat smaller cells are dividing than in the earlier stages. This is quite compatible with reaction-diffusion theory. In a symmetrical cell division, we are supposing that the division plane is specified by the crest of a morphogen wave in which the long dimension of a cell is matched by a half-wavelength of the pattern with a Dirichlet boundary at each end (Figure 3.1c) or a full wavelength with a Neumann boundary at each end (Figure 3.11d, inverted). The wavelength, however, need not stay constant through a long sequence of divisions. As discussed in Section 7.3.2, wavelength can be sensitive to precursor concentrations. If in the course of a division sequence the supply of precursor improves, the wavelength is expected to fall, so that smaller cells will indeed divide. Figure 3.10 shows the same division sequence as Figure 3.9, but instead of the system growing by a factor $2^{1/4}$ in each time step and cells dividing always at the same critical length, the cells have grown by $2^{1/8}$, and the wavelength, or critical length for division, has fallen by the same

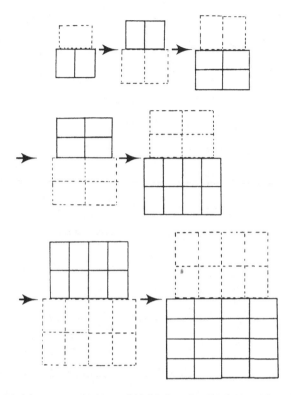

Figure 3.9. Division sequence for an idealized sequence of rectangles with ratios of $2^{1/2}$ and $2^{1/4}$.

factor inverted, $2^{-1/8}$. This gives a better match to the actual size changes between division steps in Figure 3.6.

The experimental sequence is of course distinctly more irregular than my idealized example. Even the example of two unequal rectangles as the starting configuration would have given a less regular sequence if I had not idealized to the extent of very careful choices of the shape of each rectangle and the size relationship of the two initial ones. I am not at this point trying to get a perfect fit between reaction-diffusion theory and every detail of the *Azolla* sequence. I am seeking only to show that the theory can be applied and its capability assessed at some length and in some detail, at least as deeply as any rival theory, and without having to plunge into mathematics.

The only notably asymmetric division is that in Figure 3.6c. It does not correlate well with the concept of control by some residuum of the polarities in the apical cell from which the pie slice arose. The asymmetry shown corresponds to a sequence in which face *ACD* cleaved off immediately before *ABC*, and therefore *C* should have the more "morphogen-negative" character.

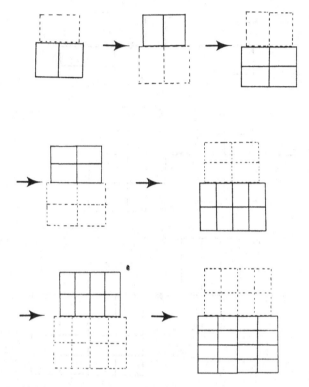

Figure 3.10. Division sequence for an idealized sequence of rectangles with ratios of $2^{1/8}$ and $2^{-1/8}$, giving a close match to the actual size changes for the *Azolla* root, as shown in Figure 3.6.

This would seem more likely to direct the division plane to left of centre. At the last stage of the sequence, Figure 3.6k, Gunning has pointed out evidence for control on the multicellular scale. Of the two side-by-side cells apparently eligible to divide simultaneously, only one does so. Further, this happens in only two of the three pie slices that have reached the same stage, and in such a way that the two dividing cells are opposite and the overall symmetry of the root is reduced from threefold to twofold.

In Section 4.3 this example is taken up again to compare the kinetic explanation with the possibilities of an equilibrium explanation.

3.3   Animal development: response to damage and grafting

3.3.1   *Teratogenesis and nonlinearities in the Gierer–Meinhardt model*

Elongated systems often show an arrangement of parts, and an ability to regulate the formation of these parts, which suggests control by an ante-

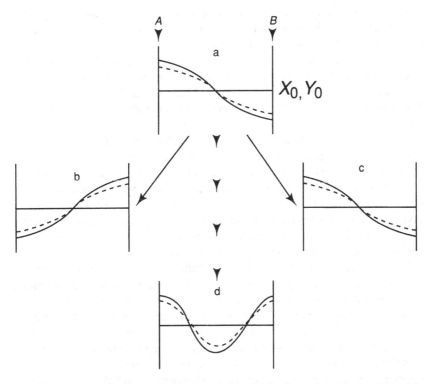

Figure 3.11. Linear Turing waves: concentration distributions of morphogens X (solid line) and Y (broken line) along an elongated system. (*a*) Initial state. (*b* and *c*) Possible responses to local morphogen removal at one end. (*d*) An impossible response to this damage.

roposterior gradient. A simple kinetic model for the establishment of such a gradient is the linear Turing model with no-flux boundaries at the anterior and posterior extremities and just half a wavelength fitting the system length (Figure 3.11a). Such a pattern could be the preferred one, that is, the pattern of highest k_g, for systems varying in length by as much as a factor of 5 (Lacalli and Harrison, 1978) (see Section 10.4).

Suppose that some damage is done to the developing system in a small region at one end, such that the morphogens are destroyed. Later, following Meinhardt (1982, chap. 8), I shall assume that in early insect development the inhibitor Y is preferentially destroyed both by ultraviolet (UV) irradiation and by puncturing the end of the embryo. In the latter case, the inhibitor escapes faster because it is always the faster-diffusing of the morphogen pair. The destruction of morphogens is analogous to the putative effect of the new cell plate, as discussed in the preceding section. For a linear Turing wave, there could be only two responses to such damage. If it were at the peak (or "activated") end of the morphogen gradient (A in Figure 3.11), and severe enough, the gradient would simply reverse (Figure 3.11b). If it were at end B, or at end *A* but insufficiently severe, the gradient would be unaltered (Figure 3.11c). Response (b) would tell us at once that *A* was the activated end. Response (c) would not answer that question.

The response which would never occur is that of Figure 3.11d, teratogenesis giving a gradient with a peak at both ends and hence a mirror-imaged embryo, with either a head at both ends or an abdomen at both ends. The Turing wave "knows" its wavelength and is not about to put one wavelength into the system where half a wavelength is preferred. But this unexpected response, always in the double-abdomen form, never the double-head, is exactly what happens in response to UV or puncture damage at a very early stage of insect embryogenesis (Figure 3.12). Meinhardt (1982) has given a carefully ordered account of the experimental evidence and of how his nonlinear model accounts for it. Here, I do not seek to repeat his account, but to give the gist of how his model works, setting this in the context of the linear model and other nonlinear models, especially the Brusselator. The comparison is taken up again, with mathematics, in Chapter 9.

The Brusselator, when taken in computations to a steady state in which concentrations have altogether stopped changing in time, seems to retain some of the properties of the linear model, almost miraculously. It still "knows" its linear wavelength. If the system is disturbed, the pattern responds as a Turing pattern would. If a pattern has been computed to steady state, and the system is stretched to a length at which the preferred Turing pattern would be more complex, the Brusselator pattern starts at once to change into that more complex one (Lacalli, 1981, and previously unpublished calculations in my laboratory; Section 9.1.5, Figure 9.4).

The Gierer–Meinhardt model behaves quite differently. As mentioned in Section 2.6, it tends to produce concentration profiles in the form of "isolated mountains in the middle of a vast plain." That is, the peaks and troughs are different in instantaneous shape and dynamic development, and the pattern does not look the same, statically or dynamically, if it is turned upside down. A Turing pattern does, except for the matter of end-to-end reversal, as between (b) and (c) in Figure 3.11. Its peaks and troughs have exactly the same shape, and at any instant the dynamics of the downward movement of the troughs exactly match those of the upward movement of the peaks. The "one-morphogen" optical-resolution model shows a similar property (Figure 2.4) continuing into its nonlinear regime (Figure 2.4c). This is because upward and downward movements on the diagram correspond to the two possible optical resolutions, as D and L. The symmetry of dynamics corresponds to a feature of the structural symmetry of the system. This is why the optical-resolution model is a useful introduction. But the symmetry of Figure 2.4 is imposed by an initial sine-wave input. The model has no permanent pattern-forming ability. Crests and troughs will develop differently if they are different in the initial input.

Brusselator patterns do not look the same turned upside down, but the

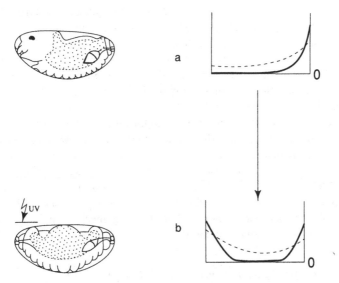

Figure 3.12. The double-abdomen deformity in early insect embryogenesis. Modified from Meinhardt (1982, Fig. 8.4), in which the experimental drawings refer to the midge *Smittia*; results of Kalthoff and Sander (1968), redrawn from Kalthoff (1976). (*a*) Normal development. (*b*) Development with two mirror-imaged abdominal ends and no head, following UV irradiation at the cellular blastoderm stage, before the structures identifying head and abdomen had started to appear. The graphs are the corresponding distributions of the Gierer–Meinhardt activator *A* (solid lines) and inhibitor *H* (broken lines). These correspond, respectively, to Turing morphogens *X* and *Y* in the linear regime. See also Slack (1983, chap. 9).

distortions from linear waveforms are only moderate and, as already mentioned, do not significantly change the regulatory properties of the patterns. The flat plains of the Gierer–Meinhardt patterns do, however, change these properties, by effectively limiting communication across the system. This is not immediately apparent from the differential equations of the model, which, like the Turing equations or Brusselator equations, have diffusion terms for both morphogens, with no spatial limitations other than the boundaries of the system. What happens in practice, however, is that a peak tends to suppress the formation of others in its immediate vicinity, but has very little interaction with other peaks beyond a certain range. Thus the concept of "chemical wavelength," as it arises from the Turing model, is replaced by the concept of the "range of the activator." This corresponds to much looser control of positions of activator peaks, such that teratogenesis can be described in terms of the formation of new peaks.

For this to happen, there must be a way in which new peaks can rise up out of the plain in response to some kinds of damage. Any explanation of this requires some mention of the special mathematical form of the nonlinearities of the Gierer–Meinhardt model. The essential feature is that there are two morphogens, activator A and inhibitor H (roughly equivalent to Turing X and Y), and the rate of growth of activator A contains a term proportional to A^2/H. This, obviously, is extremely sensitive to changes in the inhibitor concentration H when that is very low (i.e., down on the plains). Thus, if anything happens in a local region of the plain to destroy H preferentially, there can be a very rapid production of A in that region. This may be sufficient to establish a new peak.

The effects of damage to the system are thus quite different for the Turing model and the Gierer–Meinhardt. In the former, the end at which damage can bring about a change is the activated end, and the resulting change is end-to-end reversal of pattern. This was illustrated in the preceding section. For the Gierer–Meinhardt model, the damage-sensitive region is the inhibited one, and damage introduces new peaks without removing old ones. Thus it is possible for the anteroposterior-gradient-forming mechanism in insects to identify the posterior end as being the activated one.

3.3.2   *"Firing" a persistent gradient: grafting behaviour* of Hydra

Turing's objective was to show how a uniform system can self-organize into nonuniformity with nothing but random disturbances to provide rudiments for the nonuniformity. The pattern-forming mechanism decides which of these rudiments will be amplified. For a half-wave pattern like that of Figure 3.11a–c, in individual organisms using a Turing mechanism the pattern would always form, but equally often in the alternative orientations of (b) and (c).

In the developmental sequence of a living organism, a pattern-forming mechanism very often will be operating not upon uniformity, but upon a system with gradients built into it by preceding developmental events. The previously graded property may be persistent and unalterable by the new event which is about to occur, and it may lead to a gradation along the system in the initial rate of supply of activator to start the pattern-forming mechanism. Here, "unalterable" does not imply an infinite time scale. A morphogen prepattern may be envisaged as something which can form, or change from one form to another, in a period from a few minutes up to a few hours. Anything which needs more than a day to alter substantially is unalterable with respect to that morphogen prepattern. For instance, a pair of morphogens may react on the molecular scale in a pattern-forming reaction-diffusion mechanism. But the activator may then induce cells to differentiate, over a longer time period, into a form which manufactures more of the activator (and perhaps secretes it, if the morphogenetic region for the reaction-diffusion mechanism is extracellular). Thus a gradient of cell types could be set up along a tissue. There are two positive feedback loops in such a system: molecule-to-molecule autocatalysis of the activator (fast), and molecule-to-

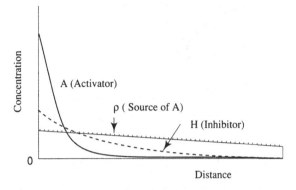

Figure 3.13. Steady-state distributions of activator and inhibitor produced along with an elongated system by operation of the Gierer–Meinhardt model upon a shallow source gradient. From Gierer and Meinhardt (1972), with permission from Springer-Verlag.

cell-to-molecule feedback leading to increased activator production (slow). The latter may be neglected for the time scale of the former, and the gradient of cell types regarded as fixed.

This picture of morphogenesis raises some important issues which are enlarged upon in later chapters: Does every pattern-forming process build upon a preexisting shallow gradient, so that in some sense the whole final shape is present ab initio? (see Chapter 5.) Are the basic "particles of the system" to which pattern-forming dynamic equations refer molecules or cells, or both together in the same mechanism? (see Section 4.4, anticipated in Section 1.3.2.) Is the observation that a particular substance induces differentiation to be regarded as evidence that the substance may be a morphogen? (see Section 10.3.1.)

Whatever may be the answers to these questions, the value of seeing what a morphogenetic amplifier can do to an initial shallow gradient is undeniable, and that was Gierer and Meinhardt's initial (1972) objective. They showed how to construct a two-morphogen model (substances A and H) which would work upon a shallow, fixed, straight-line gradient of sources of A (ρ, Figure 3.13) and convert it into the now-familiar spiky peak and long flat plain. This they called "firing" a gradient. Their models all contain a term A^r/H^s in the rate of activator production. The analysis shows that a gradient will be fired only if $r > 1$, so that if r is an integer it must be at least 2. In most subsequent work they have used the form A^2/H. The parallel between this and the Mills (1932) proposal for optical resolution is quite striking (see Mills' explanation as quoted in Section 5.3.1).

The predictions of the Gierer–Meinhardt model about the behaviour of a

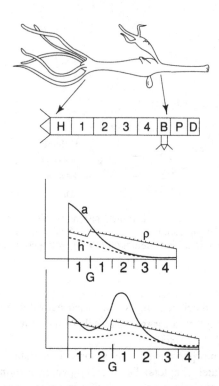

Figure 3.14. Schematic of a *Hydra* grafting experiment; *a,* activator concentration; h, inhibitor concentration; ρ, source gradient. A graft containing the region immediately below the hypostome (region 1) fails to give a second whorl of tentacles if it is grafted on immediately below another region 1. But if it is grafted below a region 2, a second whorl of tentacles forms. This illustrates the Gierer–Meinhardt model's ability to amplify the top end of a gradient behind a discontinuity exceeding some threshold value. From Gierer and Meinhardt (1972), with permission from Springer-Verlag.

system in response to disturbance may often be arrived at pictorially with the aid of a few simple rules which characterize the particular idiosyncratic behaviour of this model and are not necessarily applicable to any other nonlinear model. As for any pattern-forming model, one prediction cannot be made without mathematics: the number of peaks which will appear when the pattern first arises out of uniformity or out of a shallow straight-line gradient. This needs a calculation of the "chemical wavelength" of the model in its linear Turing regime. (I am avoiding using the word "linear" for the gradient because of the danger of confusion with the quite different major meaning of "linear" in this topic, as applied to dynamic equations.) If, however, we have good reason to believe that we are dealing with a chemical pattern with only one peak along the length of the system, predictions can be made regarding the changes upon disturbance:

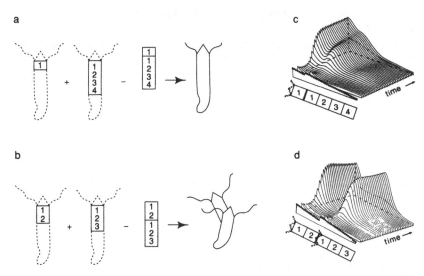

Figure 3.15. The same grafts as in Figure 3.14, differently illustrated. The block diagrams show concentrations of activator *a* only, but as functions of both distance along the organism and time. From Meinhardt (1982), with permission.

1. An existing peak is not easily destroyed by removal of the morphogens from any small fraction of its total extent.
2. An existing peak prevents formation of a new one within a certain "range of lateral inhibition"; beyond this range, positions at which new peaks may form are not precisely correlated by any "chemical wavelength."
3. Two possible initiators of new peaks are (a) local destruction of inhibitor somewhere in the nonactivated region, as discussed for insect teratogenesis in Section 3.3.1; (b) production of peaks in the source gradient, for instance by grafting of tissues into unnatural juxtapositions.

Here an additional meaning appears for persistence of a gradient. Its changes must be slow compared with the morphogenetic event which amplifies and expresses it, not only in the natural state but also after grafting, so that pieces which have been moved around and rejoined will retain their gradients. There is extensive experimental evidence for such persistence of gradients both in insects and in *Hydra*. One example from the latter serves well to illustrate the explanatory or predictive power of the Gierer–Meinhardt model, and at the same time to clarify pictorially the extent of the range of lateral inhibition mentioned in rule 2. Experimentally, if a *Hydra* is decapitated just behind the hypostome, leaving the abdomen intact (cut between H and 1, Figure 3.14), and another *Hydra* is severed part way back in the abdomen and the anterior part is retained, then grafting these two pieces together may or may not lead to teratogenesis by production of a second tentacle whorl at the graft (Figures 3.14 and 3.15). Success in forming a second whorl depends on how far back the graft is from the retained head end. Here, by contrast to the previous discussion of insect embryogenesis, the Gierer–Meinhardt model works on the assumption that the head is the activated end and that a whorl of tentacles will arise wherever the model shows an activator peak forming.

4

Structure, equilibrium, kinetics

What is the ultimate reality in terms of which scientific explanations should, at the deepest level, be written: matter or motion? Scientific philosophy has most commonly asserted that it is the latter. D'Arcy Thompson (1917) quoted instances of this starting from Aristotle's concept of the "efficient cause" and proceeding by way of Newton to a statement by F. G. FitzGerald that "all explanation consists in a description of underlying motions." But in the same chapter, Thompson indicated that an accepted axiom of biology at that time was that "function presupposes structure." And he also stated that "the overwhelming progress of microscopic observation has multiplied our knowledge of cellular and intracellular structure; and to the multitude of visible structures it has often been easier to attribute virtues than to ascribe intelligible functions or modes of action." This seems curiously modern for something written long before theories of motion had arrived at the wave nature of the electron and so made possible the structural revelations of the electron microscope.

Thompson could have had no inkling that such "overwhelming progress" was only the beginning of the possibilities of microscopy. He envisaged, rather, that there would soon be a concentration of effort on dynamics, in which "the *things* which we see in the cell are less important than the *actions* which we recognize in the cell." In the event, the infant microscopy of 1917 metamorphosed into an adult capable of finding more things in the cell than were dreamed of in Thompson's philosophy. The account from which I have quoted (Chapter 4 of *On Growth and Form*) was considered by J. T. Bonner to be so completely out of date that he omitted it entirely from his 1961 abridgement. Yet the first few pages of that chapter express very well the philosophy of kinetic theory to which this book is addressed. Despite the glories of modern microscopy, there are senses in which the cell remains as empty as it seemed in 1917, as I have pointed out in Section 3.2.2. An old book should be read to perceive, through the deadfall of a million trees of detail, the living forest of an enduring philosophy. Some of its best seedlings may be concealed in the deadfall. Do not abridge.

In Vancouver and in England, my two homes, the rain that is forecast for tonight often falls on us tomorrow; I am familiar with predictions which give the sequence correctly but get the timing wrong. I still expect that the concentration on structural studies, so prolonged by the electron microscope, will be followed by attention to dynamics, as Thompson expected, but in the twenty-first century. Meanwhile, we have to cope with the full flowering of the structural approach along with the first sprouting of the dynamic. (A somewhat parallel instance of mistiming was Malthus's indication of the perils of the population explosion, an outcome apparently invalidated but really only delayed by the remarkable technological improvements in transport which made possible the feeding of a much greater population, for a while. Exponential growth of anything, whether a chemical concentration, or the size of an organism, or a population, or the scientific literature, must always reach a limit beyond which it will turn into something else.)

Elsewhere in the same work, Thompson explored in some detail the notion that the forms of individual cells and of the partitions between cells in tissues are governed, like those of soap bubbles, by a thermodynamic drive toward minimum surface free energy, that is, toward equilibrium. Some of the crucial diagrams in that account, showing the expected shapes for successive cell divisions within a disc, were also omitted from Bonner's abridgement, but were revived by Green and Poethig (1982); see Sections 4.3 and 4.4 for modern uses of this concept in relation to both the static arrangements of plant cells and the dynamical sorting-out of animal cells.

In the remainder of this chapter, I try first (Section 4.1) to give definitions of structural, equilibrium, and kinetic theories of development. Sections 4.2–4.5 discuss particular examples, not to show that any one of them belongs exclusively to one of the three categories of theory, but rather to analyze what *aspects* of the development belong to each of the categories. Crystal growth is treated first because it most clearly and unequivocally involves all three.

4.1   Definitions of the categories

Any phenomenon of the physical universe, in material living or otherwise, may be looked at from a great variety of viewpoints, and as arising from a complex etiology of contributing factors. These may perhaps be put together in a formidable multiterm equation. But often it is possible to ignore many of these terms and concentrate upon one or two which give an essentially full account of the phenomenon in the simplest way. We tend to describe the view thus obtained as "*the* explanation" of the phenomenon. When a theory has not yet been scientifically established by the test against experiment, such application of Ockham's razor is usually necessary to permit the scientific method to proceed. It leads to a separation of categories.

4.1.1   Structure

In the most penetrating view of the ordinary matter around us, everything is mechanics: movement of particles in force fields, and their collisions with

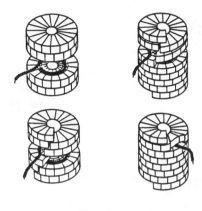

Structure

Figure 4.1. The self-assembly of the tobacco mosaic virus, illustrating the generation of form by structural fitting together of parts. From Klug (1972), with permission.

each other. In such terms we explain the formation of molecules and of the solid, liquid, and gaseous phases. But if we wish, for example, to determine the conformation to which some large molecule will fold up, we often think of the problem as a geometrical one of "how pieces fit together," rather than as a mechanical one of "how the net force on every particle may be reduced to zero." In using such concepts as atomic radius and steric hindrance, we replace the continuous variation of force with distance between objects by the concept of idealized hardness: Down to a certain approach distance, repulsive force is zero, and at that threshold it becomes effectively infinite. The objects are then "in contact." This is a crude approximation, but it serves us very well very often. Without it, we could not use in any simple way the concept that an object has a size and shape. The fitting together of parts in this approximation is what I call the *structural approach*. Forces have essentially disappeared from the discussion, which is all in terms of geometry.

At the supramolecular scale of, for example, viruses (tobacco mosaic virus, Figure 4.1) (Klug, 1972; Lehninger, 1975), this is the concept of development of form traditionally designated in biology by the term "self-assembly." Both this term, however, and the similar term "self-organization" are often used in ways which overlap the three categories of physicochemical theory which I am here seeking to distinguish.

4.1.2 Equilibrium

When a solution, cooled below a certain temperature, separates out into two liquid phases, or when a mixture of cells in culture from two tissues of an embryo sorts out into an aggregate of cells of one type enveloped by an

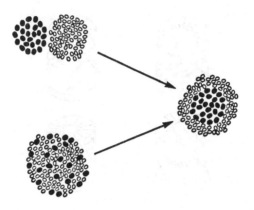

Equilibrium

Figure 4.2. An idealized drawing of two types of experiments, as used in the work of Steinberg (1970) on cell sorting (see Section 4.4.2). The upper arrow indicates an engulfment experiment in which two pieces of different tissues in culture are threaded together on one skewer. One piece moves to engulf the other. The lower arrow indicates a cell-sorting experiment in which the two tissues have been disassembled (by proteolytic enzymes) into separate cells, which have been randomly mixed. They move to form two spheres, one inside the other. The fact that for the same two tissues both experiments almost always give the same result was cited by Steinberg as an indication that that result is most probably an equilibrium structure.

aggregate of cells of the other type (Figure 4.2, Section 4.4.2) (Townes and Holtfreter, 1955; Steinberg, 1970), then we tend to think in terms of the approach to equilibrium, minimization of free energy, rather than of geo-metrical fitting or mechanical forces. Indeed, the geometry now has some indeterminate aspects. Minimization of free energy, including surface and interfacial free energies, requires that one of the two immiscible systems must envelop the other, in a specific order, and that each must have a surface in the form of a single sphere; but there is no requirement that the two spheres be concentric.

(It is, however, possible to switch back and forth rather readily between the thermodynamic description of a surface in terms of excess free energy and the mechanical description in terms of a force, the surface tension. If each is done properly, the two approaches are equivalent. This illustrates that assignment of the explanation of a phenomenon to a category is not always unique.)

4.1.3 Kinetics

It is sufficiently evident that if material is transported to and deposited at some places and is not transported to others, a shape is going to arise in the distribution of that material. This shape is an immediate result of the trans-

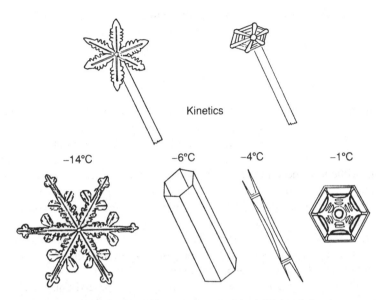

Figure 4.3. The shape of a snowflake is governed by how much its temperature of formation differs from the equilibrium temperature (about 0°C). This shows that kinetic factors dominate in generating shape. Top drawings: a needle growing at −4°C and then transferred to −14°C grows a star on its end; when changed to −1°C it grows a plate. Modified from Mason (1961), with permission.

port, which is therefore always in some sense among the causes of the shape. But very often we do not view the transport as being a really significant component of the etiology of the shape (in Aristotle's terminology, an efficient or formal cause). Similar transport processes for stones might give rise to a wall, Stonehenge, an Egyptian pyramid, or a Gothic cathedral. The essential explanation of which one will arise lies in the architect's plans and the mental processes which gave rise to them. Transport is essential to the production of shape, but in such instances trivial in the determination of what shape will arise.

I reserve the term *kinetic explanation* for one in which the transport processes and the on-the-spot production and removal processes of chemical reactions themselves intrinsically fulfil the function of the architect's mind: They determine what shape is to arise. No example is needed here because most of the chapters of this book are devoted to possible examples.

The essence of the contrast between the equilibrium and kinetic categories is that in the latter the system moves *away from* equilibrium. One of the most promising lines for conducting a "crucial experiment" is to devise a test for whether or not form can be affected by distance from equilibrium. For the shapes of snowflakes (Section 4.2, Figure 4.3), where there is only one simple phase equilibrium to think about, the test proves crucial and gives the answer that form is kinetically determined. But living things have thousands of simultaneous departures from various physical and chemical equilibria, and a similarly clear-cut test is difficult to devise.

4.1.4   Mechanochemistry

To reiterate: At a sufficiently fine level of subdivision of matter, all explanation is in terms of forces and motions. Often we avoid explicit discussion of these in one of two ways. Structurally, the biologist can explain many things quite well by stopping at the stage of static molecules and their geometrical fitting, without the need to consider molecular motions or to resolve the static molecules into moving electrons and nuclei. In both equilibrium and kinetic approaches, the macroscopic scale is so huge compared with the molecular that forces and motions can be blurred into continuum properties, such as energy, entropy, diffusivity, concentration, and kinetic rate constants. One may pursue reaction-diffusion theory a long way by knowing that the diffusivities of small molecules in aqueous solution are of order 10^{-6} cm^2 s^{-1}, and that those of membrane-bound species are of order a thousand times smaller, without seeking to explain those orders of magnitude in terms of molecular collisions and suchlike.

There are, however, some experimental phenomena in which the existence of forces and motions is brought directly to the attention of the observer, and it appears a priori that they must be included explicitly in the explanations. Most obvious among such instances are the phenomena in which a macroscopic object undergoes change because of the movement and rearrangement not of molecules but of pieces which are themselves large enough to be macroscopic and individually observable. In biological systems, these pieces are often the cells. Oster (1983) has written that "anyone who has watched time-lapse movies of developing embryos cannot help being astounded by the almost miraculous way they pulse, jerk and heave about, as they gradually shape themselves into complex geometries and tissue configurations. . . . While it is clear that these formshaping movements – which are very essence of morphogenesis – must be driven by mechanical forces, until quite recently the origin of these forces was a complete mystery."

I have been struck with exactly the same impression, as, for example, on seeing such a movie of gastrulation in the nematode *Caenorhabditis elegans*, shown by von Ehrenstein (1980) at the Second International Congress on Cell Biology. In this process, two cells on the boundary of an elongated cell mass have become visibly different from the rest (Figure 4.4a), and they move into the interior (where they start to divide to form the gut) by a series of almost imperceptible nudges and jostlings in which no cell ever seems to make a single jump anything like as large as its own size. This aspect is strikingly similar to the same aspect of the plastic deformation of crystals. The latter process occurs more than a million times more easily than it should (in terms

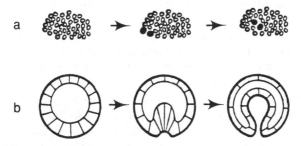

Figure 4.4. Two kinds of gastrulation, to illustrate a type of phenomenon in which the existence of mechanical forces is rather obvious, and to which the mechanochemical division of kinetic theory is likely to be relevant: (*a*) Gastrulation in nematode worms (and many other lower invertebrates), sketched from memory of the film shown by von Ehrenstein (1980). Two cells on the edge of the mass differentiate and move inward. From these, the gut then starts to form. (*b*) Gastrulation in echinoderms, *Amphioxus*, and elementary textbooks, which have some tendency to convey an erroneous impression that this idealized geometry is to be seen in the corresponding process in higher vertebrates. These are schematic cross-sections through a hollow spherical shell of one layer of cells.

of threshold shearing force and its relation to modulus of elasticity) if one molecular plane had to slide over another in a concerted movement of the order of one intermolecular spacing at a time. The theory of dislocations, originating in the simultaneous and independent papers of Taylor (1934), Orowan (1934), and Polanyi (1934), explains this in terms of a defect which is really a geometrical fault of the whole crystal but which appears to be a localized region of disorder. In a movie, one would see this defect work its way right across the middle of the crystal, with the result that the top half would finally slide one molecular spacing over the bottom half. But again, as for the nematode gastrulation, the whole mechanism would seem to involve some sleight-of-hand. No molecule ever makes an individual movement of more than a small fraction of the intermolecular spacing, and thus it does not have to overcome the energy barrier of such a jump. In the hands of metallurgists, the theory of dislocations has become firmly established and extensively elaborated over half a century. It seems to me that there are opportunites for the application of some themes from this body of theory to many of the movements within cell masses which occur in animal development.

Up to the present time, this has not been done; but Oster and collaborators (Oster and Odell, 1980; Odell et al., 1981; Oster, 1983; Oster, Murray, and Harris, 1983) have developed theories of the mechanochemical basis of morphogenesis. Although a model in this category had been devised for *Acetabularia* by Goodwin and Trainor (1985), most of this work has been relevant to phenomena in the animal kingdom which have no counterpart in plant morphogenesis. Where they refer to cells, in apparently very general terms, it is useful to remind oneself that often the word "animal" should be inserted in front of "cell."

These papers describe a variety of complex concepts, but the gist of the mechanochemical interaction as a mechanism parallel to reaction-diffusion and as another subset of kinetic theory may be shown by a simplified abstract from Oster (1983). The contractile mechanisms of cytoplasm in general, involving actin and other associated proteins, are not so well known in detail as is the actin-myosin-troponin-tropomyosin-myosin-ATPase-calcium mechanism in muscle cells, but are generally believed to be similar and are known to involve control by calcium concentration. Now intracellular calcium is subject to an autocatalytic phenomenon known as "calcium-stimulated calcium release," in which an increased calcium concentration in turn increases the rate of its release from calcium-sequestering vesicles. If mechanical stretching can deform the cell membranes so as to allow a leakage of calcium, either from outside across the plasma membrane or from the calcium-sequestering vesicles into the cytoplasm, then we have at least the equivalent of a one-morphogen mechanism for unstable pattern analogous to the reaction-diffusion mechanism for optical resolution. There is autocatalysis; and, without the need for invoking calcium diffusion, there is long-range transfer of a stimulus tending to even out the calcium to a high level everywhere, in the form of the mechanical stress. This becomes the analogue of the long-range communication by diffusion. To stabilize pattern, there is no analogue of the inhibitor in reaction-diffusion theory, but rather a strongly nonlinear behaviour of the system of mechanical stress.

In some instances, however, the formulations of the mechanochemical theory include diffusion terms. In the case of mesenchymal morphogenesis, in which migrating cells often form clusters with definite geometries and controlled spacings between them, the randomly moving "particles" for which diffusion terms appear in the equations are the cells themselves (Oster et al., 1983). In such cases, which, as these authors suggest, may include feather primordium initiation and chondrogenesis, the mechanical terms can, just as for molecular motion, disappear from explicit treatment and be replaced by diffusion parameters.

The concept of cell-as-molecule is a recurring theme in the topics of sorting-out and aggregation. The analogies are strong and useful, but also have limitations. This theme is taken up again in Section 4.4.

4.1.5 Semantics of the term "field theory," and electric fields

Morphogenesis is a field of study; as a developmental phenomenon, it occurs within the boundaries of morphogenetic fields; Goodwin and Trainor (1980) proposed a field description of the cleavage process in embryogenesis; and the work of Jaffe is largely concerned with the possible role of the electric field as a long-range communicator in morphogenesis (self-electrophoresis) (e.g.,

Jaffe, 1982). The four usages of the noun "field" in the preceding sentence are clearly different, but have sufficient overlap in meaning that possibilities of confusion arise. Of the seven numbered definitions of the uncombined noun "field" in the latest edition of the *Concise Oxford Dictionary,* just one (number 6) covers all four of the preceding usages: "area or sphere of operation, observation, intellectual activities, etc., . . . region of electric, gravitational, magnetic, etc., influence, presence of such influence, force exerted by it on standard object."

As an item in the terminology of physical theory, the word "field" is intimately related to the concepts of "action at a distance" and "delocalization of matter." General acceptance of the notion that objects not connected to each other could exert forces on each other arose principally through the work of Kepler and Newton, and that notion was a main feature of the Scientific Revolution in the seventeenth century. This concept inevitably entails that of a dual nature for a particle: as a highly localized object, and as an influence extending through the whole of space. This duality is not something new to the quantum mechanics of the twentieth century (wave–particle duality). Physics has had to live with it for three centuries, and keeps on finding new resolutions for the apparent conflict. In the present century, two contrasting views have arisen: a completely particulate view, in which interparticle forces arise from exchange interactions between them which quite replace the "field" concept; and "field theory," in which everything is fields and there really are no localized particles. In the words of Dodd (1984):

In the most sophisticated form of quantum theory, all entities are described by fields. Just as the photon is most obviously a manifestation of the electromagnetic field, so too is an electron taken to be a manifestation of an electron field and a proton of a proton field. Once we have learned to accept the idea of an electron wavefunction extending throughout space . . . it is not too great a leap to the idea of an electron field extending throughout space. Any one individual electron wavefunction may be thought of as a particular frequency excitation of the field and may be localised to a greater or lesser extent dependent on its interactions.

Physicists such as Trainor are not seeking to introduce quantum mechanics into the field (that word again!) of biological development. But it is in the context of that fashion of twentieth-century physics that the term "field description" is being used by Goodwin and Trainor. Their strategic objective is to try to set up theories of morphogenesis *in the first instance* in terms of the mathematical properties of a very generalized abstraction, a "field" of undefined physical nature, and to proceed step by step toward what that field might be in more concrete terms. In the first instance, they appeared to envisage fields in which a process is operating for which a mathematical minimization is appropriate, such as the minimization of free energy. Thus the concrete representation of the field would involve the free energy, and ultimately the structure, of the cell surface – something not far away from the soap-bubble analogies of D'Arcy Thompson (1917), as revived by Green and Poethig (1982). More recently, via personal communications and in unpublished lectures, Goodwin and Trainor have suggested that reaction-diffusion and the mechanochemical theory might come under the all-sheltering umbrella of field theory. To my mind, we should not be using the same term to include theories of pattern formation as *approach to* equilibrium and theories of pattern formation as *departure from* equilibrium (kinetic theory). If the term "field theory" is going to obscure that distinction, its use is not conducive to progress in the interaction between theory and experiment, and it should be avoided. Most of the theory which the advocates of this term have been using could be restated very similarly except for the omission of the term.

Self-electrophoresis is, like reaction-diffusion and the mechanochemical theory, a division of kinetic theory. It has not been under development as long as reaction-diffusion, nor pursued by so many workers. But the whole of kinetic theory remains so generally foreign to developmental biologists that the relative popularity of its various branches cannot yet be taken as any index of eventual promise.

The concept of self-electrophoresis, or a role for the electric field in pattern formation, starts from the experimental knowledge that cell membranes are often furnished with two kinds of structure relevant to the transport of simple ions, such as Na^+ or Ca^{2+}, across them: passive leaks and active pumps. Second, both in some fully developed structures (frog skin) and in a variety of developing systems, both animal and plant (Figure 4.5), the pumps and leaks are spatially segregated, often to opposite ends of a cell. This pattern of distribution leads to the possibility of closed loops of electric current, which have been observed. A current arises because of the existence of a potential gradient (voltage gradient), otherwise known as the quantitative measure of the electric *field*. It is evident that where all this is known to exist, there is a possibility of directional movement of various charged species toward or away from particular locations. This is self-electrophoresis. But do we have in all of this a pattern-creating mechanism? That is, are the observed nonuniform distributions of ion pumps and ion leaks self-organizing, with the electric field playing some role in the process? Or are the distributions of pumps and leaks set up by some quite different mechanism, such as reaction-diffusion, mechanochemistry, self-assembly, or whatever, which does not invoke self-electrophoresis as a process capable of setting itself up ab initio?

The answers to these questions are not yet known. There is, however, some information relevant to important features of self-electrophoresis as a possible self-organizer. First, are the fields big enough to act as the principal long-range communicator, replacing diffusion or mechanical stress? The problem here is that the important self-organizing region, the cell membrane, is usually in contact with a highly conducting solution. A measurable current in the *solution* may correspond to a field which is tiny indeed (Jaffe's vibrating

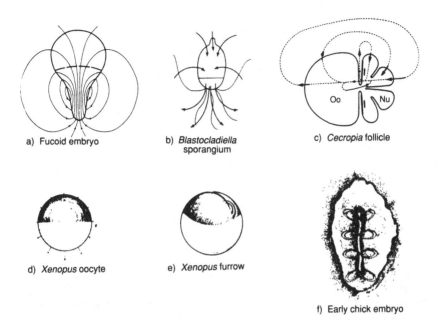

a) Fucoid embryo

b) *Blastocladiella* sporangium

c) *Cecropia* follicle

d) *Xenopus* oocyte

e) *Xenopus* furrow

f) Early chick embryo

Figure 4.5. Some patterns of electric current observed during early development in a wide variety of organisms. From Jaffe (1982), © Wiley-Liss, with permission from Wiley-Liss, a division of John Wiley and Sons, Inc.

electrodes will detect 10^{-9} V cm^{-1}) and which is unlikely to bring about significant movement in the more sluggish charged particles attached to the membrane. But Jaffe (1982) gives a number of instances of fields up to 100 mV mm^{-1} in developing systems.

Second, can an electric field move charged particles in such a way as to enhance itself? Usually, one expects the opposite. An electric field set up by positive charges in one place and negative charges in another is so directed that each kind of charge will move away from where it now is, leading to diminution of the field. This is simply the discharge of a capacitor through a conductor connected to both plates.

There are, however, plausible ways in which the expected flows in a simple capacitor-conductor system can be reversed in the much more complex systems here considered. As one example, Jaffe (1982), in his Figure 7 and its related text discussion, refers to a model by McLaughlin and Poo (1981) for simultaneous self-electrophoresis and electroosmosis. In this model (Figure 4.6), an electric field along the outside of a membrane can move positive ions in the exterior solution in one direction and set up a countercurrent of water in the opposite direction. A large membrane-bound particle may have an exterior appendage which then will act as a kite. The particle will move, impelled by the water current, toward the positive end of the electric field, more or less

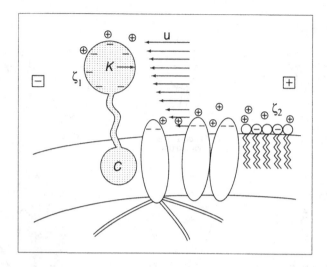

Figure 4.6. Cartoon showing the joint effects of electroosmosis and electrophoresis on a particle C within a cell membrane, but mobile and moved by its "kite" K. The movement of water (u) causes the negatively charged K to move away from the net positively charged region on the right (i.e., in the direction opposite to that of simple electrostatically generated movement in a vacuum). This is therefore a possible mechanism for self-organization of membrane-attached species. From Jaffe (1982), © Wiley-Liss, with permission of Wiley-Liss, a division of John Wiley and Sons, Inc., modified from McLaughlin and Poo (1981).

regardless of the charge, or lack of it, on the large particle. If this particle is part of the mechanism for setting up the positive end of the electric field, one then has a possible mechanism for the field to be self-organizing.

At the end of his 1982 review, Jaffe has a comparison of the electrical theory with reaction-diffusion theory which to my mind totally misses the most significant point. He enquires whether pattern formation is likely to be "exclusively chemical" or to involve electrical forces. I do not understand his concept of "exclusively chemical." Most modern chemists, especially analytical electrochemists, but including increasing numbers of organic chemists who find that important oxidations and reductions sometimes may best be done at an electrode surface, will tend to switch back and forth quite readily between charged and uncharged species. In particular, the practitioner of polarography and other voltametric methods knows that a fine line separates the conditions in which current is diffusion-limited and independent of field (even though fields are there and the species are charged) and the conditions unwanted in such experiments in which field-governed transport dominates. Such a worker will see diffusive and electric forces as two aspects of a given experimental system, not as different worlds.

I am inclined to believe that the important thing about reaction-diffusion, the mechanochemical theory, and self-electrophoresis is that they all involve fundamentally the same kind of principles of self-organization. To be capable of generating pattern, any of these three must include the attributes of self-enhancement (positive feedback, autocatalysis), long-range communication (whether diffusive, mechanical, or electrical), and, for formation of stable pattern, something equivalent to inhibition. All three together are branches of kinetic theory, and involve movement away from equilibrium. In this they stand together, and they stand opposed to theories of equilibrium and structural self-assembly.

This opinion should not be construed as an attempt to subsume the forces of Oster and Jaffe under the banner of the Turingians. The *field* has no need of a supreme commander, but we do need to recognize that we are all on the same side. Neither should it be supposed that the object of the battle is to occupy all the territory of the structuralists and the equilibrists. It would be surprising indeed if nothing in morphogenesis turned out to need their kinds of explanation. The point is not that the territory of the kineticists should be the whole world of morphogenesis, but that it is not yet generally recognized as being a vast territory, which I believe will one day happen.

In the remainder of this chapter, I try to indicate what are the structural, equilibrium, and kinetic aspects of some phenomena mostly chosen because they have all three.

4.2 Developmental control of the shapes of crystals

4.2.1 *Structural aspect: just a few symmetry elements*

The shapes (or *habits*, to use the crystallographer's term) of snow crystals (Figure 4.3) have, when seen together, two striking aspects: first, their diversity; second, that all of them (except, perhaps, some of the needles) clearly show a single sixfold axis of symmetry. This feature correlates, if we go from the macroscopic to the molecular scale, with the symmetry of the hexagonal *lattice* in which the water molecules are arranged. (Actually, the hexagonal system is defined by a unique threefold axis, but in simple cases it often appears as sixfold.)

Likewise, sodium chloride has a lattice of Na^+ and Cl^- ions of cubic symmetry (essential elements: four threefold axes mutually at the tetrahedral angle of 109° 28′) and normally crystallizes in a cubic habit. If, however, the salt is crystallized from a solution also containing urea, it forms octahedral crystals (Figure 4.7a,b). Both cubic and octahedral habits reflect the symmetry of the lattice, as also would an intermediate shape with parts of both cube and octahedron faces exposed (Figure 4.7c). Thus the correspondences between crystal habit and crystal lattice are rather limited, being confined to a few symmetry elements. On the basis of the regular habits of crystals it was speculated for centuries that they contained orderly arrangements of minute

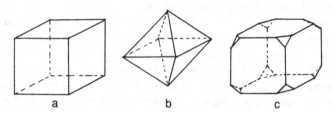

Figure 4.7. Three contrasting crystal habits all compatible with the self-assembly of any lattice of cubic symmetry. Only two types of faces are used: (*a*) (100); (*b*) (111); (*c*) both of these. The "morphogenetic" question is, What determines which of these shapes (or many other possibilities) a crystal will adopt; and in case (*c*) what determines the ratio of areas of cube and octahedron faces? For the small use of Miller indices made here, the reader unacquainted with them need only know that they relate the orientation of a plane (or a line) to a set of rectangular coordinates x, y, and z; (100) specifies a plane which intercepts the x axis but is parallel to the y and z axes, e.g., the right and left faces of the cube shown in (*a*). Faces "of type (100)" include (010) and (001), i.e., all six faces of the cube. A (111) face makes equal intercepts on all axes. In two dimensions, a face becomes a line, or edge. The designation "of type (12)," which includes (21), will specify all lines with slopes of $\frac{1}{2}$ or 2, i.e., a factor of 2 between its intercepts on the x and y axes.

particles. In 1611, Kepler, in an essay on *The Six-Cornered Snowflake,* published with English translation in 1966, almost anticipated modern concepts of close-packing. Later in the seventeenth century, Robert Hooke got even further with stacks of cannonballs. And toward the end of the eighteenth century, de Häuy invented the word "molecule" for what would today be called a unit cell. But the evidence from habit could give no detailed information on the arrangements of particles within crystals, beyond a few symmetry elements, until the twentieth century produced X-ray diffraction. For crystals, the very broad correspondences in symmetry represent the limit of self-assembly in accounting for shape. Everything more detailed has some other explanation.

4.2.2 Equilibrium shapes: surface free energy and Wulff's theorem

When crystals are grown close to equilibrium conditions, that is, from a liquid solution or vapour with a very low degree of supersaturation, they most usually have rather simple shapes, for which the surfaces are only one or two types of crystallographic planes. It seems reasonable to expect that such shapes may often correspond to equilibrium, that is, to the minimization of free energy. Surfaces usually have excess free energy over the same material in the bulk. This excess, per unit area, is different for different kinds of crystal planes. Commonly, it is lowest for low-index planes, such as the cube faces, or (100)-type planes, of sodium chloride. This is the reason why such planes are commonly found as the only bounding surfaces of many crystals.

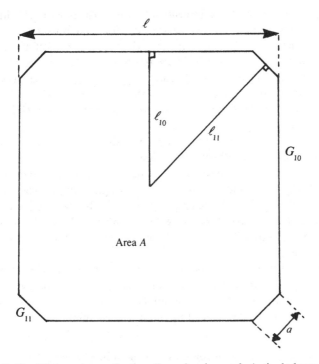

Figure 4.8. Equilibrium shape of a two-dimensional crystal. A single layer of molecules forms a simple square lattice. For a crystal of finite size, a (10)-type edge (horizontal or vertical) has excess free energy G_{10} per unit length. A (11)-type edge, at 45°, has excess free energy G_{11} per unit length. For a crystal of a fixed number of molecules, area A is fixed. (Thus, if the shape changes because of increased or decreased bevelling-off of corners, both l and a must change.) If $G_{11}/G_{10} = n$, and all edges other than those of the (10) and (11) types have such high free energy that they need not be considered, then the equilibrium shape is given by $l_{11}/l_{10} = n$. (Problem 6.5.2 asks for proof of this, and the solution is given in Section 6.6.)

Two different kinds of low-index planes, such as the cube and octahedron faces, (100) and (111), may often have rather similar surface free energies. If the free energy of the latter is somewhat, but not much, greater than that of the former, one can see the possibility that free energy might be minimum for a shape such as that in Figure 4.7c, in which cube corners are bevelled off to expose pieces of the octahedron faces. But how much of each corner will thus be "chopped off"?

The answer is given by Wulff's theorem: "In a crystal at equilibrium, the distances of the faces from the centre of the crystal are proportional to their surface free energies per unit area." For a discussion, including a proof for two-dimensional systems, see Burton, Cabrera, and Frank (1951). This simple theorem makes it quite easy to solve the bevelling-off problem, provided that one knows the excess free energy per unit area of both kinds of face (G_{oct}

and G_{cube}). The distance from the centre of a cube to a corner is $3^{1/2}$ times the distance from the centre to a cube face (and the direction from centre to corner is perpendicular to the octahedron face which might cut off the corner). Therefore, if $G_{oct}/G_{cube} > 3^{1/2}$, there is no thermodynamic advantage to appearance of the octahedron faces. But if $G_{oct}/G_{cube} < 3^{1/2}$, there is such an advantage. As G_{oct}/G_{cube} decreases below this threshold value, the size of the octahedron faces should increase relative to the size of the remaining parts of the cube faces.

Quantitative calculation of the shape is most simply illustrated by the same situation reduced to two dimensions (Figure 4.8). The corners of a square crystal will be bevelled off to such a depth that the distance l_{11} from the centre to bevelled edges is n times the distance l_{10} to square edges, where n is the ratio of excess edge free energies per unit length for bevelled and square edges. Proof of this is asked for as the first exercise in the use of calculus in Chapter 6 (Section 6.5, Problem 6.5.2) and is a proof of Wulff's theorem for this special case.

The structuralist will have noticed that the equilibrium shapes of crystals depend upon a property related to structure: the excess free energy for different kinds of crystal faces. But there is still no way of avoiding calculus to find how the shape will be determined by those free energies.

4.2.3   Kinetic aspect: the diverse shapes of snow crystals

Notwithstanding F. C. Frank's remark that "the average photograph of a snowflake is more symmetrical than the average snowflake," the stellate crystal must have some remarkably good control mechanism for the production of a shape far more complex than any predicted by equilibrium concepts. If we idealize the crystal as two-dimensional, a single layer of molecules in hexagonal array, the simplest representation of a stellate crystal (Figure 4.9) shows two kinds of edges exposed. In the rhombic indexing of the two-dimensional hexagonal lattice, horizontal edges and ones at 60° to the horizontal (i.e., what would be the edges of a simple hexagonal plate) are (11) edges. Vertical edges and others at 30° to the (11) edges are (12) edges. This labelling is used in Figure 4.9, but no knowledge of Miller indices is in fact needed to follow the argument.

The shape is accounted for, in general kinetic terms, if molecules are added to (12) edges much faster than to (11) edges. As the six arms grow out, with (12) edges on their ends, they can remain narrow because they are bounded on the sides by slow-growing (11) edges. Here, the principle of equilibrium embodied in Wulff's theorem cannot help. The sides of the growing rays are clearly at varying distances from the centre of the crystal, according to what point along the ray one looks at. Thus the rays do not satisfy Wulff's theorem and are not part of an equilibrium shape.

Whereas equilibrium thermodynamicists are thus confounded in trying to

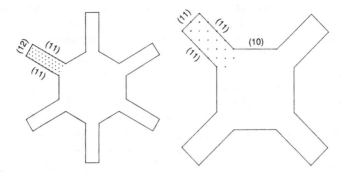

Figure 4.9. Basic stellate shapes, in two dimensions, for a hexagonal lattice and a square lattice. Both have two kinds of faces (or edges, in two dimensions) exposed. The type of face is labelled with Miller indices for rhombic and square indexing, faces such as (10) and (01) all being labelled as type (10), and so forth. Thus a change of shape because of differential growth rates for different types of faces is possible. But the six-pointed star can grow long thin rays if (12) grows faster than (11), because (12) faces occur only at ray ends; the four-pointed star will grow fatter rays because the side faces are of the same type as the ends. Knowledge of indexing is not needed to appreciate the difference in face types. For the six-pointed star, the spacings of points on ends and sides of a ray are clearly different, whereas for the four-pointed star they are the same.

explain the shape, structuralists may see a valid extension of their realm. The crystal can develop rays, in the geometry of the six-pointed star, because the sides of rays, together with the boundaries of the central hexagonal plate, are structurally different from the growing ends of rays. By contrast, a square crystal of a substance crystallizing in a simple square lattice could not send out rays as in Figure 4.9 because the sides of the rays would be the same kind of edges as their ends and should grow at the same rate. Much is thus explained by looking at structure, but only because structure determines rate.

Rates of processes usually depend on some measure of how far a system is away from equilibrium. Hallett and Mason (1958) considered two possible mechanisms for the contrast in rates of growth: (1) different rates of direct condensation from the vapour phase onto different faces; (2) different rates of migration of adsorbed water molecules across different faces of the crystal surface. They were able to devise an experimental test to distinguish these possibilities. For (1), the appropriate measure of distance from equilibrium should be the degree of supersaturation of the vapour; for (2), it should be the temperature. In the laboratory, using an atmosphere of water vapour at controlled pressures and temperatures, they could control the two variables separately and show that the shapes of snowflakes depend directly on temperature, not on degree of supersaturation. Thus (2) is the phenomenon which shapes the crystal (see Section 6.5, Problem 6.5.3.). There is current interest among physicists in the dendritic crystallization of supercooled succinonitrile, which is thought to have some analogy to formation of ice (Saito, Goldbeck-Wood, and Müller-Krumbhaar, 1988). Dr. Müller-Krumbhaar (personal communication) assures me that Mason's much earlier work on ice remains valid, and that the process is still not understood in detail beyond the generalization in this paragraph.

What is the significance of all this to a student of biological morphogenesis? First, this discussion of crystal growth is offered as a general conceptual introduction to the modes of thinking of a physical chemist in approaching the problem of shape-generation. Particularly, such a scientist would have no aversion to switching rapidly between concepts of structure, equilibrium, and kinetics as various manifestations of shape-production might seem to indicate. Biologists in general seem to have much more reluctance to switch back and forth rapidly between the concepts of self-assembly and those of kinetic control.

Second, a specific feature of the evidence on the stellate snowflake is that the shape of a three-dimensional structure is controlled by rates of transport across its surface, tangential to the actual growth. This may sometimes have close analogues in the control of biological growth. It draws attention to the probable importance of events at the cell surface, including transport along the plasma membrane.

Third, and somewhat more likely to make the biologist a little despondent, the discussion has described a type of experiment in which we see the possibility of two measures of displacement from equilibrium being important, and we separate them experimentally so as to vary them one at a time. The strategy is important to bear in mind, but much more difficult to put into practice in biological systems, because they inevitably contain thousands of assorted displacements from equilibrium. Most of these cannot be adjusted experimentally independently of all the others. This is what makes it so difficult to devise the "crucial experiment" to decide whether control is or is not kinetic.

4.2.4 Biological control: the echinoid spicule

Inoué (1982), reviewing the role of (structural) self-assembly in morphogenesis, presented "a graded series of examples that . . . illustrate the extent to which self-assembly can specify the product formed and the degree to which kinetic, or cellular, features govern or modulate the generation of biological structures by self-assembly." Last in this graded series, and so by implication showing the greatest kinetic component, he placed his own observations (Okazaki and Inoué, 1976; Inoué and Okazaki, 1977) on the skeletal spicules of sea-urchin larvae. These have complex shapes entirely failing to display the symmetry of the calcite lattice. But polarized light suggests a uniformity of crystal-lattice orientation throughout the spicule, and X-ray

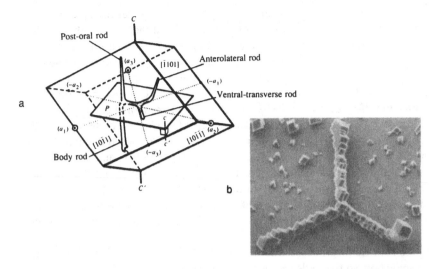

Figure 4.10. (*a*) Shape of a sea-urchin spicule, surrounded by an indication of the orientation of the calcite lattice within it, the whole spicule being a single crystal. (*b*) A triradiate rudiment of a spicule removed from the larva and grown further in saturated calcium carbonate solution, showing the preferential growth of low-index faces, and equal growth of all crystallographically equivalent faces, which does not occur in vivo. From Inoué (1982), with permission. Originally published in *Development, Growth, & Differentiation,* the Japanese Society of Developmental Biologists (1976).

diffraction confirms that the spicule is a single crystal of calcite (Figure 4.10a). Growth of a spicule in saturated calcium carbonate solution takes place by preferential advance of some low-index faces, as one would expect for the approach to equilibrium. The crystal then begins to assume the appearance of a set of connected blocks all in the same orientation (Figure 4.10b). The change from out-of-equilibrium shape determination in the living system to a movement toward equilibrium in the inanimate solution is very striking.

This evidence is of great philosophical interest as an example of the kind of thing that different individual scientists are likely to interpret quite differently, according to their *preconceptions*. Inoué seems to classify this example as still belonging to the category of self-assembly, but with some fine-tuning or modulation provided by kinetics. A vitalist could take the contrast between Figures 4.10a and 4.10b as a striking illustration of how living things defy the principles of chemical behaviour in vitro. To my mind, the vitalist would be right if it were not that physical chemistry provides mechanisms for movement away from equilibrium, including the breaking of those basic symmetries which are inevitably maintained when growth control is by self-assembly. To me, therefore, the kinetics form the basis of the growth mechanism, not just the fine-tuning.

Spicule shape, in vivo, is controlled by clusters of mesenchyme cells with pseudopods fused to form a syncytium. Okazaki and Inoué found that as few as three mesenchyme cells had the capacity to generate the unique spicule shape; they suspected that even one cell might be enough. From further details of the evidence, they concluded that "healthy mesenchyme cells must be suppressing the growth of calcite crystal faces. While permitting the calcite lattice to self-assemble, the cells regulate the exact pattern of calcium carbonate deposition to meet the genetically dictated form." I am quite happy with the implied characterization of genetics as a dictator, provided that it is recognized that dictators and their ilk commonly employ artists and architects to control the design of elaborate structures intended to aggrandize the dictator, and that here a mechanism within the mesenchyme cells is the architect. In this sense, I see the calcite lattice of the spicule as having no more and no less significance than the same lattice within a piece of marble which is being carved by a sculptor.

No specific theoretical model has yet been developed for this phenomenon, to show how so few cells may act as so talented a sculptor. But kinetic effects of mesenchyme cells are numerous and varied. Oster's mechanochemical theory of mesenchymal morphogenesis is outlined briefly in Section 4.4. It requires quite large numbers of cells which move independently and divide. This leads to controlled clustering in definite geometrical patterns. In the echinoid larva, this mechanism might be relevant to the determination of where in the larva the syncytial clusters form and hence where the spicules originate (i.e., the stage of morphogenesis preceding spicule growth).

4.3   Division of plant cells: Is control kinetic, thermodynamic, or mechanical?

Kinetic control of shape is easy to prove when the equilibrium shape is known. But for many intracellular events, no such clearly defined structure is formed. Then, rather than separating structural, equilibrium, and kinetic aspects, one is left to speculate on alternative mechanisms which may be quite disparate in respect of which aspect is dominant in shape determination. Kinetic theory of plant cell division was exemplified at length in Section 3.2. A well-known form of equilibrium theory was presented by D'Arcy Thompson in his 1917 and 1942 editions, though it was not given fully in Bonner's 1961 abridgement. It was revived by Green and Poethig (1982), and something similar is implied, in connection with cleavage of animal egg cells, in Goodwin and Trainor's "field description" (1980).

The equilibrium principle used by Thompson is that cell surfaces are likely to have excess free energy and that the shapes of multicellular aggregates produced by cell division should be such as to minimize this free energy. Roughly, this means that cell surfaces should assume shapes similar to those of assemblies of soap bubbles. A well-known property of such films is that they tend to avoid a quadruple junction, such as that in the middle of Figure

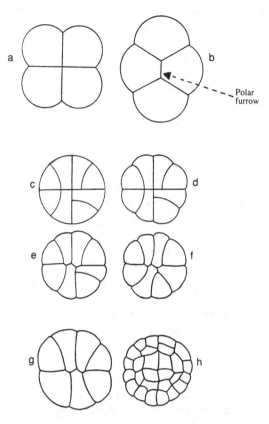

Figure 4.11. (*a*) Unstable equilibrium of dividing surface between soap bubbles. (*b*) Stable equilibrium to which the former inevitably changes (being an example of symmetry-breaking which arises from equilibrium considerations). (*c–f*) Equilibrium divisions within each cell readjusting to equilibrium of the 8-cell assembly. (*g,h*) Sections of actual assemblies of cells in a developing moss. From Thompson (1917).

4.11a, replacing it with two triple junctions with angles close to 120° separated by a "polar furrow," this being the vertical line in Figure 4.11b. (The polar furrow arises again in a discussion of how animal cells may rearrange in sorting-out; see Section 4.4.4.)

In many instances of the formation of sheets of cells by repeated division in the plant kingdom, quadruple junctions indeed seem to be avoided most of the time, but not completely. Figure 4.11c–f shows Thompson's concept of how a disc-shaped cell might divide into eight by divisions at first idealized to the situation within each dividing cell, but with readjustments to a stable equilibrium for the whole assembly. Figure 4.11g,h, also from Thompson, shows sections through early developmental stages of a moss. Gunning (1982) indicated a similar fairly general but incomplete avoidance of the quadruple junction in his observations on the *Azolla* root.

Lintilhac (1984) proposed a model for the orientation of the preprophase band which could be regarded as a problem for the student of this chapter. To which of my categories, structural, equilibrium, or kinetic, does it belong? (Or, for a much more detailed answer, how would you start to set up this model mathematically? It is clearly at the stage of an interesting and well-presented suggestion, and yet it remains somewhat undefined until one tries to express it mathematically.)

The model starts from cellular deformation arising from both internal turgor forces and external forces. In response: "A dimension-seeking mechanism within each cell would therefore be the most direct method of determining the directions of principal stresses and strains, effectively transforming each individual cell into a three-dimensional, multi-axial strain-gauge." Lintilhac proposes that this mechanism involves stiff, noncontractile elements such as microtubules, randomly oriented but spanning the cell, with attachments to the plasmalemma on both sides. In response to extension of the cell in any direction, elements running along that direction will be broken. New linkages are then assumed to be formed randomly to restore structural continuity. "Thus, a period of extended dimensional change will result in the gradual depletion of structural elements along the axis of maximum dimensional change and the accumulation of elements in the orientation which is the most protected from shattering forces. This . . . will be transverse . . . to the direction of maximum dimensional change" (Figure 4.12).

The short answer to the problem posed earlier may be arrived at by applying a well-known test for whether or not a final state is an equilibrium state: Does one reach the same state from different starting points, or is the state path-dependent? If the former is answered affirmatively, the final state *may* be an equilibrium state; if the latter, it *definitely is not*. Let us suppose that the final cell shape, corresponding to Figure 4.12e or 4.12j, is a cube. In which of the three mutually perpendicular planes bisecting this cube will the "preprophase band" form? In Lintilhac's mechanism, this is governed by which dimension has been altered the most to achieve the cubic shape. The symmetry of the *change* governs the symmetry of the *product*. The model is kinetic.

Section 3.2 and the present section are, I believe, together sufficient to show that despite all that has been discovered about the cytoskeleton and other contents of the cell since Thompson wrote, yet in terms of surely known or generally accepted developmental mechanisms the cell is as empty today as it was in 1917.

4.4   Animal morphogenesis: rearrangement, deformation, and proliferation of cells

4.4.1   Some phenomena, and the cell-as-molecule concept

Molecules move randomly, being propelled passively, that is, from without, by their collisions with other molecules. They attract each other by forces

a b

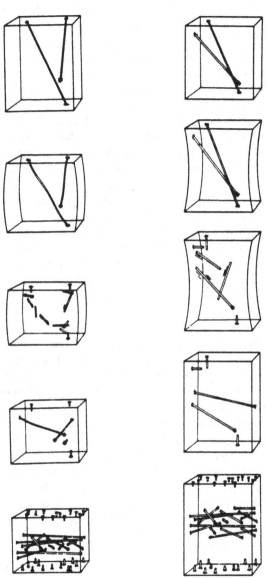

Figure 4.12. Lintilhac's diagram of the determination of a plane of future cell division by the breaking of an array of elements such as microtubules chiefly in the direction of maximum dimensional change, followed by random re-formation: (a) for cell shortening; (b) for cell lengthening. In both cases, the "pre-prophase band" develops perpendicular to the direction of maximum change. From Lintilhac (1984), with permission.

mainly requiring a quantum-mechanical explanation. And they take part in reactions, including in some instances autocatalytic proliferation.

Some of the behaviour of cells as units of an assembly seems analogous to these properties of molecules. Animal embryogenesis frequently involves movements of cells which lead to rearrangements (e.g., the gastrulation process in nematodes and many other lower invertebrates, as sketched in Figure 4.4a). In an assembly of cells in which all cells are dividing at regular intervals, their total number increases in the same manner as a concentration increases in a first-order autocatalysis. Taken together, these two properties seem to indicate that cells may sometimes have the essential attributes for them to serve as the "molecules" in a reaction-diffusion formulation.

Many instances of animal morphogenesis involve an epithelial sheet lined on one side with a basal lamina of extracellular-matrix materials, and an adjacent layer of mesenchyme cells. These last display the attributes of independent mobility and proliferation. In some instances, experimental evidence suggests that the positions of particular epithelial structures (e.g., the regular hexagonal arrays of chick feather primordia) are first established as clusters of mesenchyme cells. One is then left to wonder if a mesenchyme cell can sometimes be identified with an activator morphogen "molecule" X. The theory of Oster et al. (1983) (see Section 4.4.4) leads to that interpretation in certain limiting conditions.

There are, however, important properties of the motion of cells which limit the validity of the analogy between that and molecular motion. This is seen particularly in epithelial sheets, which often change their shape by deformation of individual cells precisely orchestrated over certain areas of the sheet (e.g., the gastrulation process in echinoderms and higher animals, Figure 4.4b). Some components of this long-range organization can be destroyed, so that others can be studied by themselves, by taking a tissue apart into separate cells. This was done in the in vitro sorting-out experiments of Steinberg and co-workers (Phillips and Steinberg, 1969; Steinberg, 1970, 1975; Steinberg and Wiseman, 1972). When cells thus obtained from two different tissues of a chick embryo are mixed randomly together in culture, they sort out into two separate aggregates, one enveloping the other, as in Figure 4.2. This requires movement of cells relative to each other much as molecules must move when a solution separates into two phases, as a result of change in temperature or other conditions. But cells would take most of eternity to do this if each could be regarded as a molecule moving only with the kinetic energy appropriate to the temperature. To move quickly enough, cells must be propelled from within by the contractile machinery which can bring about shape changes. This is active movement, quite unlike the internal motions of molecules. But it can still be directionally random. An assembly of rearranging cells can then still usefully be thought of as rather like a set of molecules, but with a mean kinetic energy appropriate to some fictitious temperature enormously greater than the real temperature of the system.

Such in vitro experiments were used to establish the concept that differential adhesion exists between various types of cells, and that this can be a factor in morphogenesis. This is equilibrium theory, and the in vitro phenomena were carefully shown by Steinberg (1970) to be in that category (Section 4.4.2), and analogous to phase separations arising from differential attractions between molecules. To what extent is it a factor in living pattern formation? These workers have been careful never to claim it as a complete model for this, and to stress that the order of envelopment of one tissue by another makes no anatomical sense. When sorting-out seems to be a component of morphogenesis in vivo, it is usually evident that there is some long-range organizer of the cell movements which may or may not take the system toward an equilibrium. Cells, for instance, often move as coherent aggregates, or form into such aggregates as they move. Steinberg showed that when pieces of tissue from chick embryos are excised and placed adjacent to each other on a skewer, they move so that the final conformation is the same as that in the sorting-out experiments, in respect of order of envelopment of one tissue by another. Here, equilibrium still dominates. The path by which it is reached is a secondary matter.

By contrast, the cellular slime mould *Dictyostelium discoideum,* at one stage in its life cycle, forms an elongated mass of nonfeeding, nondividing cells nowadays usually called a "slug." This differentiates into two cell types, the pre-stalk cells occupying about the anterior quarter of its length, and the pre-spore cells the remainder (Figure 4.13a). Both the formation of the slug and the following stage, culmination, in which the slug is transformed into a fruiting body of spore cells and stalk cells, are relevant to the present topic, except that the organism does not seem to own allegiance to either the animal or the plant kingdom.

In both developmental stages there is cell movement. For the slug stage, it remains uncertain whether a spatial prepattern is laid down by some chemical mechanism which then commands the differentiation of the cells, or whether the cells differentiate randomly and thereafter sort out spatially; see Mac-Williams (1982) and earlier references there cited. But the shape of the slug does not correspond to the end of Steinbergian sorting to minimize the free energy of differential adhesion. One cell mass does not envelop the other. If sorting is dominant, it must be by some such means as chemotaxis. Pre-stalk cells are known to produce cAMP and to move toward high concentrations of it. Anything along these lines would be a kinetic mechanism, and a proper mathematical treatment of it would probably have resemblances to reaction-diffusion, with pre-stalk cells as a "morphogen." See Section 10.3.1 for further discussion and two other morphogens.

Culmination (Figure 4.13b–e) shows even more striking evidence for mechanokinetic organization. The pre-stalk cells, instead of forming a stalk in the geometrically simplest way on the side of the spore mass where they are already located, first move anteriorly (now upward), but at the tip form a

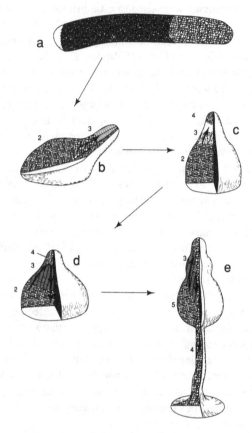

Figure 4.13. Two morphogenetic stages of the cellular slime mould *Dictyostelium discoideum*. (*a*) The fully developed pattern in the slug (grex, or pseudoplasmodium), with the pre-spore cells (dark) and the pre-stalk cells (light) distinguished by the periodic acid–Schiff staining method. (*b–e*) Stages between the slug and the fruiting body, which is a spherical mass of spores atop a narrow stalk with a basal disc at the bottom. Pre-spore cells (2) mature into spore cells (5); pre-stalk cells (3) mature into stalk cells (4). As the stalk forms by downward movement of the stalk cells, it acquires a heavy cellulose sheath. Arrows in the pre-stalk and stalk cell masses show approximate directions of the reverse-fountain flow, which is clear evidence for a system of mechanical stresses highly organized throughout the structure. Modified from Bonner (1967), with permission; Sections *b–e* adapted from *Scientific American 201* (December 1959), 152–62.

narrow file which then moves right through the spore mass in the opposite direction to form the stalk. The whole flow path of cells is essentially a reverse fountain, highly organized and not at all like random sorting, and leading to a structure which would be most unlikely as an equilibrium shape.

Morphogenesis is a continuous sequence of processes in which the occurrence of one stage is often dependent upon the prior completion of another event. For two such successive stages, it is not always clear which should be accorded the greater credit for determination of form. This is particularly the case when there is good experimental evidence that one event requires the presence of a preexisting and persistent gradient in some property along the morphogenetic field of the event. Does one take the gradient for granted as an experimental fact, or does one maintain that the crucial question is how the gradient was set up in the first place? The former view has been taken, at least for the time being, in some important accounts of kinetic theory, especially Gierer and Meinhardt (1972), as well as equilibrium theory (Steinberg and Poole, 1981). These latter workers showed that the pronephric duct of the axolotl (*Ambystoma mexicanum*) grows across the flank mesoderm in a dorsocaudal direction in response to an adhesive gradient in the mesoderm. While the duct may be regarded as seeking equilibrium, it is trying to hit a moving target, because there is evidence that the gradient was not fully set up previously and then became static, but is itself moving as it guides the duct. Description of the gradient therefore needs kinetic theory.

The two kinds of gastrulation used here as examples, for nematodes and echinoderms, both suggest the presence of some prepattern before the onset of cell migration or deformation. Oster's (1983) mechanochemical description of the onset of invagination in the echinoderms gives a good account of how the signal to deform may be passed from cell to cell, and how the resulting concerted deformations may produce the observed change of form. But it presupposes that the cells on one side only of the blastula are somehow prepared to receive the mechanical signal. This is a prepattern, and its establishment is the fundamental symmetry-breaking between blastula and gastrula, before the mechanical deformations begin. It was this demarcation of the blastula into two disparate sides which Turing had in mind as the role for his kinetic mechanism when he remarked that the spherical blastula, in the absence of such a mechanism, "could not give rise to a horse, which is not spherically symmetrical."

Nematode gastrulation looks superficially very different, leaving one to wonder if nature has found two separate ways to form a gut, that is, in topological terms, to form a sheet of cells into a non-simply-connected surface, the topology of a doughnut. Yet, in the sorting-out process which looks so random if one gives detailed attention to the jostlings of a few cells, the resultant cell movements suggest a system of long-range forces organized just as precisely as those in the echinoderm gastrulation. This suggests that there may here also be a hidden prepattern which prepares the ground for organization of forces by an Oster mechanism.

In this view, the two kinds of gastrulation may in fact be rather similar at the stage at which organized forces arise to move and deform cells. The processes look different because the starting shapes of the assemblies of cells are different, and the prepatterns are different. Comparative study of morphogenesis in different organisms requires first a careful analysis of the morphogenetic event into stages, so that one may then try to pick out which stages are similar and which are different between various examples. The initial analysis requires recognition of complex sequences of the making and breaking of symmetry, which is the topic of Chapter 5. The question of linking morphogenetic processes in sequence is taken up again in Sections 9.1.4, 10.1, 10.2, and 10.3.3.

For the purpose of the present chapter, the operation of systems of organized forces to shift cells around and to change their shapes is clearly in the realm of kinetic explanation. One cannot say much that is definite about the probable chemical prepatterns until more is known about them experimentally, but such distributions of material are quite difficult to account for other than kinetically.

The last example in this section is the account of morphogenesis of the mouse submandibular salivary gland given by Bernfield et al. (1984). This seems to be a phenomenon which has everything that is kinetic, and very little that is not. The dominant feature seems to be establishment of nonuniform distributions of chemical substances along the boundary of the developing multilobate form (Figure 4.14). There seems to me to be something here which is far more complex than, but yet strategically similar to, the morphogenesis of the stellate snowflake: One's attention is continually brought to the directions tangential to the advance of the developing outline, and to how disparities arise along those directions because of differential rates of processes.

The system is one of three layers: an epithelial sheet, bounded outside by a basal lamina, with a layer of mesenchyme cells outside that. Bernfield and associates give an extensive catalogue of the known aspects of the behaviour, cellular and histochemical, which gives rise to morphogenesis. This includes almost every kind of behaviour which can form part of a kinetically organized process. For instance, the extra surface area of the enlarging and branching lobes is provided by proliferation of cells. This has been shown to involve substances, probably classifiable as mitogens, transmitted from the mesenchyme cells. What is missing is any model for how these cells go about providing this chemical signal more at the lobe tips than elsewhere. From the description, it would appear that this is more a matter of nonuniform distribution of the chemical signal than of clustering of the mesenchyme cells. Possibly, then, a reaction-diffusion mechanism along the mesenchymal layer could be involved. Also, there is a nonuniform distribution of extracellular matrix on the epithelium. There is less basal lamina (e.g., type IV collagen) on the lobule tips than in the notches, and the latter have an additional layer of type I collagen. The basal lamina is produced by the epithelium, but some modifications of it are due to the mesenchyme. The control of lobule branching needs not only a mechanism for an increase in area but also a shaping influence, probably involving the cytoskeletal mechanics which can change the shapes of

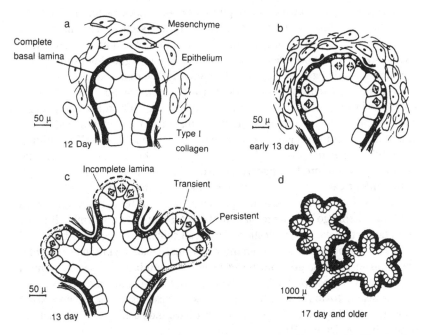

Figure 4.14. Morphogenesis of the mouse submandibular salivary gland. Changes in the extracellular matrix (removal at active areas of lobe growth, and the laying down of extra type I collagen in the notches), as well as cell proliferation and shape change at the growing and branching lobe tips, involve chemical interactions between epithelium and mesenchyme. This implies the formation of chemical prepatterns to control lobe formation and branching. How are these prepatterns formed? From Bernfield et al. (1984), © Wiley-Liss, with permission of Wiley-Liss, a division of John Wiley and Sons, Inc.

cells. But those authors mention (in a general context, rather than with specific evidence from this example) that changes in cell shapes can affect rates of DNA synthesis and cell proliferation. This is a feedback linkage which is quite appropriate to the differential equations for kinetically controlled development.

This instance well exemplifies the general state of the field. There is quite a wealth of cellular and chemical information which gives strong indications of the main players in the game of morphogenesis. But there can be no proof that these various behaviours can indeed be responsible for the observed shape changes short of a mathematical analysis of their kinetic interactions. Everywhere, their account stops short of that, and for good reason. In doing all that work, those authors had enough on their hands without embarking on that equally time-consuming but very different part of the work. A main purpose of this book is to try to encourage more people into activity in that side of the work.

Table 4.1 is a checklist of the equilibrium and kinetic factors likely to be involved in the various examples described in this section. The three terms used (Yes, ?, and No) should be taken as indications of relative probability, as I see it, rather than absolutes. It would be possible to argue about some of the affirmatives and negatives. The three columns on the right, under the heading "Cells as units," must be understood in a very specific sense. "Yes" does not merely indicate that mechanical strain, or chemotaxis, or proliferation occurs and has some role in the process. It means that there is a high probability that one of these factors has an essential role in the self-enhancing, self-inhibiting, cross-enhancing, or cross-inhibiting interactions which confer on a mechanism the power to produce pattern or form. Thus the word "Self-organizing" in the title of each of those columns is vital to its meaning.

The items have been listed in a particular order, to form a series in which the "Yes" items go in a diagonal band from top left to bottom right. It is my expectation that when the question marks can someday be replaced by "Yes" or "No," this pattern will be even more obvious. Thus, the instances, in order from top to bottom, go from those in which equilibrium adhesion is most important, via those probably involving kinetic self-organization on the molecular scale, to those in which the cells themselves are in a sense morphogens. For the two instances in which I have indicated "Yes" to reaction-diffusion, see Section 10.3.

4.4.2 Equilibrium aspects: differential adhesion

Conceptual advances in science frequently become firmly established through the recognition of a new property which can be quantified: the velocity of light, believed infinite up to the seventeenth century; the mechanical equivalent of heat, essential to establishment of the concept of energy; and the statistical ratios of Mendelian genetics, which paved the way toward molecular biology.

Sometimes the recognition of a quantifiable property occurs when a set of objects A, B, and C are shown to have a property (represented by the same letters) which, by *qualitative* test, obeys either (1) the law that if $A = B$ and $B = C$, then $A = C$, or (2) the so-called transitive law that if $A > B$ and $B > C$, then $A > C$. Here, the signs for "equals" and "greater than" acquire their usual mathematical meanings only after the qualitative comparison has been used to assert the possibility of, and then set up, a quantitative scale. Case (1) is exemplified by the zeroth law of thermodynamics, being the starting point of an approach to the subject which efficiently extracts the essence of its beginnings by mincing and boiling its historical details. Thermal equilibrium between two bodies can be defined, without the concept of temperature, as the relationship between them when all their properties have ceased to change as a result of contact (e.g., size, pressure, electrical conductivity). The law is as follows: If two bodies are each in thermal equilibrium with a third body, then

Table 4.1. *Physicochemical factors likely to dominate specific examples of form determination*

Experimental phenomenon	Equilibrium: adhesion	Kinetics				
		Molecules as units		Cells as units		
		Prepattern (unknown origin)	Molecular reaction-diffusion	Self-organizing mechanical strain	Self-organizing chemotaxis	Self-organizing proliferation (cells as X)
Sorting-out in vitro (chick embryo cell types)	Yes	No	No	No	No	No
Pronephric duct growth (*Ambystoma*)	Yes	Yes	?	?	?	?
Nematode gastrulation (*Caenorhabditis elegans*)	?	?	?	Yes	?	No
Echinoderm gastrulation (the phylum in general)	No	Yes	?	Yes	No	No
Axolotl heart formation (*Ambystoma mexicanum*)	No	Yes	?	?	No	No
Slime-mould slug pattern (*Dictystelium discoideum*)	No	?	Yes	?	Yes	No
Slime-mould culmination (*D. discoideum*)	No	?	?	Yes	Yes	No
Mesenchyme clustering (chick feather primordia)	No	?	?	?	?	Yes
Mouse submandibular gland morphogenesis	No	Yes	?	?	?	Yes

they are in thermal equilibrium with each other. It leads to the recognition of a property which can be given the same value for the entire set of bodies in thermal equilibrium with each other: the temperature.

The transitive property (2) is exemplified by a replacement series. If Zn replaces Cu from solution, and Cu replaces Ag, then Zn replaces Ag. This implies the existence of a hierarchy $A > B > C$, and hence of a quantifiable property corresponding to order in the hierarchy. This is of course the reversible electrode potential: oxidation potential for Zn > Cu > Ag, or reduction potential for the reverse order. It is often arbitrary in a hierarchy which direction is defined as increase.

Hierarchical ordering is commonly to be expected for equilibrium properties. Kinetic pair-interactions are idiosyncratic: If a snake can kill a man, and a man can kill a mongoose, it does not follow that a snake can kill a mongoose.

Steinberg (1970) established the existence of a transitive property and hence of a hierarchy in sorting-out and engulfment experiments, as shown schematically in Figure 4.2. He used six tissues from chick embryos, taken at sufficiently early stages (1–8-day embryos) that each tissue usually contained only one cell type:

A: back epidermis, 8-day
B: pigmented epithelium of the eye, 5-day
C: heart ventricle, 5-day
D: liver, 5-day
E: chondrogenic cores, fore- and hind-limb buds, 4-day
F: posterior neural tube, 36-hour

For these 6 items, there are 15 pair combinations; with each, both a sorting-out experiment and an engulfment experiment were done. The former involves taking a tissue apart into cells, mixing them in culture, and observing how the aggregate rearranges. The latter involves skewering two bits of tissue together on a glass fibre in culture, and observing which crawls over the other. The work is difficult, and the amount of work increases roughly with the square of the number of tissues. One might therefore ask: Why six tissues and two kinds of experiments?

Steinberg's arguments were meticulous. If the same result is achieved by two different paths, that is a strong indication, though not a proof, that the result is determined by equilibrium. As to the number of tissues, if one has only 3, A, B, and C, then of the 8 possible sets of results of pair-competition experiments, 6 correspond to hierarchies. That is, there is a 75% chance of "proving" the existence of a hierarchy when the pair-competition results are in fact random, with no difference in property driving them (Table 4.2). For 6 items, the chance of finding a hierarchy by such an accident is reduced to 2.2%.

Table 4.2. *Pair competitions between three items:*
the eight sets of results and six hierarchies,
showing 75% chance of finding a hierarchy
"by accident" from random results

Possible data sets			
A/B	B/C	C/A	Hierarchy
$A > B$	$B > C$	$C > A$	None
$A > B$	$B > C$	$A > C$	$A > B > C$
$A > B$	$C > B$	$A > C$	$A > C > B$
$A > B$	$C > B$	$C > A$	$C > A > B$
$B > A$	$B > C$	$C > A$	$B > C > A$
$B > A$	$B > C$	$A > C$	$B > A > C$
$B > A$	$C > B$	$C > A$	$C > B > A$
$B > A$	$C > B$	$A > C$	None

If we take the sign ">" to mean "goes inside," for the final configuration of one spherical tissue inside another, the experiments give the hierarchy $A > E > B > C > F > D$. Of the 15 pair competitions, only one gives a mixed result ($F > D?$).

The final configuration is just that which would be assumed for two immiscible liquids put in contact with each other. Suppose that at the liquid/air interface, liquid A has surface tension γ_A and liquid B has surface tension γ_B. Where the two liquids are in contact, there is an interfacial tension γ_{AB}. Surface tensions are excess free energies per unit area. A system will tend to adopt the geometrical configuration which minimizes the product of surface tension and area, summed over all surfaces. In the engulfment of one liquid by another, the liquid/liquid interface is a complicating factor. γ_{AB} may be an idiosyncratic property of the AB pair, not simply related to γ_A and γ_B, and therefore capable of destroying hierarchical ordering. But if the two drops of liquid placed together have equal volumes, then the liquid which goes inside will form a sphere of the same size regardless of whether A goes inside B or vice versa. The contribution of γ_{AB} is then the same in either configuration. The order of engulfment will then be determined by the excess free energy at the air/liquid interface, the surface of the larger sphere. For minimization, the liquid of higher surface tension will go inside the other. The sign ">" was taken earlier to mean "goes inside" in anticipation of this correlation with a surface-tension-like property for an aggregate of cells. Steinberg and his co-workers did not stress the matter of relative volumes in their writings, but in fact seem usually to have taken rather similar volumes of the two tissues to be compared.

Phillips and Steinberg (1969) performed an experiment to measure the surface tension of an aggregate of cells of one kind by the sessile drop method. A drop of liquid sitting on a flat plate assumes a shape close to spherical if it has a high surface tension, and tends to flatten if it has a low surface tension. The experiment can be made quantitative by precise measurements of the shape. Phillips and Steinberg did not try to represent the results numerically, but showed that aggregates centrifuged on a flat plate assumed flatter drop shapes if the cell type was lower in the engulfment hierarchy; specifically, in ordering according to drop shape, $E > C > D$ (Figure 4.15).

The surface tension of a liquid is a manifestation of the energy of cohesion between its molecules. Using the cell-as-molecule concept, Steinberg attributed the gradation in a surface-tension-like property as described earlier to differential adhesion between different kinds of molecules. At the level of adhesive molecules attached to the cell surface, two general possibilities are distinguished: homophilic adhesion, in which a molecule on the surface of one cell adheres to a molecule of the same kind on another cell; and heterophilic adhesion, in which the adhesive molecules on the two surfaces are different. For the homophilic case, cells of different kinds are supposed to have different fractions of their surfaces occupied by adhesive molecules, to account for the differential strengths of adhesion. If for cells of type A this fraction is a, and for type B it is b, then the strength of A-A adhesion is proportional to a^2, and that of B-B adhesion is proportional to b^2. Unlike cells (A-B) should then adhere to each other with a strength proportional to ab, being the geometric mean of the A-A and B-B adhesions. It is a curious coincidence that this is the same relationship proposed by Berthelot in 1898 for the attractions between unlike molecules in a mixture of imperfect fluids; see Hildebrand and Scott (1962) for an account of the theory of mixed fluids. On this basis, immiscibility of fluids, being the molecular analogue of cell sorting-out, should require the difference $|a - b|$ to exceed some threshold value. That threshold, however, is related to the entropy of mixing. For the statistics of molecules, entropy is a powerful opponent of unmixing. For the small-number statistics of cells (a million is a very small number for this purpose), the entropy of mixing is negligibly small, if indeed such a quantity can be defined at all. Cells are individually distinguishable from each other. Molecules are not, and that indistinguishability is essential to the calculation of statistical entropy. Thus, any threshold value of $|a - b|$ as an immiscibility criterion for cell types may be expected to be quite minute.

Heterophilic (lock-and-key) adhesion was proposed by Marchase, Barbera, and Roth (1975), with a speculative molecular mechanism. They envisaged cell coat glycoproteins with side-chains composed of 10 assorted monosaccharides. Each of these would be attached in turn by a glycosyl transferase specific to recognize the partly completed chain and attach the next monosaccharide. But if that unit was not supplied, the enzyme would stay attached to the partly complete chain indefinitely, in lock-and-key fashion.

Such ideas remain useful. The molecular aspects of adhesion are not yet all known. But there is now one known class, called cell-adhesion molecules

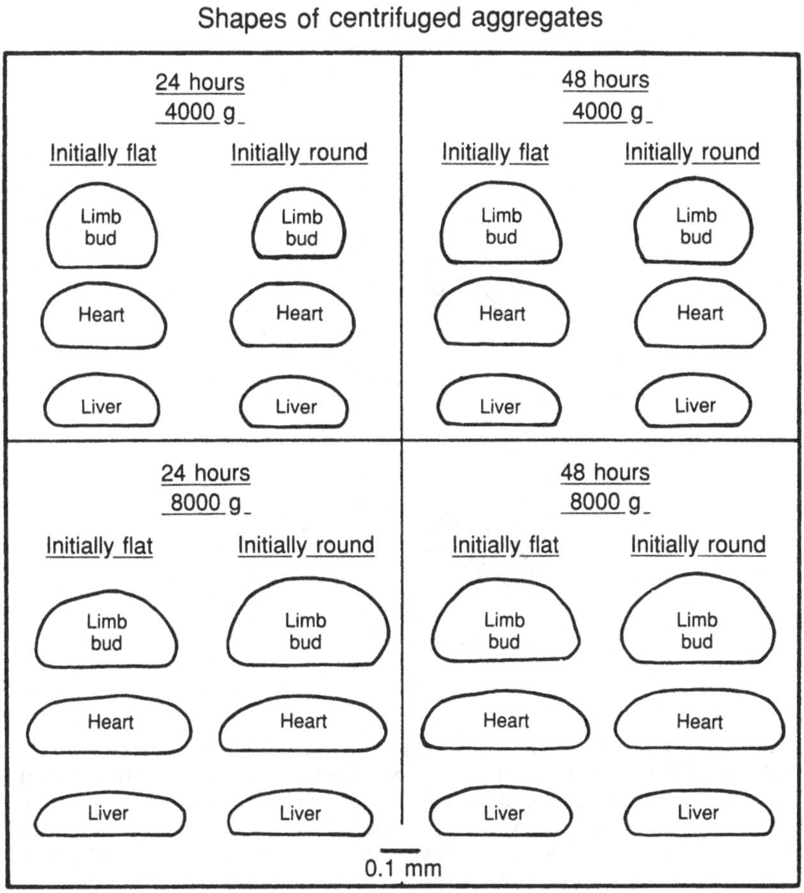

Shapes of centrifuged aggregates

Figure 4.15. Schematic representation of centrifugation experiments on aggregates of cells from a single embryonic tissue, analogous to the "sessile drop" method of measuring surface tension of a liquid. The tissues were dissociated with trypsin. The cells were suspended in culture medium and spun down into flat sheets. Some pieces of these were used as starting material. Others were rounded up on a gyratory shaker. The bottoms of the centrifuge tubes were coated with agar. From Phillips and Steinberg (1969), with permission.

(CAMs), about which much is known. They are in the extended family of immunoglobulins (Edelman, 1984, 1988).

4.4.3 From equilibrium to kinetics: incompleteness of adhesive-gradient theory

The preceding section concerned the interaction of two cell types differing in cell-to-cell adhesive strength. There is substantial evidence for a much more

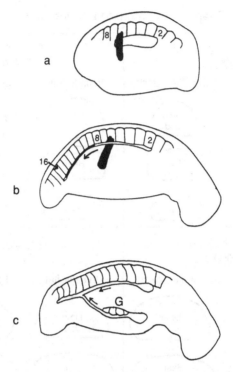

Figure 4.16. Pronephric duct development in the salamander or axolotl, used by Steinberg and Poole (1981) in a series of experiments showing that the duct is guided by an adhesive gradient in the flank mesoderm. (*a*) Stage 22 embryo. The structure below somites 2–7 is the beginning of development of the duct, and the dark mark across it is a vital dye. (*b*) The same embryo at stage 32, showing how the dyed section of the duct has moved. (*c*) A similar embryo to which an extra duct primordium has been grafted at G, showing that the extra duct is guided to join the normal one and then travel with it. From Steinberg and Poole (1981), with permission of The Company of Biologists Ltd.

complex situation involving such adhesion, that is, for a continuous gradient in adhesive properties across a tissue. Steinberg and Poole (1981; Poole and Steinberg, 1981) found evidence for an anteroposterior gradient of increasing adhesion along the flank mesoderm of the salamander or axolotl, *Ambystoma maculatum* and *A. mexicanum,* during embryonic development (Figure 4.16). Their experiments included transplanting a portion of this tissue to a more anterior location, where, instead of fusing into the surrounding tissue, it rounded up in the manner of a strongly adhesive aggregate surrounded by a more weakly adhesive one. Most particularly, they performed a series of experiments on the directional control of the development of the pronephric duct from an anteriorly placed primordium posteriorly into an elongated structure. The experiments were designed to distinguish between guiding by an adhesive gradient and other possibilities, such as chemotactic signalling, and the results showed definite indications of control by adhesive gradient.

From such evidence as this, I am left with little doubt that adhesive gradients exist and that they are important in morphogenesis. But I have serious doubts that they can be regarded as the whole story, or even the greater part of it for a single morphogenetic event. As is so common in this field, the crucial event seems to have hidden itself one level behind what experiment is revealing. How is the adhesion gradient set up and maintained? Its maintenance is, in fact, neither temporally nor spatially well separated from such events as the migration of the pronephric duct, because this occurs during a period of increasing size of the flank mesoderm region, and the tip of the duct follows close behind the leading edge of the sequentially forming somites (Figure 4.16, from a to b).

An adhesion gradient, operating within a tissue without a strong polarizing influence from the surroundings, will not organize itself monotonically from high at one end to low at the other. Just as in the sorting-out experiments on two cell types the more strongly adhesive ones went inside, so in an assembly of cells with a range of adhesiveness the drive toward minimization of free energy will lead to the most adhesive cells ending up somewhere in the middle. The details of the geometry may be complicated, according to the shape of the tissue and the relative abundances of cells of various adhesive strengths. Two simple examples are shown in Figure 4.17.

The lock-and-key model runs into different (but strategically related) difficulties. One of its special areas of application is in the joining of the optic nerve to that part of the brain surface known as the optic tectum, in the lower vertebrates. Grossly simplified, but I believe in a manner which retains the essence of the morphogenetic problem, the process is as follows: The tectum may be idealized as a square of a million cells in a single epithelial sheet, one thousand cells on a side. About an equal number of retinal axons form the extremity of the optic nerve which joins the tectum. They do so in such a way that visual information goes from the eye to the brain with a precision of about one part in a million for its geometrical ordering. This can be and commonly is taken to mean that every axon finds exactly the correct tectal cell to join up to, although it is difficult to confirm this by direct histological examination.

Diverse theories have been proposed for how such precise connection may be directed. Willshaw and von der Malsburg (1976, 1979) proposed a model with Turing-like features on a cellular rather than a molecular scale. A directional information model is presented in the form of a problem for readers to discuss for themselves, in Section 4.5. Sperry (1965) put forward a hypothesis, usually called neuronal specificity, wherein every neuron is supposed to be in some way chemically labelled differently from every other. This does not necessarily mean that a million substantially different molecules are involved. Sperry mentions gradients, and the hypothesis is formulated suffi-

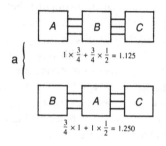

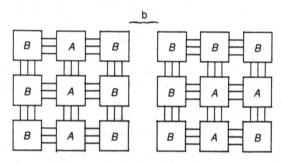

Figure 4.17. Self-organization of assemblies of cells of diverse adhesiveness, if guided by the equilibrium criterion. (*a*) A row of three cells with homophilic adhesive energies determined by the fraction of each cell surface covered by adhesive molecules, which, for cells *A*, *B*, and *C*, respectively, is $a = 1$, $b = \frac{3}{4}$, $c = \frac{1}{2}$. The favoured arrangement (*high* number, because these are *binding* energies, which should strictly be given a negative sign) is the one in which the most adhesive molecule, *A*, is in the middle. (*b*) Three strongly adhering cells *A* and six weakly adhering cells *B* in a 3 × 3 square array. If one calls the *AA*, *AB*, and *BB* binding energies large, medium, and small and counts the number of interactions of each kind, one may draw all possible arrangements of the cells and conclude that the two arrangements shown are equal in energy and are favoured over all other possibilities. It is not necessary to use a precise rule for how the *AB* energy is related to the *AA* and *BB* energies, except that it lies between them. This case illustrates both that the strongly adhering cells tend to go to the middle and that the equilibrium geometry can be ambiguous.

ciently loosely that, as Hope, Hammond, and Gaze (1976) pointed out, the same experimental results can be seen by some people as supporting it, by others not.

The lock-and-key model of Marchase et al. (1975) may be regarded as one interpretation of neuronal specificity which minimizes the need for an unlikely multitude of different molecules. They consider that one-in-a-million specificity in a two-dimensional array could be achieved by two crossed gradients, at roughly right angles to each other, each with a specificity of one part in a thousand. Suppose that the substance concerned were a glycoprotein, that three of its oligosaccharide side-chains each had 10 monosaccharide units if complete, and that, by the presence or failure of the supply of each monosaccharide at the right time, each could be in any of its 10 possible states of partial completion. This gives a thousand different chemistries. (Their models also included gradients in concentrations of the various glycosyl transferases needed, but they could find no experimental evidence for such gradients.)

This model is a useful concept in relation to the achievement of molecular diversity at cell surfaces. But to my mind, as a model for self-organization of neuronal systems it has two serious areas of incompleteness, one of them specific to the kind of molecular diversity proposed, the other a much more general problem of the need for precise definition in setting up any lock-and-key model.

The first is the matter of where all these incomplete side-chains came from, each with exactly the right degree of incompleteness for its spatial location. Just as much as the homophilic gradient, though for different specific reasons, the heterophilic gradient is not self-organizing, and the real organizer must be hidden at least one level deeper. What is needed seems to be a supply of the monosaccharides for production of the side-chains, these substrates being themselves in spatially graded distributions so that there is enough concentration for reaction in some places and not in others. This seems to transfer the burden of self-organization back to kinetics again.

The second problem lies in how far the locks and keys are able to move around on the surfaces in order to maximize the linkages from one cell to another. It is implied, but not usually explicitly stated in descriptions of the lock-and-key concept, that the adhesive molecules can move around freely on one cell surface to achieve this optimization, but cannot jump from one cell to the next in the epithelial sheet. To see the need for mobility, consider what would happen if there were none, the locks and keys on any cell being in random arrangement but in fixed positions (Figure 4.18). A gradient from all locks at one end to all keys at the other across the tectum is represented by its two end cells and the middle one. If the incoming axons arrange themselves in the same gradient reversed, then the two end cells can join axons with the use of all their adhesive molecules (12 in the diagram). But the middle cell is quite nondirecting if all molecules are fixed. It will make 6 molecular connections to each of the incoming axons shown. In order to make 12 connections to the middle axon, and that one only, the locks and keys must be able to move around on each cell surface to find each other.

The adhesive molecules must not, however, be able to diffuse from one cell surface to the next, or their diffusion would destroy the gradient, unless it were continuously kinetically maintained. The discontinuity between epithelial cells shown in Figure 4.18, however, is not actually present. The tectum, as is common for epithelial sheets, is covered by a continuous basement membrane (the preferred term today is basal lamina), and the adhesive molecules must be above that to interact with the incoming axons. The paper of Marchase et al. (1975) is followed by several pages of published discussion

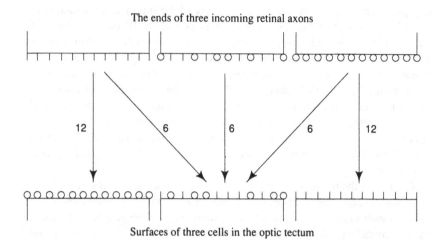

Figure 4.18. The lock-and-key model for the retinotectal junction, to illustrate the need for the adhesive molecules to move on the cell surfaces. Circles indicate "lock" molecules; vertical bars indicate "key" molecules. Numbers (6 and 12): the number of lock-and-key connections made from each axon to each tectal cell if the adhesive molecules do *not* move, illustrating lack of discrimination in the central region.

in which, especially on pages 329 and 331, Brenner, Wolpert, and Roth zero in on this point. It is important, and still easy to overlook today.

There have been several treatments of the shaping of epithelial sheets in which they have been regarded as visco-elastic solids or elastico-viscous liquids (e.g., Jacobson and Gordon, 1976). In these, both deformation of the shapes of cells and rearrangement of cells are considered to be related to the mechanical stresses in the sheets. In the visco-elastic model, a cell can be displaced a certain distance without losing its position in the assembly (elastic behaviour), but greater stress causes it to lose its position and start moving past other cells (viscous or plastic flow). In the elastico-viscous model, cells have no threshold strain for flow in response to an in-plane shear stress, but can show elastic properties for out-of-plane bending of the sheet.

Full consideration of models of these kinds is quite beyond the space available for equilibrium models in a book intended to expound chiefly the kinetic models. But they are mentioned here because they come very close to the border between the two fields. It should be evident from the general nature of these models that they may be set up in terms of approach to equilibrium, mechanical and chemical, but that they could also contain features which would create feedback loops in which the mechanical forces would take on the role of long-range communicators, and the system would self-organize by moving away from equilibrium. Models of that kind are described in the next section.

4.4.4   Kinetic aspects involving mechanical forces

Most cells contain cytoskeletal structures capable of generating mechanical forces on a supramolecular spatial scale. They include microtubules, contractile arrays of microfilaments of actin and other proteins, and, in animal cells, intermediate filaments which, after dissociation of their ends from the cell surface, have an almost miraculous capacity to find their way back to the same point of attachment. They are responsible for cytoplasmic streaming, the complex sequence of events in mitosis and cytokinesis, and most deformations of the shapes of cells during development.

The cytoskeleton is the great hope of the structuralists, because it appears capable of structural self-organization and then of bringing about orderly movements of material on a large spatial scale. Such mechanisms are undoubtedly important. But if I believed that structure could be the great organizer of pattern and form, with dynamics serving only as its handmaids and porters, I should not have written this book. I believe that where mechanical forces are clearly present, and a morphogenetic event is occurring, one should a priori keep an open mind to the possibilities that the mechanical force may be (1) irrelevant to the shaping event, (2) part of the mechanism of the event, but in a subsidiary role, or (3) an integral part of the shape-forming dynamics.

Consider, for instance, the early events of insect embryogenesis. Repeated division of the fertilization nucleus produce a multinuclear syncytium. The daughter nuclei move outward and form a single layer close to the surface. There, plasma membranes form, and the syncytium is transformed into the cellular blastoderm. Apparently at the same time as this sequence of events, determination of the anteroposterior gradient and the segmentation is taking place. Clearly, the outward movement of the nuclei to their places in the blastoderm is a shaping event which involves patterned forces, everywhere directed toward the cell surface. But the formation of anteroposterior gradients may involve both type I (see Section 10.2) and type II (see Section 3.3.1) morphogen behaviour in reaction-diffusion models without mechanical forces. Such forces may be quite irrelevant both to gradient formation and to periodically repeating pattern formation in segmentation, though both of these occupy time intervals overlapping that of the nuclear migration.

Mechanical force as an essential but subsidiary player in morphogenesis (a construction worker, not part of the mind of the architect) may be illustrated by the requirements of sorting-out in assemblies of cells. In the analogy of a mixture of molecules separating into two phases, the necessary molecular motion is thermal. This is a collective property of the assembly, in which every molecule receives its kinetic energy through buffeting by others. For cells, thermal energy is quite negligible as a source of locomotion. Cells can, however, move relative to each other in random directions, so that the motion can be called diffusion, by means of deformations caused by their internal contractile mechanisms. Steinberg and Wiseman (1972) discussed the relative roles in sorting-out of this kind of force and of adhesive forces at the cell surface. They pointed out that a cell cannot locomote unless it adheres to nearby objects, and they raised most of the questions that a theoretical modeller of rearrangements in an assembly of cells needs to worry about, as follows: ". . . a cell might extend a filopodium, attach it to a nearby object, and then contract it, pulling the cell toward the object. The cell still provides all the energy for movement, but here it is especially obvious that certain adhesive relationships must exist for the described movement to occur. The object must be rooted strongly enough, the filopodium must adhere to the object strongly enough and have sufficient tensile strength, and the force must be exerted in such a way that the filopodium neither pulls loose nor snaps and that it is the cell rather than the object that moves."

They pointed out also that rearrangements could be propelled exclusively by the adhesive forces at cell surfaces: "If sufficient inequalities in adhesive energy exist among the various regions of contact, the greater energy of the stronger contacts will cause these to 'zip up' at the expense of the weaker ones, which will tend to yield." In the operation of both internal and surface forces together, ". . . a cell might extend and attach a filopodium to a nearby object, but the filopodium, instead of contracting, might adhere to the object so strongly as to produce a zipping up of the two. . . . this would result in the pulling of the cell toward the object. . . ."

In considering such possible mechanisms for motion, one should not lose sight of the fact that they are means for speeding the approach of the system to its equilibrium geometry, but they do not determine that geometry. Sorting-out remains an instance of pattern determination by minimization of free energy.

Computations on sorting-out may, however, tackle the problem of the equilibrium configuration either directly, without mechanistic detail, or by following a postulated kinetic path. For example, the former approach was taken by Goel and Rogers (1978; Rogers and Goel, 1978). Figure 4.17 is sufficient to indicate that finding the equilibrium configuration of a number of cells with a range of adhesive strengths is not a trivial problem, and needs a computer for anything more than a very small number of cells. Goel and Rogers used a three-dimensional array of cubic sites. Their programs examined pair-exchanges between sites and allowed those that were energetically favourable.

A kinetic approach was taken by Matela and Fletterick (1980). It is based on contact exchanges in groups of four cells in a way analogous to the treatment of changes in bonding between small groups of atoms in the transition-state theory of chemical kinetics. As mentioned in Section 4.3 (especially Figure 4.11a,b), if the surfaces of such a group of cells have appreciable surface tension, the equilibrium configuration, in respect of minimizing surface free energy, will not have a four-cell junction, but two three-cell junctions and a two-cell junction called the "polar furrow." Matela and Fletterick picture (Figure 4.19) the elementary contact-exchange step as a switch of the polar furrow from contact between cells A and B to contact between C and D,

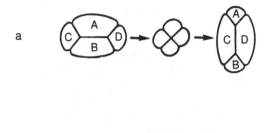

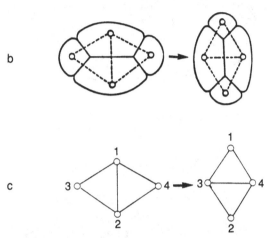

Figure 4.19. The four-cell analogue of a chemical reaction occurring via an unstable transition state, as used by Matela and Fletterick in computations on the kinetics of cell sorting. (*a*) The cell surfaces. (*b*) Transformation to the representation of each cell by a "vertex" (V), and each cell-to-cell contact by a line. (*c*) The transformed symbolism by itself, without the cell surfaces, as usually shown in results of computations. From Matela and Fletterick (1980), with permission of Academic Press, London.

by way of an unstable intermediate state (the analogue of the chemical transition state) which has a four-cell junction. The step is equivalent to a reaction:

$$C + AB + D \rightarrow ABCD ^{\ddagger} \rightarrow A + CD + B.$$

Irregular clustering of cells of assorted polygonal shapes, according to their contacts with neighbours, can be represented. To do this, the cell surfaces are not drawn. Each cell is represented by a small circle (black for strongly adhesive cells, white for weakly adhesive ones), with lines joining any two cells sharing a contact surface. The elementary step of contact exchange is allowed if it improves adhesion energy. As mentioned earlier in connection with Figure 4.17b, when there are only two types of cells, it is necessary only to specify interactions as strong, medium, and weak; numerical values need not be assigned.

Initial

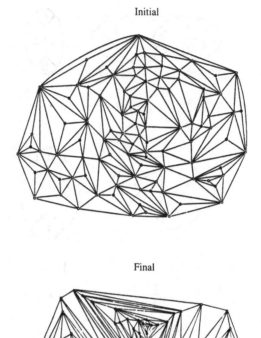

Final

Figure 4.20. Initial and final configurations in a sorting-out computation of Matela and Fletterick. The proper equilibrium configuration is approached, but the distortions of shapes of some cells are enormous, especially near the interface between the clusters of the two cell types. The symbolic representation of cells and contacts is that of Figure 4.19c. From Matela and Fletterick (1980), with permission of Academic Press, London.

The result reported by these authors for a sorting-out computation is shown in Figure 4.20. All the strongly adhesive cells end up clustered inside the array of weakly adhesive cells, but some rather alarming shape changes have taken place, as indicated by very long lines joining cells which supposedly still have surfaces in contact.

This model has the virtue of seeking to represent the kinetics of cell rearrangement, but one may question whether it is at all close to a correct representation. Shape change of cells is unrestricted, except for the use of the polar-furrow concept of contacts. Much more experimental work, such as close examination of the details of time-lapse photographic sequences on cell sort-

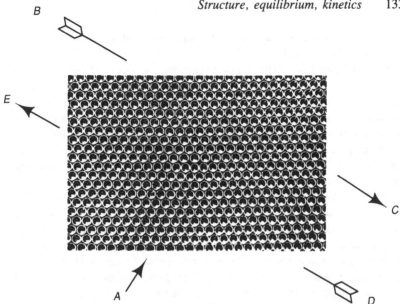

Figure 4.21. A bubble-raft analogue of a two-dimensional crystal containing a dislocation. The defect is close to the centre of the raft. It appears to be localized disorder. But if one looks along the page at a glancing angle, along the arrow *A*, one can see that the defect is in fact a much-longer-range displacement of the lines of bubbles. The presence of such defects can make plastic deformation up to 10^7 times easier than it would be if a whole plane of atoms had to slide simultaneously over another. In this case, the movement facilitated would be of the top right region relative to the lower left, in response to the shearing force represented by arrows *BC* and *DE*. All elementary displacements of the bubbles are much smaller than those of Figure 4.19a. Bubble rafts were invented by Bragg and Nye (1947). This picture, by Lomer, is from Moore (1972), with permission.

ing, needs to be done to establish the details of shape changes. But in the present state of knowledge, it seems to me that two kinds of processes are more likely than the polar-furrow exchange. One is the involvement of contractile filopodia, as mentioned by Steinberg and Wiseman in the quotations given earlier. The other is some analogue of plastic deformation of crystals, in which, as mentioned in Section 4.1.4, the presence of the type of defect known as a dislocation enables molecules – or in the present case cells – to move past each other without any elementary readjustment of more than a fraction of the size of a cell (i.e., nothing nearly as big as the polar-furrow exchange step). Figure 4.21 illustrates the subtlety of this defect by showing just one of them in an otherwise orderly assembly of bubbles on a soap-solution surface, a model which has been used to simulate the mechanical properties of a crystal of copper metal. An array of cells in a developing tissue would contain many such defects.

Up to this point, kinetics and mechanical forces are still employed as obedient labourers in the work of pattern formation. The models of Oster and co-workers would promote them to the managerial office of the architects. Two aspects will be outlined.

The first, from Oster (1983), is a mechanism for concerted deformation of cells in an epithelial sheet, capable of accounting for such shape changes as occur in gastrulation, by invagination of one side of a spherical sheet of cells. The positive feedback loop which makes this self-organization possible, in this model, is as follows: an increase in calcium concentration in the cytogel stimulates production of mechanical stress by contractile microfilament systems; increased mechanical stress stimulates an increase in calcium concentration. The long-range communication needed for self-organization can involve either side of this feedback loop. Calcium may diffuse from a high concentration in one cell to a lower concentration in a neighbour, so that the latter concentration rises and the contractile machinery of the second cell is activated in concert with that already active in the first cell. In that case, the morphogenetic mechanism is basically reaction-diffusion, with mechanical force acting as the final labourer converting the orders of the chemical prepattern into a shape change. Alternatively, in the absence of any significant cell-to-cell calcium transport, the mechanism allows for the forces deforming one cell to deform also, to a lesser extent, attached neighbours. This can switch on an increase in calcium concentration in the neighbours which will lead them to complete the job with their own contractile mechanisms. In that case, the fundamental pattern-forming mechanism is mechanochemical, and does not depend on diffusion.

In either case, the protagonist here is a mechanism known as the calcium trigger, which Oster takes from Jaffe (1980). It causes the cell which receives a small-to-moderate signal, either of force or of diffusing calcium, to amplify it into a big change in the same direction. The amplifier resides in the surface properties of calcium-sequestering vesicles. These are known to respond to changes in external calcium very little up to a certain point, and then by a rapid increase in calcium outflow rate from the vesicle over a small range of external concentrations (Figure 4.22, curve R). If this were an ideal switching mechanism, the curve would be a step function; the rate would jump from zero to its maximum suddenly at a concentration near to x_2. Once the phenomenon has been qualitatively demonstrated, I think most biologists will readily accept, without full quantitative demonstration, a sketch like curve R of a switching function with nonideality in the form of rounded corners to the step.

This raises an important philosophical question of plausibility. If biologists are asked to accept as a working hypothesis that the rate of increase of a concentration C is controlled autocatalytically, being proportional to C^2, many will be quite unhappy proceeding from this starting point. But the initial part of curve R could quite well represented by this dependence. What is happening on the vesicle surface may be quite a complicated process, but the essence of its kinetic consequence is the autocatalysis commonly used as the basis for reaction-diffusion theories.

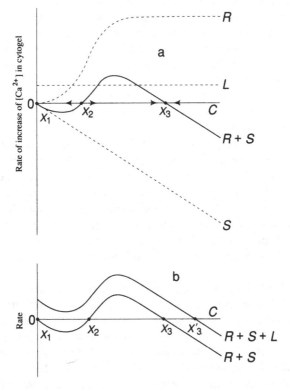

Figure 4.22. The calcium trigger. (*a*) Broken lines represent contributions to the free-calcium concentration, $[Ca^{2+}] = C$, in the cytogel, from various processes: R, release from calcium-sequestering vesicles; S, resequestration; L, leakage, either into the cell from an external supply of calcium or from the vesicles. The latter type of leakage is strain-induced and can increase with increasing mechanical forces exerted on the cell by its neighbours. The solid line is the sum of the release and resequestration terms, without the leakage term. It is of the general type of multiple-steady-state curves, often involving cubic terms (Section 6.4). There are two stable steady-state concentrations of calcium, x_1 and x_3, and an unstable steady state x_2. This value is a threshold: If $C < x_2$, then C will sink toward x_1 (which is zero if there is no leakage term). If a brief pulse of calcium puts C momentarily above x_2, then C will continue to rise until it reaches x_3. (*b*) C can be switched to the upper state not only by a pulse but also by an increase in the long-term leakage rate L. Curve $R + S$ is the same as in (*a*). Curve $R + S + L$ is the same with the value of L as shown in (*a*) added. This shifts the whole curve upward so that there is only one steady state, x_3'. Because L can be increased by forces from neighbouring cells, it is possible for these neighbours to "pull the trigger" on the increase of C in the cell under discussion.

If calcium can be resequestered by a process which is not the thermodynamic reverse of its release, but a simpler process at a rate proportional to external concentration (curve S), then the balance of release and resequestration is a sigmoid curve of the basic type often used to discuss multiple steady states, which are in this case x_1, x_2, and x_3. The remainder of the operation of the calcium trigger is given in the legend to Figure 4.22. If the theory of self-organization is to be mechanochemical rather than reaction-diffusion, leakage L from the vesicles must be promoted by mechanical strain transferred from neighbouring cells. What category of theory we have depends on where L comes from.

To use this mechanism for gastrulation, Oster required two additional assumptions. First, the disposition of contractile elements in a cell was pictured as what the structural engineer would call a truss (Figure 4.23a), but with only one element of it becoming active through the self-organizing process. Second, the cells had to be in a condition to take part in the propagation of this self-organization only on one side of the blastula. Does this asymmetry mean that another pattern-forming mechanism has already done its work on the blastula (which is one of the structures Turing had in mind when he first devised reaction-diffusion), or is it a direct descendant of the animal-vegetal asymmetry of the oocyte? This matter of where symmetry-breaking is needed and where symmetry is already broken is taken up as a major topic in Chapter 5.

Oster et al. (1983) published a paper entitled "Mechanical Aspects of Mesenchymal Morphogenesis" showing how, by kinetic self-organization, mesenchyme cells adjacent to an epithelium might group themselves into clusters in periodic patterns with a quantitatively controlled spacing between them (Figure 4.24). They applied this theory to feather-follicle primordium formation and to the formation of the cartilaginous rudiments of the vertebrate skeleton. Obviously, such a theory is promising for many other morphogenetic phenomena, such as the glandular morphogenesis shown in Figure 4.14.

Like the preceding example, this theory contains both forces and diffusive movements as possible means of long-range communication. But the diffusing species in this case is the mesenchyme cell, not a molecular species. Also, the kinetic self-enhancement which is important in determining the spacing between clusters, at least at early stages, is not a process on the molecular scale, but the mitotic rate r of the mesenchyme cells. To my mind, one of the most interesting features of this model is not that it concerns "mechanical aspects" but that, at least in the limit of the early stages of pattern formation, it reduces to reaction-diffusion with the cell-as-molecule concept. From their paper: ". . . when patterns first appear the spacing between the cell aggregations is given approximately by the expression

$$\text{Spacing} \sim 2\pi [D_2/rN]^{1/4}.$$

However, the precise pattern that eventually evolves can only be determined by numerical simulation of the model equations." (D_2 is diffusivity for the cells, r is mitotic rate, and N is cell density.)

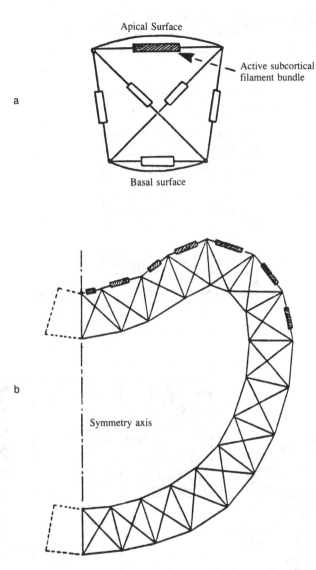

Figure 4.23. The truss-like arrangement of contractile elements in each cell, and the assumptions as to which ones are active, and in which cells, as used in Oster's theory of gastrulation. From Oster (1983), with permission.

4.5   A few problems (without solutions)

A person indoctrinated into the standard ways of thinking of physical chemistry tends to be very sensitive to a particular kind of looseness in the discussion of a phenomenon, and upon detecting that problem, will at once say:

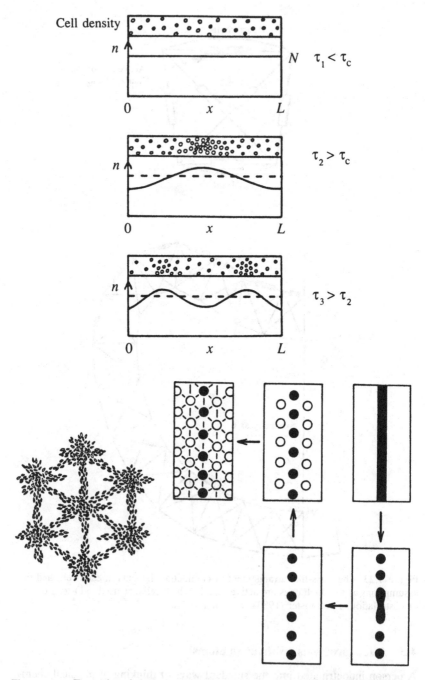

Figure 4.24. Examples of clustering of mesenchyme cells as predicted by the theory of Oster et al. (1983). From Oster et al. (1983), with permission of The Company of Biologists Ltd.

"Now, wait a minute. Are we dealing with equilibrium or kinetics here?" The intent of Chapter 4 has been to try to indoctrinate other kinds of scientists with this attitude, because I believe the adoption of it to be essential to acceleration of progress in solving the problems of morphogenesis. Here are a few things to think about in order to exercise that way of thinking. It is not expected that anyone, physical chemists included, will find clearly unequivocal answers to all these problems.

Problem 4.5.1

A stack of dinner plates on a shelf consists of two sizes of plates. The convenient stack has all the smaller ones together on top of all the larger ones. Initially there are two plates on the shelf, arranged as one small plate on top of one large plate. The stacker takes unassorted plates one at a time, does not look at the plate in his hand but only at the stack, and inserts the next plate always into the present junction between smaller and larger ones. The stack will develop as desired. This could be an analogy for mechanisms of biological ordering in which the plates are analogues of molecules, or cells, or other biological structures. Discuss the structural, equilibrium, and kinetic aspects of any hypothetical biological process for which this could be an analogy.

Problem 4.5.2

Hope et al. (1976) published a model for retinotectal specificity in which the tectum supplies not positional information but only "directional information," such that incoming axons can find from local information on the tectum which is the rostrocaudal direction, and which its rostral sense, and similarly for mediolateral. The analogy is of a parade ground, in which soldiers know which is the front side toward which they face, and which are the right and left sides. They form up in rows, facing the front. In the words of these authors:

Suppose that there is a row of soldiers lined up side by side, facing the same way. The order of the soldiers is initially random with respect to their height. The problem is to give a set of instructions which, if obeyed, will cause the soldiers to be lined up with the tallest man on the far left, the next tallest to his immediate right, and so on, until the smallest is on the far right.

The set of instructions which defines the arrow model, and which is to be obeyed by each soldier in turn, is as follows: Choose either of your neighbours. If you have chosen your left-hand neighbour, then if he is taller than you, stay put; if he is shorter than you, change places. If you have chosen your right-hand neighbour, then if he is taller than you, change places; if he is shorter than you, stay put.

Discuss (1) whether the existence of the local "directional information" on an epithelial sheet does or does not imply the existence of a hierarchy $A > B > C \ldots$ for some kind of property of the cells in any row across the sheet and (2) the structural, equilibrium, and kinetic aspects of this model.

Problem 4.5.3

The establishment of a quantitative measure of length for any part of a biological system, including a macromolecular fibre, is a form-determining event which potentially could have structural, equilibrium, and kinetic aspects. For a protein, the process is entirely structural. But consider a cellulose microfibril in a plant cell wall. Current evidence indicates that some, and quite probably most, such fibres are formed by the action of a rosette-shaped array of a few cellulose synthetase molecules in the plasma membrane, and that each fibre stops growing at approximately the same length. What might be the structural, equilibrium, and kinetic aspects of this length determination?

Problem 4.5.4

Computer programs intended to find the equilibrium configuration of an assembly of two types of cells of different adhesiveness by examining pair-exchanges for whether or not they are energetically favourable are notoriously prone to poor performance. What are the structural, equilibrium, and kinetic aspects of this difficulty? (Consider, for instance, whether the equilibrium configuration is ambiguous, as in Figure 4.17, or whether it is unambiguous but there are numerous only slightly suboptimal configurations. Some of these latter might be much more easily reached kinetically. For example, in precipitations of a solid from solution, the equilibrium state of the precipitate is always just one crystal, but usually large numbers of small crystals are formed. They then move toward, but rarely reach, the equilibrium configuration, by "Ostwald ripening," in which some small crystals redissolve while others grow. It is instructive to write a simple microcomputer program for cell sorting and compare its performance with that of a simple reaction-diffusion model of optical resolution, for which a program is given in Section 10.5.)

Problem 4.5.5

Edelman (1984) asked his reader to

> . . . imagine a stream of water running down a mountainside and striking a submerged boulder whose temperature is below freezing. At first the flow of water will be influenced only slightly by the boulder and the stream will remain a single stream. In time, however, as water freezes onto the boulder, the enlarging structure may suddenly become a barrier causing the stream to split in two and assume a new shape as it runs down the mountain. All subsequent shapings of the stream will be influenced by the effect of the original freezing. Rivulets downstream may break into a variety of new and intricate patterns . . . one can imagine that the cellular counterparts to the driving force of the gravitational potential on the mountain might be the processes of cell division, cell death and cell movement. The counterpart to the freezing of the water would be the attachment of cells to one another by cell-adhesion molecules. . . .

Consider both the biological analogue as Edelman gives it and the alternative possibility that the analogue of the flow of the stream could be the metabolic flow of material in homeostasis, either in a single cell or in an assembly of cells in the same differentiation state. For the latter analogue, what might be the equivalent of the boulder and its enlargement which at some threshold size would lead to a switchover of homeostasis? For any level, cellular or molecular, of biological interpretation of the features of the model, discuss the structural, equilibrium, and kinetic aspects of it. A living cell can have some analogies to a molecule, but is always an out-of-equilibrium flow system. In light of this, are the structural, equilibrium, and kinetic aspects the same or different for molecular analogues and for cell-as-molecule analogues?

Problem 4.5.6

Paul Green, in reviewing the manuscript of this book, threw me a curve in a notation at the end of Section 4.3, which I shall paraphrase as follows:

A balloon is being inflated and is increasing in size. This is nonequilibrium. But the shape of the balloon is essentially spherical at all times. This is a shape of minimum free energy, corresponding to equilibrium.

Discuss the kinetic, structural, equilibrium, and mechanical aspects of the shape of the growing balloon. (Hints: It may emerge in your discussion that equilibrium and mechanical descriptions address the same aspect in different ways. The concept of a quasi-equilibrium, in which various aspects of a system keep pace with a slow change, is very commonly used, both in physics and in chemical kinetics.)

PART II

Pattern-forming processes

Events are ordered rather than structures.

—Thurston C. Lacalli (1973)

Part II deals with the kinds of processes capable of forming pattern. Here the word "process" implies a time sequence of events with essential properties which can be understood only by considering times and rates of change. It is contrasted to "structure," in which parts may have to be fitted together in some definite succession, though it doesn't really matter when; any time intervals, regular or irregular, long or short, can be used between successive steps. The same distinction is implied in this book between the terms "self-organization" (involving processes) and "self-assembly" (involving structural fitting).

To describe processes and their intrinsic pattern-forming capabilities beyond what is given in Part I requires more of the mathematical language of the infinitesimal calculus. In Part II, this is kept as elementary as the subject will allow, and some attempt is made to instruct the beginner in the terminology of this branch of mathematics.

Often the formation of a pattern can usefully be thought of as a gain or loss of symmetry; so Part II commences with a discussion of symmetry. This leads into the instance of symmetry-breaking which is the simplest for mechanistic and mathematical exposition of the principle of symmetry-breaking by chemical kinetics: optical resolution. As mentioned in the Preface, a beginner in this topic who is interested in developmental biology but who doesn't feel especially interested in the old problem of spontaneous optical resolution should not pass over this section; there is no other equally easy example. In Chapter 5 this topic is discussed without mathematical formulation. The example serves to define how far one can go in explaining development by direct simulation of models, with or without a computer, but without equations. Chapter 6 introduces the mathematics needed for reaction-diffusion theory step by step and at one point returns to optical resolution to show the greater power of the mathematical analysis. Alone of the chapters in this book, Chapter 6 has, in textbook style, a set of problems at the end. They are not difficult, but are not intended for the veriest beginner, who should seek routine practice in an elementary calculus text. Under their disguise, my problems are philosophic essays, like most of this book. The discussion in chapter 6 is in terms of kinetic equations, but it allows some exposition of how the second law of thermodynamics is satisfied, so that this topic may then be set aside and the kinetic approach pursued for the rest of the book.

Chapter 7 is the main introduction to reaction-diffusion models for the generation of stable pattern. Turing's model is presented, first without equations, and then with them. We are then equipped to cope with some of the basic questions about what reaction-diffusion can do, and what place it may have within the overall framework of developmental mechanisms, kinetic and otherwise. This is the topic of Part III.

Part II will have achieved my purpose if the experimental biologist understands it but is left in a state of desperation, having realized that it is essential to spend long periods with pen and paper doing algebra in order to study development, and that this activity is just as much a part of the "real world" as is maintaining a culture or purifying an extracted active substance; but together, all these seem to need the 48-hour day.

5

The making and breaking of symmetry

In a review seeking to draw the attention of physical chemists to morphogenesis (Harrison, 1981) I wrote: "Morphogenesis is the creation of a complicated shape out of a simpler one by chemical processes in living organisms. To the physical scientist, the essence of it is symmetry-breaking." Almost simultaneously, Meinhardt (1982) wrote: "In most biological cases, pattern formation does not involve symmetry breaking (although the proposed mechanism can perform this). . . ." Here we have an apparent contradiction between two people who usually see eye to eye about the general strategy of the explanations of morphogenesis. Each of us then proceeded to mention antecedents such that the precept "asymmetry begets asymmetry" would not be violated; but they were different in the two accounts. From mine: "Natural disturbances, which are continually present everywhere, contain adequate asymmetry to serve as antecedent for any shape, however complex. But how do living organisms go about making a rather precise selection of what parts of the available asymmetry to amplify?" From Meinhardt's: "The asymmetric organism forms an asymmetric egg and the orientation of the developing organism is therefore predictable." In the first quotation, I stressed the inheritance of a selection mechanism to extract order out of chaos; Meinhardt emphasized the direct inheritance of asymmetry. Both are aspects of macroscopics, and the contradiction between the two attitudes is only apparent. To see this, however, it is necessary to consider rather precisely what we are doing when we observe a living organism and perceive symmetry, or the lack of it, in what we see.

5.1 Symmetry is in the spline of the beholder

It is a commonplace that the human body displays a certain semblance of bilateral symmetry. This idea is so built into our minds that it is easy to perceive symmetry when it does not exist at all. I once came upon a textbook which illustrated bilateral symmetry with a drawing of one of those very formal and orderly looking ancient Egyptian statues of a standing man. In accord with the artistic convention of the period, the statue had its left foot a little farther forward than the right. This displacement destroys, totally, the

145

symmetry of the statue. A living, flexible human body in fact is rarely close to showing external bilateral symmetry except when it is standing at attention.

In common with other well-known properties of geometrical figures, such as the constant sum of the angles of a triangle, symmetry is an attribute of a mathematical idealization, not precisely attained by any real physical system. This idealization is an array of points in rigidly fixed relative positions. For spherical or cylindrical symmetry, it is an infinite array of contiguous points. In looking at any living organism, we make some kind of approximation to extract this idealized structure out of it. There are various kinds and levels of approximation commonly used. One was illustrated earlier. We tend, usually without formulating the problem to ourselves in precise terms, to define the symmetry of a flexible object as the *maximum* symmetry attainable by any rearrangement of its parts allowed by its flexibility.

Even when the symmetric possibilities are thus maximized, to perceive symmetry we must ignore many structural features. Looking only at the outline of the body, and not contending with the disposition of the internal organs, to see symmetry we still have to pass over such things as fingerprints. The perceived symmetry is that of a smoothed-out envelope drawn through these small-scale irregularities. In visualizing it, we do, mentally, what any scientist does mechanically to experimental data by drawing a smooth curve through a set of somewhat scattered points. For nonlinear plots, this fitting often used to be done on a drawing board with the aid of a flexible strip of wood or metal or plastic which was held in any desired curved configuration by a set of weights; it was called a spline. Nowadays, the scientist is more likely to put the points into a computer and summon up a genie with some such name as SPLINEFIT to do the job. This deprives the operator of seeing in an obvious, concrete, Aristotelian way that he has imposed a long-range continuous organization upon the set of points.

An egg cell is often spherical in outline, to a good approximation. This is a legitimate symmetry to describe as such. At other levels of approximation, we perceive the animal–vegetal polarity, and see that neither contents nor surface are spherically symmetrical in detailed chemical composition. If our description goes down to the molecular level in spatial scale, no real system is spherical, because spherical symmetry is a property of an infinite set, and all real systems have finite numbers of molecules. But the idealized spline curved through the outline is spherical. When the egg first divides, that symmetry is broken. As it divides repeatedly, if it is, for instance, an echinoderm egg, a new spherical symmetry arises, that of the blastula. This is perceived as spherical at a different level of approximation, by another spline-fitting. Turing referred to this level of spherical symmetry and pointed out that a system with spherical symmetry would, except for the presence of natural disturbances, remain spherically symmetrical forever, even if it did have reaction-diffusion amplifiers in it. He wrote: "It certainly cannot result in an organism such as a horse, which is not spherically symmetrical." (Evidently he had an

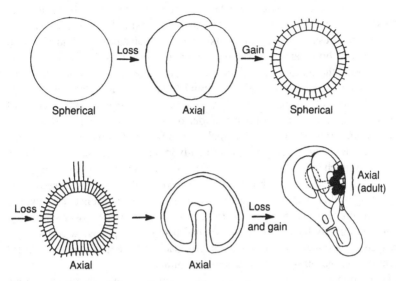

Figure 5.1. A generalized sequence of makings and breakings of symmetry typical of early echinoderm development. Both losses and gains of symmetry occur, at first in the whole organism and later in its parts. Each is seen only by using an appropriate spatial scale of approximation, which is not the same at all stages and for all aspects of development. Each, however, points to the existence of a morphogenetic field, or spatial range of operation of some morphogenetic control process. The echinoderms are the highest animals in which the blastula is clearly spherical and the early gastrula is of cylindrical symmetry (in having one infinite-fold rotation axis).

impression, which some elementary textbooks of embryology seem quite anxious to convey, that spherical blastulae are more widely distributed in nature than they are.)

Thus, while nature is breaking a symmetry as seen to some approximation at the single-cell level, it may be on the way to making another symmetry as seen at the multicellular level with a grosser spatial scale of approximation. Figure 5.1 shows a typical sequence of such changes in symmetry taken from the early stages of echinoderm embryogenesis. Such complex sequences can be perceived throughout development. Whenever one recognizes such a change, one has applied a new spline-fit to the developing shape. Each time, the question arises: Has nature used a spline?

Whenever one figuratively reaches for the spline to put it through the developing shape in a new way, one has seen an event which is in some sense always of the same kind; though sometimes the system has gone from a higher symmetry to a lower one, and sometimes from lower to higher, the significant thing is that it has changed from one state of organization to another. The spline signals that the process is organized over whatever spatial scale the spline is used on – that is the sense in which nature has used a spline.

Symmetry-breaking should not be confused with reversion from order to chaos. It can be quite the reverse. For instance, if a simple chemical system in an elongated vessel consists of a mixture of two substances, A and B, and if they are somehow caused to separate, so that A occupies the top half of the vessel and B the bottom half, then a symmetry-breaking has occurred on the spatial scale of the vessel and substantial fractions of its total size. But on the molecular scale, the system has become more orderly. The relation of macroscopic complexity of form to order on a very much more microscopic scale is not straightforward. This makes it rather dangerous to try to put the physical science of development continually in the language of entropy. There is just too much chance of setting up an incorrect description in that language.

None of the foregoing, however, fully resolves the question raised at the beginning of this chapter. We perceive symmetry in the blastula at one level of spline-fitting. But when it goes on to show development to a lower symmetry, might we already have seen this if we had had a more precise description of the system, and traced the lower symmetry back to the egg? Or is a new event necessary at the blastula stage to break symmetry?

These questions appear superficially to present clear alternatives. One must, apparently, answer one in the affirmative and the other in the negative for any particular developmental phenomenon one is considering. To my mind, the greatest philosophical difficulty in the attempt to describe and explain development from the molecule to the organism is that such a distinction is nonsense. Both questions must always be answered in the affirmative. At one level of approximation in spline-fitting, new shapes and symmetries are appearing sequentially de novo. At another, they are all present in apparent molecular chaos. It all depends how hard one looks before bending the spline.

The essence of understanding self-organization is that the external human observer should picture himself vanishing into the system and becoming part of it. In this instance, the question becomes, How hard can the system look at itself? Or, more precisely, from what rudiments can the system start to build a new macroscopic shape? In arriving at this question, I have departed significantly from Turing's attitude toward the problem of antecedents for pattern. His opinion was that there is no problem. He used the analogy of an electrical oscillator. Any time one switches it on, connected to a loudspeaker, one expects to hear the same note, apparently immediately, on the time scale which people can follow. One simply does not worry about what gave the electrons in the circuit the "kick" to start the oscillation. From the resistance, capacitance, and inductance of the circuit one can calculate an oscillation frequency. One expects it always to be expressed.

If, however, one looked at the behaviour of an oscillator in the first few milliseconds after switching it on, one would probably get a quite different impression, of a complex and confusing signal gradually changing into a simple and precisely controlled kind of behaviour. For biological development, it is my impression that the corresponding time interval is from tens of minutes up to some days. Experimental evidence which must be used to prove or disprove kinetic theory refers to patterns which are not in the infinite-time form, equivalent to the pure note from the electrical oscillator. It is therefore necessary to ask what rudiment a system contains within it for a particular pattern. The final pattern may depend only on the mechanism and not on the starting point (except for orientation, which may or may not be important); but the time taken for its establishment may be quite sensitive to the rudiment.

Themes introduced in this section recur, with more extensive development, in various parts of this book. There is more about splines operated from within the system in Section 5.2. An example in which a human observer may fail to find a hidden pattern, though the most primitive reaction-diffusion mechanism can recognize it easily, is discussed in Section 5.3. The question of spline-fitting, when it is done from within, without an external intelligent being to work the spline, is almost identical to the question of selective amplification of the great range of available inputs.

5.2 Open and closed traverses: the accuracy of self-organization

Since I began to study morphogenesis I have continually been struck by the contrast between the orderliness of the overall shapes of organisms or their parts and the comparative disorder of the microscopic parts of which they are constructed. The first example which I encountered, and still one of the most striking, is Lacalli's transmission electron micrograph of a developing semi-cell of the desmid alga *Micrasterias rotata* (Figure 5.2). The regularity of the outline greatly exceeds that of the organelles within, and when the living cell is observed under the optical microscope, its contents are seen to undergo rapid and rather irregular cytoplasmic streaming. It is possible to maintain that both the outline and the streaming cytoplasm are dependent upon a cytoskeleton which imposes the overall order. But most pictures of the cytoskeleton which I have seen show increasingly disorderly tangles as the magnification increases; the more detail, the less order. The same is true in animal tissues for the collagen fibrils of the extracellular matrix, which un-doubtedly plays some role in morphogenesis, but surely not by a strict struc-tural conveying of short-range geometry to long-range geometry. Also in *Micrasterias* and other plants which form both primary and secondary cell walls, it is striking that the orderly morphogenesis occurs with formation of a disorderly array of primary-wall microfibrils from a disorderly array of poly-merase rosettes, whereas the nonmorphogenetic secondary wall, which later passively rigidifies the already formed outline of the cell, is regular and is formed from a regular hexagonal array of rosettes (Staehelin and Giddings, 1982). (Yet more paradoxically, the instances in which small structures stack up most precisely into orderly arrays are not concerned with the construction and maintenance of fixed shape, but with the mechanisms of motility: flagella, and the actin-myosin contractile mechanisms of muscle cells.)

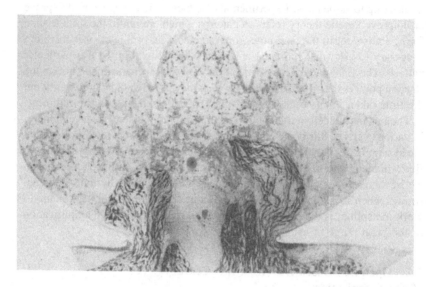

Figure 5.2. Electron micrograph of a developing *Micrasterias rotata* semicell. From Lacalli (1973), with permission.

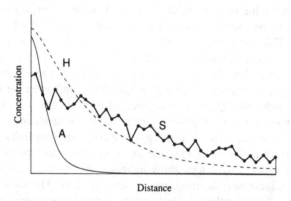

Figure 5.3. Computation by Gierer and Meinhardt illustrating a reaction-diffusion mechanism smoothing out irregularities in a system: S, source gradient, with persistent irregularities; A, steady-state concentration of activator; H, steady-state concentration of inhibitor. From Gierer and Meinhardt (1972), with permission from Springer-Verlag.

In all these instances of rough materials and polished form, I tend to see, rather directly, nature at work with the spline. An example of reaction-diffusion doing this, in a computation, was given in the first paper of Gierer and Meinhardt (1972). A source gradient that contained random irregularities of substantial amplitude was worked upon by the kinetic model and converted into almost exactly the same "fired gradient" that would have been obtained from a smooth, straight-line source gradient (Figure 5.3).

In case the reader has had enough of splines, a different analogy comes from the work of the land surveyor on the ground, with theodolite and chain. In the days before one could do everything with photographs from the air or from outer space, intervisibility of distant points was a serious problem. Let us suppose that a set of points on the terrain to be surveyed (*A, B, C*, etc.) is such that *A* and *C* are both visible from *B*, so that the angle *ABC* can be measured, but there is no longer-range visibility. For a simple illustration of error accumulation, let us suppose that all the points are actually in a straight line (Figure 5.4a). Because each direction is measured relative to the previous one, errors in direction accumulate. The last segment may be badly displaced in the survey from its true position on the ground (Figure 5.4b). This procedure, with internal short-range reference only, lacking both long-range controls and external reference, is called an "open traverse." In navigation, it is "dead reckoning." Error accumulated in the well-known statistical fashion. The measure of error in position or direction at the end of the traverse (e.g., the standard deviation of either quantity, applicable to a large number of repeats of the whole traverse) increases with the square root of the number of segments. The total length, however, goes up in proportion to the number of segments; hence fractional (or percentage) error in *total* distance drops as the number of segments becomes larger. But direction does not accumulate like total distance, and hence it is the error in direction which is clearly seen to get worse and worse as the traverse lengthens. The shape of the traverse has, of course, aspects of both length and direction.

Two things have commonly been done to improve accuracy. One is to "close the traverse" by ensuring that points *A* and *Z* are intervisible. When the direction from *A* to *Z* has been ascertained, the necessary correction can be spread over all the segments, reducing but not eliminating error (Figure 5.4c). This shows the use of a long-range control in a more active way than the previous example of a spline. When the long-range control has been established, the intermediate points are actually moved in conformity with it. In the simplest spline-fit, drawing a straight line through a scatter of points with a ruler, this is similar to finding the best line and then moving each point to a nearby position on the line, thereby removing the scatter. In a living system, there may or may not be an obvious spline, or traverse-closing observation, independent of the intermediate points. The points themselves may have to represent the line. In that case, the sophisticated long-range interactions which are the main topic of this book are probably involved. In other cases, whether or not such mechanisms are involved, some smoothing may occur in a more trivial way, with something closely analogous to a spline which, however, actually pushes the points into line (e.g., an envelope with appropriate balance of flexibility and flexural rigidity).

The other way for a surveyor or navigator to avoid accumulation of errors in direction is to use an external frame of reference by replacing the theodolite with a compass and measuring every direction with respect to magnetic north instead of measuring angles such as *ABC*. Clearly, living systems sometimes

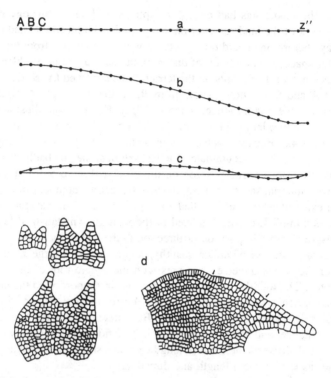

Figure 5.4. Accumulation of errors in spatial sequences of structures. (*a–c*) Surveying a traverse from point to point. (*a*) It is assumed that the true map is a straight line. (*b*) No error has been deliberately put into the length of each leg of the survey. Directional errors have been put in by tossing a coin four times for each leg. Heads are given the value 1, and tails −1, so that the values −4, −2, 0, 2, and 4 were obtained in a 1 : 4 : 6 : 4 : 1 probability distribution. Each leg was offset in direction from the previous one by this error, taken directly as degrees (negative is clockwise). (*c*) The open traverse of (*b*) has been closed by a determination of end-to-end direction, assumed to have given the correct result. The correction has been distributed along the traverse. (*d*) Four stages in the development of a leaf of the liverwort *Lophocolea bidentata*. From Bopp (1984), after Bopp and Feger (1960). Whether the regularities in overall shape (and, not shown, its reproducibility from leaf to leaf) are statistically compatible with a two-dimensional open traverse or require its closure on a scale longer than the cell-to-adjacent-cell basis is an open question. This is intended to suggest that the error-suppressing ability of every kind of pattern-forming mechanism suggested should become a major field. There is enough in it for many workers. See Lacalli and Harrison (1991).

do this (e.g., in the negative geotropism of plants, wherein the gravitational field provides an external frame of reference). Not many developmental events, however, involve a frame which is not self-generated by the organism as part of its self-organization. The theoretical modeller, especially when using a computer, must be continually on guard against putting into the models something external which the living system does not have. For instance, "cellular automaton" models usually generate patterns in a rectangular array of sites which is larger than the developing pattern and can be arbitrarily expanded indefinitely as the pattern needs more space to grow into. A living organism has to manufacture the coordinate grid on which it grows as an integral feature of its growth, and never ahead of the spatial limits it has reached at any time.

Figure 5.4d shows the assembly of cells at four stages in the growth of the leaf of a liverwort. Plant cells do not sort out. Each cell may change its size and shape, but they remain in the spatial sequence in which they are formed. They are therefore analogous to the successive stages of a survey, as shown in Figure 5.4a–c. Morphogeneticists should continually be asking themselves this question: Does the final shape show the amount of dead-reckoning error which one would statistically expect, or is it controlled over the long range better than an open traverse? Thin filaments in the plant kingdom often do have irregular changes of direction just like diagram (b), and clearly are not keeping control along the whole length. Leaves, on the other hand, seem to grow to species-characteristic shapes, despite marked irregularities in sizes and shapes of individual cells.

It is not intuitively obvious at the pictorial level whether Figure 5.4d is the two-dimensional analogue of the open traverse of (b) or the closed traverse of (c). There is scope for statistical analysis and computations here. Meanwhile, nature itself supplies some of the statistical work. Anyone wanting to photograph a wide variety of stellate snowflakes may take a selection out of a single snowstorm. To get an equal range of branching patterns for maple leaves, or diatom patterns, one would not go for a selection of leaves from one tree, or cells from one species in culture. One would look for a range of *species*. As discussed in Chapter 4, even the snowflake does not grow statistically randomly. Without some long-range kinetic control, it would not be dendritic at all. But its degree of control of detailed branching pattern is surely primitive compared with the leaf or the diatom frustule. The identical molecules of a snowflake might seem to be better starting material for building a well-controlled shape than the very variable cells shown in Figure 5.4d; but it is the latter kind of unit which is the building block for species-specific shapes. In summary:

1. Every description of a shape or symmetry in a living system is an idealization, or spline-fit.
2. For every such idealization by the external observer there is within the system a pattern-forming event.
3. The relation between the large-scale order of the shape and the degree of spatial disorder of its parts gives an indication of the controlling power of the pattern-forming event, in the following hierarchy:
   A. Least powerful: The overall pattern or form may be generated by

repeated small-scale activity only, with no long-range control. There must then be an appropriate statistical match between the overall form and small-scale irregularities.

B. Intermediate: Simple mechanical interactions, such as a moderately flexible envelope, may impose some smoothing on the form so that its irregularities are less than statistical expectation for the accumulation of small-scale irregularities.

C. Most powerful: Where long-range order seems to have overridden small-scale irregularities very markedly, the presence of a powerful kinetic organizer, such as reaction-diffusion or mechanochemical action, is strongly indicated.

4. The indications in (3) are for completely self-organizing systems. More orderly shape may arise from less internal organizing power whenever a system uses an external frame of reference for some aspects of its form.

A very remarkable instance of approximate symmetry occurs in a few non-living systems, the quasicrystals. These, to a good approximation, have icosahedral symmetry and are three-dimensional analogues of the "Penrose tilings" which, in two dimensions, show a good approximation to fivefold rotational symmetry, an element of symmetry which is mathematically forbidden in two- and three-dimensional arrays with translational symmetry (i.e., crystals). The approximation to a forbidden symmetry is a very sophisticated instance of a spline-fitting. Penrose (1989, p. 565) points out for such structures that "their assembly is necessarily *non-local*." Models of achieving the structure by successive addition of one atom at a time simply cannot work. Penrose suggests that there must be a quantum-mechanical ingredient to such an assembly process. His suggestion goes far beyond any concept of action across long distances that I use for biological self-organization. Everything I discuss involves strictly classical concepts of long-range communication by diffusion or mechanical force to establish approximate long-range symmetries. Are these enough, or will biological theory some day involve grappling with the concepts of long-range quantum interactions which are not yet well understood by physicists?

5.3   The simplest reaction-diffusion mechanism: optical resolution

The preceding section was intended to allow one to visualize oneself in the midst of a somewhat disorderly system acting as a perceiver, corrector, and amplifier of whatever long-range order is there to be found, that is, first as the operator of a spline, then as the surveyor calculating corrections after closing a traverse, but finally as something remaining to be discussed. In the arrange-

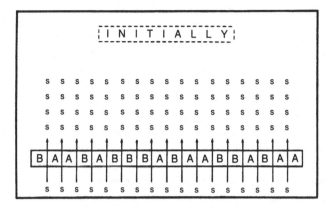

Figure 5.5. The starting point for a simulation of the Mills model for optical resolution, with diffusion added so that it becomes a simple reaction-diffusion model and permits transient pattern formation; *s* is a substrate in solution. It can be converted autocatalytically into either of the enantiomers *A* and *B*. They are shown attached to sites (arrows) on a rigid catalyst bed. For illustration of concepts throughout Chapter 5, the initial arrangement contains an almost imperceptible rudiment of long-range order. The working of the mechanism is described in Section 5.3.2 and Figure 5.6. In relation to biology, the catalyst bed with its attached activator molecules is the analogue of the living system, and the flow of *s* represents intake of food and excretion of waste products. This flow is also the communication between system and surroundings which makes the increase of order in the system thermodynamically possible (Section 6.4).

ment of *A*'s and *B*'s in Figure 5.5, can the reader who has not already glanced at the later diagrams perceive any long-range order? A reaction-diffusion mechanism not only can perceive it, but will make haste to get rid of everything else.

How spontaneous optical resolution occurred during biochemical evolution has been an important philosophical question since Pasteur (1848) sorted out a pile of crystals of sodium ammonium tartrate into separate heaps of right- and left-handed crystals. A fundamental question arises of random chance versus sequential cause, the stochastic versus the deterministic. On this basis, three kinds of theories may be distinguished:

1. Theories linking spontaneous resolution to the intrinsic universal asymmetry of the so-called weak interaction in the theory of nuclear structure, and the phenomenon of nonconservation of parity. [For a clear account of this universal asymmetry for the nonphysicist, see Gardner (1982).] This is the only type of theory which is completely deterministic: Living material has a twist in it because the universe has a twist in it. On this basis, if we were to discover life which had originated independently on a number of

planets, and with the same chemical basis as on earth, the amino acids should be L and the sugars D in all cases.

2. Theories linking spontaneous resolution to preexisting asymmetry in some inanimate system. Two examples: first, the possibility that early life interacted with asymmetric quartz crystals which happened to have an imbalance of the two types; second, possible interaction with circularly polarized radiation likewise unbalanced. In this kind of mechanism, the processes envisaged within life are completely deterministic, but the question of the ultimate origin of the asymmetry is begged. There is no explanation of how the inanimate system, crystals or radiation, came to have a chiral imbalance, and no prediction as to whether one would expect to find the same or variable chirality for life on different planets.

3. Theories linking spontaneous resolution to a chemical reaction system with the property of self-enhancement (positive feedback, autocatalysis) on a stereospecific basis. Such a system could act as an amplifier of asymmetry and use as antecedent nothing more than random disturbances, which are always present everywhere. If the significant reactants are diffusible, the mechanism becomes the simplest form of reaction-diffusion and is likely to produce transient unstable pattern as an intermediate stage in resolution. On this basis, life on different planets should display both possible chiral systems with statistical distribution, just as if the choice had been made for each planet by tossing a coin.

For collections of papers on the various theories, see the symposium volumes edited by Thiemann (1974) and Walker (1979).

The following account relates to theories of type 3, which I consider by far the most probable. Theories of type 1 would require an amplification from energies of interaction of order 10^{-14} eV to chemical bonding energies of order 1 eV. This seems unlikely; if it does happen, then more interest should attach to the amplifier than to the initiator, and the amplifier could be a mechanism of type 3. Theories of type 2 transfer the question to behaviour of inanimate matter, which is less likely than living material to contain sufficient self-organizing capability.

Nevertheless, I have noted in informal conversations that many among the most objective of scientists yet seem to have an emotional antipathy to concepts which assign a dominant role to chance at some point in what is otherwise a strictly deterministic sequence. This produces a predisposition against theories of type 3. I do not share that antipathy. It does not bother me that on some other planet, life may have arisen just like ours, but with D-amino acids and L-sugars. What would be the same on both planets would be the fundamental nature of the amplifier which would allow resolution to occur in either case, and that is what I am about to discuss. Einstein, as quoted by Wald (1957), explained the dominance of one chirality in the words "it won in the fight." What we are studying, then, is the strategy of the fight.

5.3.1 The mechanism of W. H. Mills

Mills (1932), in a review talk on stereochemistry at a regional meeting of the British Association, seems to have clear priority for the concept of the asymmetrizing power of an autocatalysis needing two product molecules. This later appears in the crucial step of the Brusselator pattern-generating mechanism of Prigogine (1967) and is implicit in the A^2/H term in the growth rate of activator A in the Gierer-Meinhardt model (1972). It is not quite so obvious that the Turing model effectively involves the same concept, but it does (see Sections 6.3, 7.1, and 9.2.2).

Mills first pointed out that Pasteur sought the origin of the optical activity of life in dissymmetry of the universe (not, in the modern way, via the minute scale of nuclear processes, but the obvious large-scale dissymmetry). He continued:

. . . it may be profitable to enquire whether the property of growth which is characteristic of living matter may not necessarily lead to its dissymmetry. . . . Let us now consider the growth of a tissue which is not completely optically inactive, that is, a tissue in which the d- and l-systems are not present in equal quantities. Let us suppose, for example, that there is twice as much of the d-system as of the l-system . . . in the process of growth . . . the d-system will increase at a relatively greater rate than the l-system. The complex dissymmetric components . . . will be built up . . . by chains of synthetic reactions, and the rates of formation of the end-products will be controlled by the velocity of the slowest link in the chains. If we consider a case in which, as must frequently happen, this slowest link is an interaction involving two dissymmetric molecules and . . . assume that, as in a simple bimolecular reaction, the reaction velocity is proportional to the second power of the concentration, then the rate of formation of the d-component will be four times that of its enantiomorph. If this applied to every dissymmetric constituent of the new growth, then, whereas there was twice as much of the d- as of the l-system in the old tissue, there would be four times as much of the d- as of the l-system in the new growth.

It will be clear that, even though the reactions of living matter may be less completely stereospecific than I have, for simplicity, assumed, and though the velocities . . . may increase more slowly with the concentration than according to the second power, yet as long as they increase more rapidly than according to the first power . . . any excess of one system over the other in the old tissue will become greater in the new growth. . . . There is an *a priori* probability that an optically inactive growing tissue would be, as regards its optical inactivity, in a state of unstable equilibrium. . . . From this point of view the optical activity of living matter is an inevitable consequence of its property of growth.

(He goes on to discuss the original bias and to show that the statistically expected scatter about equal numbers of d and l molecules would be adequate to start the asymmetrization in something about the usual size of a cell, 30 μm in diameter.)

This earliest proposal for the asymmetrizing mechanism is striking in that it lays stress on a simple general property of something envisaged from the start as a very complex reaction system. When the topic is developed mathematically, it is usually idealized into a couple of equations relating to the behaviour of a small number of substances. All such idealized treatments should be approached in the spirit of Mills' generalizing vision. The model immediately following is such an idealization, but it also adds the feature of diffusion, which was not in Mills' proposal, and hence allows pattern formation as an intermediate stage on the way to optical resolution.

5.3.2   A model without mathematics

In Figure 5.5, the eighteen boxes represent a solid surface having sites capable of being activated for catalysis of a reaction by adsorption of its product. The reaction is $s \rightarrow A$ or $s \rightarrow B$, the alternative products A and B being the two optical enantiomers of the same molecule. The catalysis is bimolecular, corresponding to Mills' model. A site is represented by the boundary between two boxes which represent attachment positions for A or B. The substrate s is envisaged as in solution and flowing upward across the catalyst bed. The reaction rule is that an s which passes between two A's is converted to an A; an s which passes between two B's is converted to a B; and an s which passes between an A and a B is unchanged. At the outset, A and B have been adsorbed at random onto the attachment positions. But randomness cannot be represented unequivocally by an arrangement of any finite number of objects, and numbers so small as nine A's and nine B's are bound to show some kind of order in their arrangement. In an infinite random array, one-quarter of the catalytic channels would be active for production of each of A and B; here, six out of seventeen are active, three to make A and three to make B (Figure 5.6a).

After some reaction has occurred, diffusion is simulated in the solution by choosing at random to leave each A or B unmoved or to move it one space right or left. Then, again by random choice, some of the molecules in solution are exchanged with ones attached to the catalyst surface. This having been done, the continual upward washout which would be occurring in a flow system is simulated by replacing all the A's and B's in solution with s's (Figure 5.6b–d). Of course, all these steps should be envisaged as going on simultaneously.

A substantial increase in activity of the catalyst has now occurred. In Figure 5.6d, eleven channels are active (six for A and five for B), and the activities for A-production and B-production are clearly becoming segregated into wide blocks on the catalyst surface. A number of stages of the same sort of thing have been omitted before the next reaction-diffusion sequence shown (Figure 5.6e,f). After one more complete sequence, the system has sorted out into a pattern of two equal parts (Figure 5.6g). This will not happen inevitably. In a number of repeats of the same simulation, the pattern at this stage will often

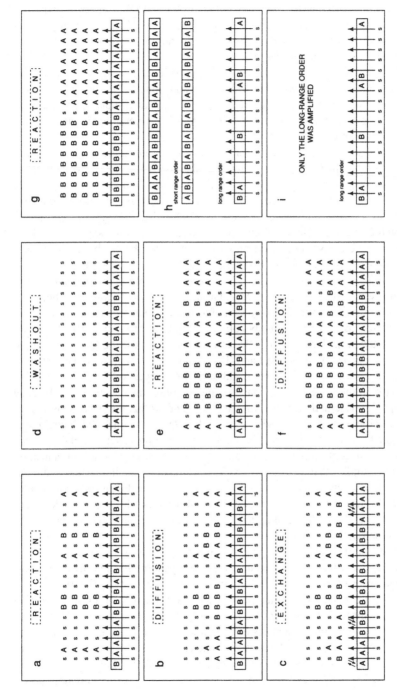

Figure 5.6. Some stages from a simulation of the reaction-diffusion mechanism for optical resolution.

consist of somewhat unequal assemblies of A's and B's, but always with the B's on the left and the A's on the right, if the identical starting configuration was used. If that was also made subject to random variation, the two-part pattern will sometimes have A's on the left.

This system has not yet reached its inevitable destiny. It has gone as far as I could take it in half an hour with paper and pencil. The next stage takes longer, and it is useful to put the model on a small computer, as described later, to pursue the development to the end. But several significant features already emerge.

First, the pattern that has arisen had an antecedent in the original arrangement of adsorbed A's and B's, however disorderly that may have seemed at first glance. In the simplest terms, the left-hand half had an excess of B's over A's (5 to 4), and the right-hand half had the same excess reversed. Figure 5.6h shows the same imbalance in a different way. The starting arrangement is compared with a strict alternation of A's and B's, a short-range order which would leave the system forever totally inactive for the catalysis, since there would be no adjacent AA or BB pairs. The six discrepancies, three in the left-hand half and three in the right, form the same pattern in the two halves, but with reversal of A's and B's. Only that long-range order was amplified.

This provides a simple illustration of a concept addressed at length and with sophisticated mathematics by Nicolis and Prigogine (1977, part III, "Stochastic Methods"). They wrote: "In essence, the purpose of the macroscopic analysis carried out in [parts I and II] was to test the *response* of the system to an *external disturbance,* which, although weak, was assumed to be present initially in the system with a macroscopically observable amplitude. The view adopted in fluctuation theory, on the other hand, is that the system has to generate spontaneously this deviation from the average regime, which could then trigger further evolution to a dissipative structure provided it contains a non-Poissonian contribution." And later: ". . . the onset of a self-organization process implies that there exists a volume element within the system of dimensions much larger than the characteristic molecular dimensions but smaller than the total volume of the system, within which fluctuations behave coherently, and that they add up to a sizable result and subsequently modify the macroscopic behaviour."

These authors are saying that it is not the random (Poissonian) aspect of fluctuations which serves as antecedent for pattern, but the incipient long-range order (non-Poissonian) contained in them. Elsewhere (Harrison et al., 1984), in connection with time and space scales for pattern formation, I used Poisson statistics to estimate the initial amplitude from which pattern starts to grow. That was for an analysis of experimental data on the kinetics of pattern switchover after rapid temperature change, for the temperature-sensitive whorl spacing in *Acetabularia.* Mills used the same kind of estimate in considering the size of a system in which optical resolution could readily occur.

Are these estimates legitimate, in the light of Nicolis and Prigogine's comments?

For a pattern of just two parts, which is a situation of widespread biological importance, simple Poissonian statistics indicates a high probability that random fluctuations will provide a substantial rudiment. Suppose, for instance, that we divide an elongated system into left- and right-hand halves. The average concentration of a morphogen is such that each half should contain 10^6 molecules of this species. The Poissonian standard deviation of this quantity is its square root, 10^3. Without further ado, we can make an order-of-magnitude estimate that imbalances between the numbers of molecules in the two halves on the order of a few hundred to a thousand molecules will commonly arise, and that that is the amplitude of the rudiment for two-part pattern. More precisely, the expectation value is $N_L - N_R = 0$, but its standard deviation is $\sigma_{L-R} = (\sigma_L^2 + \sigma_R^2)^{1/2} = 2^{1/2}N^{1/2}$, where N is the average value of N_R or N_L. If we are not bothered about which way round the pattern appears in the system (i.e., we are interested in $|N_L - N_R|$ rather than the value with sign attached), then the root-mean-square value of this is σ_{L-R}. This is about 1,400. One could take half of it, 700, as a good estimate of initial pattern amplitude. This kind of estimate is good for patterns of two or three parts, but probably overestimates initial amplitudes for patterns of greater complexity. These require coherence, or in-phase character, between fluctuations in different parts of the system; it is this non-Poissonian behaviour which a reaction-diffusion amplifier will tend to pick up. In the remarks of Nicolis and Prigogine quoted earlier, they were thinking not only of biological pattern but also of such things as nucleations of phase changes (e.g., precipitations, in which thousands of events such as formation of small crystals may be occurring simultaneously in the system). This is a different order of complexity from a two-part pattern.

Second in the list of features which may be extracted from the model before we let it go on operating is the absence of anything commonly recognizable as mathematics in this presentation of it. To see this fully, it is essential to try using the model first with paper and pencil only, abstaining from the use of any kind of electronic computer. In this way, I made enough pictures for an animated film shown at the fortieth annual symposium of the Society for Developmental Biology (Harrison, 1982), from which Figure 5.6 presents a few extracts. No numbers are used; this is an *analogue* computation, although it is straightforward enough to program it for a digital computer, as described later. Nor need any words be written down. Three or four abstract symbols have to be manipulated according to fixed rules. Random choices can be made by guesswork, or with less risk of bias by using a coin or dice or playing cards.

I recall reading in a Vancouver newspaper a few years ago of the death of an old street gambler who was adept at manipulating a pack of cards, but who had otherwise never learned to read or write. I expect he was not too good on partial differential equations either, but his thought processes must have worked quite well on playing complicated games according to rules. This is all one needs to follow the model presented here.

Third, and arising directly from the previous point, there is a variety of things one may do with a computer to find out whether or not a model works:

A. *Computer experiments:* These are necessary when one does not have a complete understanding of the behaviour of the model from mathematical analysis, which is nearly always the case for nonlinear dynamics. One programs the model into the computer and determines its properties in ways which have the style of an experimental project on a process in a living organism. Two levels of description of the model for insertion into the computer may be distinguished:

1. *Simulation without equations:* The computer program is set up directly from the verbally expressed rules of the model, as in the instance just described. Figure 5.6 is such a simulation, but was done without even using a computer, because the number of operations was small enough to do it by hand.

2. *Simulation by numerical solution of dynamic equations:* To do this, one converts the rules of the model into differential equations, as described beginning in Chapter 6. One then uses standard arithmetical methods of solving differential equations on a digital computer. This is a long step in abstraction from direct operation of the model. It can require careful consideration of the relationships of the continuous variables in the differential equations to two kinds of discontinuities: (a) the finite-difference procedures used in the computer; (b) the structural discontinuities in the living system being modelled. (See Figure 9.6 for a very coarse-grained computation intended to match the small number of rows of nuclei in a *Drosophila* blastoderm.)

B. *Numerical calculation of analytical solutions:* If one has analytical solutions, one knows already how the model behaves, and the computer is being used merely to generate illustrations. This is rare for nonlinear dynamics.

To return to the *A*'s and the *B*'s: For gaining understanding of the dynamics, there are virtues in doing the simulation by hand, and different virtues in putting it on a computer, so that it can be repeated many times with the same or different random inputs. Figure 5.7 shows a typical complete run. In the first 23 steps, the mechanism quickly produces a patch of *B*'s which floats around in the middle to left region and eventually settles down as the left-hand half of the pattern. Repeat computations with this same input (which is the same as that in Figure 5.6) usually produce some variant of this behaviour. Here, the mechanism emulates the human observer who has done some counting and realized that there is an ambiguity as to whether there is a rudiment of pattern with B's on the left and A's on the right or of pattern with A's on both sides and B's in the middle.

As the model continues to run from the two-part pattern onward, the interest is in what happens to the boundary between the B's and the A's. In the next step, one of four things can happen: (1) the system can remain unchanged, (2) the system can take a step backward in order from the two-part pattern, (3) the sharp boundary can become "fuzzy," and (4) the boundary can move one column to the right or the left. Only this last possibility is significant in the further development of the system over a large number of steps. Possibilities (1) and (2) could serve only to delay the approach of the system to its final destiny. If (2) occurs, it will soon be reversed to restore a two-part pattern. If we concentrate upon (3) and (4), we perceive that the boundary has become an entity which is executing a random walk in one dimension (the horizontal); it is jumping to left or right, one step at a time, from whatever its previous position was, just like a diffusing molecule. Eventually it must reach either the left-hand or the right-hand edge of the system. One of the enantiomers, A or B, will then have disappeared from the system, and it will never reappear. Optical resolution has occurred, spontaneously.

What is going on in this stage of the development is a version of the classical statistical problem known as the gambler's ruin, which is discussed in texts on probability theory (e.g., Feller, 1968). Of two gamblers, one starts with a capital of z dollars, and the other with $a - z$. A game in which the first gambler wins or loses a dollar with probabilities p and q is played repeatedly between the two, until the first gambler's capital reaches zero or a; one of the gamblers is then ruined. The classical problem concerns the probability of the gambler's ruin and the probability distribution for the duration of the game.

In our case, each B or A may be taken as a dollar in the possession of gambler B or gambler A. The game is simply a toss of a coin, giving equal probabilities $p = q = \frac{1}{2}$ that the boundary will move left or right. It is evident in this symmetrical case, with $a = z = 8$ in Figure 5.5g, that both gamblers stand equal chances of ruin. The average duration of the game (expectation value of the number of repeats before one player is ruined) is $D = z(a - z)$. In our case, $D = 8 \times 8 = 64$ repeats. Feller points out that a game can be very long. If each player started with 500 dollars, the average duration would be 250,000 tosses of the coin before the ruin of one player. (In our case, we have overlooked the steps in which the boundary did not move, which will add to the length of the game.)

Because of this long duration, it is easy to overlook the fact that the ultimate ruin of one or the other of the gamblers is inevitable; that is, in our model, optical resolution is inevitable. (What is "intuitively obvious" in statistical situations of this kind varies quite a lot from one individual to another.) Several things may help to throw some light on various aspects of this:

1. A simple computer program for the gambler's-ruin problem may be run

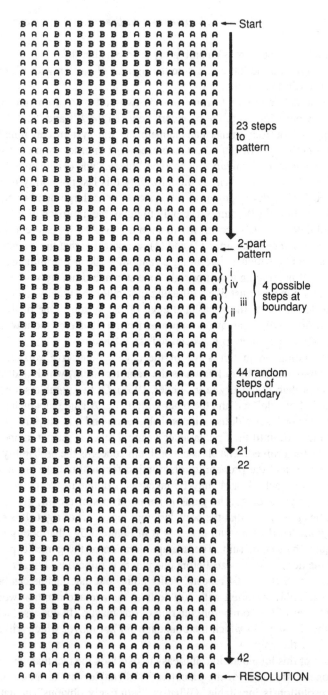

Figure 5.7. Computation of a larger number of stages of operation of the same model as in Figure 5.6, going from the same input through pattern to resolution. The stages of reaction, diffusion, exchange, and washout are computed in an abbreviated manner

repeatedly, with the same starting capitals, while the observer watches the changing scores. After a number of games, the average length of a game should be calculated. It is very instructive to give one gambler only 1 dollar and the other 100. Half the games must end on the first toss, but the average length is 100 tosses.

2. One should recognize that in a sequence of a longer series of coin tosses the imbalance of heads over tails, or vice versa, often keeps the same sign for very long sequences. This was stressed by Feller (1968, chap. III), especially in his Figure 4, being a record of a computer simulation of 10,000 tosses of a coin, reprinted in Mandelbrot's 1982 book on fractals. He found that even experts on statistics could have incorrect intuitions about this. In this book, Figure 5.4b, the example of surveying an open traverse, is essentially the same thing. Notice that the map has strayed away from the true path in the downward direction throughout, though positive and negative errors of angle are selected with equal probabilities.

3. The computer program for Figure 5.7 (Harrison and Green, 1988) should be run repeatedly, both with the same starting sequence and with different ones, which may be selected at random or prearranged as one wishes. The paper cited gives programs and suggestions for two-dimensional computations.

4. One should consider the analogy between these various random sequences and the random walks of diffusion. For a sequence of coin tosses it is easy to get the impression that over a long sequence the numbers of heads and tails will approach equality and the ruin game will never end. The approach to equality is true on a fractional or percentage basis, but on an absolute basis the discrepancy between total heads and total tails will, on the average, grow with the square root of the number of tosses of the coin. This is equivalent to the observation that a coloured material placed in a small region of a colourless solvent *always* spreads; on the average, every molecule moves away from its starting point in proportion to the square root of the time, and every one will at some time hit the wall of the container.

The dependence of the length of the game on the square of the size of the system is rather significant in relation to optical resolution during biochemical evolution. The games in Figure 5.6 and 5.7 are being played in regions about 100 Å long, if the symbols A and B represent molecules adsorbed on sites a few angstroms apart. Unit steps of reaction, diffusion, and exchange probably

Caption to Figure 5.7 *(cont.)* which does not make them available for separate display. Each row of the output is the set of A's and B's on the catalyst bed after a sequence of these processes. See Harrison and Green (1988) for a discussion, with computer programs given, of this computation and similar computations in two dimensions, in which the movement of the boundary between two oppositely resolved regions is much more complex. To try some such computations is of pedagogic value to students of pattern formation contemplating the transition from stochastic to deterministic behaviour, or from small-number to large-number statistics. From Harrison and Green (1988), with permission.

have to be assigned times from 10^{-6} to 10^{-2} s, and the complete optical-resolution game needs at the very least 10^{-4} s up to 1 s or even more. This gets us well away from having to assume that chirality arose because all life is derived from a single molecule (a mechanism which I consider so unlikely that I did not even list it among the possibilities), but still requires a microscopic scale for optical resolution; Mills (1932), without detailed analysis of the time scale, was thinking of something like the size of a single cell, a few micrometres, to have enough starting imbalance. From the foregoing estimate, our game would go to completion in something from an hour to a couple of weeks when played in a region 10 μm long; but for 10 m, it might need the total time that life is believed to have been present on earth.

The game has a feature, however, which could cut these times drastically. This is the possible interaction with long-range disturbances at the stage at which pattern has reached whatever I just meant by long-range. A disturbance could be, for example, a failure of the food supply (stoppage of, or reduction in, the supply of reactant *s* in the model) over only one part of the system. If this affected, say, most of the left-hand half of Figure 5.6g, then the disturbance would effectively have handed the game to *A* with no need for much further play. As every experimental biologist knows, mysterious variations in the health of a living system can easily affect regions as small as adjacent cells in a carefully maintained culture dish. The growth of transient pattern in our game sensitizes the system to disturbance on that scale in an hour or a week. Evidently a number of jumps of this kind could successively take increasingly large reefs in the time scale. I see, therefore, no essential difficulty in this kind of mechanism having operated to produce optical resolution reasonably quickly in a system some metres or even kilometres across.

I tend to regard spontaneous optical resolution as being, in respect of the essence of the mechanism, a solved problem. We may never know precisely what molecules were involved, nor whether the crucial autocatalysis was a simple bimolecular step or some behaviour of a complex reacting system which, overall, gave self-enhancement kinetics in some order exceeding the first. But the general strategy of the mechanism is so much in accord with the self-replicating property of life, so powerfully asymmetrizing, and so capable of making use of the asymmetry of a great variety of disturbances on a wide range of spatial scales that the strategy must surely be accorded a very high probability. Frank (1953) also regarded the problem as thus solved.

The relevance of similar concepts to other aspects of biological development is much more complex and speculative. This game has not directly addressed the question of shape development, because it was played with a fixed frame of reference, the set of sites on the catalyst bed. The preceding discussion of random walks is of course applicable to the increasing error limits in the later steps of an open traverse (Figure 5.4b) and hence to the consideration of whether or not small-scale disorder can produce observed long-range order (Figure 5.2) without a long-range organizer.

This mechanism has the territorial ambition of an imperial expansionist of unlimited greed. It amplifies disturbances more rapidly the longer they are, without limit, until the longest takes over the system. This is shown mathematically in Chapter 6. What is lacking is the stabilization of a particular finite size of pattern repeat. The addition to this kind of model needed to do that was first devised by Turing (1952) and is described in Chapter 7.

This principle of optical resolution has been discussed by a number of workers, some of whom rediscovered it independently without knowledge of the earlier work. Frank (1953) gave a mathematical discussion including consideration of the game in two dimensions. He pointed out that a boundary between regions resolved as *A* and *B*, which can be a curve in two dimensions, will advance in the direction toward which it is concave. Therefore, a region which has become surrounded has lost the battle. This feature resembles the strategy of the Japanese game of Go. For other discussions, see Seelig (1971a,b), Decker (1973, 1974, 1979), Harrison (1973, 1974), and Harrison and Lacalli (1978).

This kind of mechanism, as well as everything else about kinetic determination of form in this book, has possible implications in regard to biological evolution as well as development. The reader may find it interesting to reflect on whether or not the system in Figures 5.6 and 5.7 has evolved. It has become more orderly; it is better adapted to handling the *s* supply from the environment, because by the stage of the two-part pattern all of the system can assimilate *s* (except the central boundary), whereas only part of it (one-half, on a statistical average) could at the outset. But its complexity in terms of macroscopic parts increases from Figure 5.6a to Figure 5.6g, and then *decreases* as one enantiomer is annihilated.

In regard to the excessively popular definition of natural selection as "the survival of the fittest," it is astonishing that of two exactly equally matched adversaries, one must inevitably be annihilated totally by the other.

5.4 Asymmetry begets asymmetry

5.4.1 Trivial and significant antecedents

Developmental mechanisms are amplifiers, and form arises from what is most efficiently amplified. The example in Section 5.3 of one of the simplest possible developmental amplifiers has shown that what is amplified may cover a wide range of spatial scales. That particular mechanism favours the longest spatial scale it can find in any asymmetric antecedent. An organism which inherits, by the genetic code, the ability to manufacture the molecules called *A* and *B*, or 0 and 1, has inherited this property of long-range selection. Even if, in a primitive form, it contained both those enantiomers, it would have inherited the property of optical resolution; that is, it must inevitably lose *A* or *B*. That is clear without any consideration of the nature of the asymmetric antecedents on which the mechanism works. But which enantiomer is lost, and how long the whole procedure takes, can be strongly dependent on the antecedents. If at the stage of Figure 5.6g a gradient in concentration of s is somehow set up from left to right across the system, it will determine the winner and accelerate the process. Now the distribution of s may not be relatable back to the genetic code on the time scale of one generation. The asymmetric organism may form an asymmetric egg (i.e., the s gradient may be inheritable as a long-range template outside the genetic code). The model therefore resolves not only optical enantiomers but also the philosophical paradox propounded in the first paragraph of this chapter. It contains both an antecedent-independent, genetically determined inevitability (so long as there is some antecedent, it does not matter what) and a genetic-code-free, organism-scale sensitivity (so long as the genetic code has provided the amplifier, different templates can be amplified).

This model is, of course, very indeterminate in relation to the precise patterns which may form at intermediate stages; the pattern of two equal parts does not necessarily appear. More sophisticated kinetic models (beginning in Chapter 6) make finite pattern determinate. The quantitative measure of distance between repeated parts in a pattern is then linked to the chemical nature of the morphogens and hence is attributable to the genetic code. But for an asymmetric pattern of different parts, the matter of "which way round" is still dependent on long-range templates.

There is no doubt, then, that kinetic amplifiers will respond to polarities. The question arises whether the polarities are a significant part of development or just an antecedent where any antecedent would do. This question surely has different answers for different cases. Let us consider first a polarity provided by an external frame of reference, such as the gravitational field. It is significant to a plant that it should grow upward. But if a *Xenopus* egg is rotated through 90°, by an osmotic technique of collapsing the vitelline membrane onto the egg (Kirschner et al., 1980), the orientation of the dorso-ventral axis can be completely reversed. This gravitational effect appears completely trivial, of no possible use to the organism; but it is there, and it shows that an insignificant polarity can be amplified. In this case, the amplifier is probably of quite a different kind than the reaction-diffusion systems discussed here, at least for the first step of rearranging the cell contents in the gravitational field. The cell's later response to that rearrangement, leading to the establishment of the dorso-ventral axis, could, of course, involve kinetic mechanisms.

The energetics of the coupling of a cell to gravity are interesting in relation both to how small an energy can be amplified and to how a response to an unnecessary polarity may arise. In Section 5.3 I cast doubt on the possibility of significant interaction of living material with the nuclear weak interaction because it involves an amplification from 10^{-14} to 1 eV. But gravitational interaction requires a response, in a cell 1 mm high (*Xenopus* egg), to the difference in gravitational potential between a structure at the top of the cell and one at the bottom. Per atomic mass unit, this is 10^{-10} eV. Amplification to chemical energies simply requires that the cell contain some organized structure of 10^9-10^{10} daltons which moves as a single unit in the gravitational field and has a density significantly different from that of the surrounding cytoplasm. Such a structure needs to be about 0.2 μm across in each dimension. Because organized structures of at least this size are common in all cells, an accidental gravity response is not at all unlikely. In an oocyte, it could be a gradient of density through the whole contents of the cell which is involved. This kind of amplifier involves the approach to equilibrium (Chapter 4) rather than a kinetic departure from equilibrium.

A polarity can, on the other hand, be of great significance, especially in ensuring that a developing part of an organism has the correct orientation relative to other parts. I expect that this is what Meinhardt had in mind in suggesting that whereas developmental mechanisms are capable of symmetry-breaking, actually they usually amplify asymmetries already present. Examples abound in the embryology of vertebrates. The specificity of attachment of retinal axons to the optic tectum is an example for which a variety of models have been proposed, including one which uses self-enhancement and self-inhibition on the grand scale, the units being entire synapses and their overall electrical activity, rather than individual molecules (Willshaw and von der Malsburg, 1976, 1979). The various models all need some form of positional or directional information, the extent of which varies from model to model. Experimentally, it is known that changing the region of contact between the tectum and the thalamus induces errors in orientation. This kind of polarity sensitivity is nontrivial; the proper orientation of eye-to-brain messages depends upon it.

This last example points to another aspect of the model described here. If one considers A and B to represent molecules, one must, as always, be somewhat suspicious of any attempt to illustrate large-number statistics with a small-number example. Features of enormous significance can be absent from or inadequately represented in the small-number situation (e.g., entropy). But our A's and B's may sometimes represent whole cells, perhaps at the onset of differentiation into two types. Then to the list of essentially synonymous terms "autocatalysis," "self-enhancement," "positive feedback," "assimilation," and "reproduction" we must add another: "homeogenetic induction," the phenomenon in which differentiated cells cause neighbouring cells to differentiate in the same way. The numbers of cells across a system at crucial stages in pattern formation often can be as small as the eighteen attachment sites used in the model, and these small-number statistics are then quantitatively appropriate.

When we move up in spatial scale while considering optical resolution as illustrative of a fundamental property of life, surely we should return at last to Pasteur. He believed in an interaction involving the large-scale asymmetry of

170   *Pattern-forming processes*

the universe, as we perceive it so readily in the arrangement of the starry skies at night. Einstein used a device known as the thought-experiment (Gedanken experiment) to illustrate succinctly the gist of some theoretical topics. Here is a Gedanken experiment.

5.4.2 Life on the planet Gedanken

The planet Gedanken is exactly spherical and is covered with an ocean of uniform depth which initially contains a "primeval soup" of primitive life in which the organisms (or large self-replicating molecules) are present everywhere in both enantiomeric forms. The planet orbits a sun. It has a magnetic field, with an axis which coincides with its axis of rotation, and this axis lies in the plane of its orbit. This arrangement minimizes the possible large-scale asymmetries in the system. The planet also has an atmosphere.

In this arrangement, the interaction of radiation from the sun with the magnetic field and with the atmosphere has enough asymmetry to encourage separation of the enantiomers spatially into two separate hemispheres (north and south). This happens because unpolarized light can be resolved into right- and left-hand circularly polarized components. These are chiral (i.e., they have right- and left-hand screw-thread asymmetry) and therefore absorbed to different extents by the opposite enantiomers of a chiral substance (the Cotton effect). This can potentially lead to a difference in rates of destruction of the two enantiomers by photochemical action. For this to happen, one of the circularly polarized components must reach the surface of the planet in greater intensity than the other. The magnetic field can make this happen, provided that the planet has an atmosphere. Wherever the magnetic field has a component parallel to the direction of the light, it makes the refractive indices of the two circularly polarized components different. The one of greater refractive index will be deflected downward more by the atmosphere, because its refractive index is greater in the denser lower layers. Thus, more of one chiral component of the light will reach the surface of the planet.

The effect is a maximum when the axis of the planet points toward the sun. We need not trouble ourselves as to which way round the effect is. Half a year later, the planet will be on the other side of the sun, with the other end of its axis pointing to the sun. The entire effect will be mirror-imaged. Whichever enantiomer of the primitive life was encouraged in one hemisphere, the other will be similarly encouraged in the other hemisphere (Figure 5.8). Thus, in a time-average over a full year, the planet receives from the celestial spatial scale an asymmetrizing antecedent for a pattern in which one hemisphere is occupied by life of one chirality, and the other hemisphere by life of the opposite chirality. Classical properties of radiation and of the magnetic field are used in this model, but the physics of the weak interaction and nonconservation of parity does not come into the model at all.

Large-scale separation of enantiomers could, of course, take place without

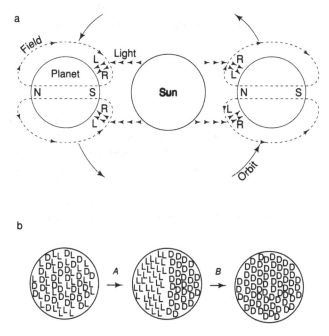

Figure 5.8. Life on the planet Gedanken. The magnetic field, together with the refractive-index gradient through the planet's atmosphere, causes right- and left-hand circularly polarized light (L and R) to be deflected downward to varying extents. The effect is exactly mirror-imaged at half-year intervals, if the orbit is circular. This can encourage two-part pattern formation in two hemispheres (arrow *A*) by the photochemical effects of L and R light on D and L enantiomers. A large-scale external disturbance (arrow *B*) can encourage loss of L, and resolution.

such an antecedent, because random fluctuations in the ocean and its primeval soup contain some rudiment of long-range pattern. But, as indicated earlier, if one depended on those, one might have to wait billions of years for the separation to get up to a spatial scale of a few metres. The interaction with asymmetries on a vaster scale is capable of providing a bigger antecedent and speeding the process up.

As to the eventual loss of one enantiomer, there are two possibilities for an asymmetrization which would accelerate the process much above the rate which a random walk of the boundary would achieve. One, as described by Mörtberg (1974), from whose account this is adapted, is that the arrangement of the axis of the planet and the axis of its orbit is not as symmetrical as suggested here, and the planet has an elliptical orbit, not a circular one. This gives enough asymmetry for the two hemispheres to get unequal treatment. The other is simply that an accident on the large scale, such as collision with a meteorite, or a burst of destructive radiation from some particular direction, may destroy much more of the life on one hemisphere than on the other. Such antecedents need no knowledge of modern particle physics or statistical thermodynamics for their appreciation. They are big things, that we can all see or visualize.

5.5   The paradoxical nature of symmetry

One might think, a priori, that if anything is a straightforward manifestation of static aspects of geometry, it is the symmetry of an object. Yet whenever one studies symmetry in nature, somehow one soon seems to be concentrating on motion. In the quantum mechanics of atoms and molecules, symmetry and angular momentum, the measure of circular motion, go hand in hand. Here, motion has entered in quite a different way, but has been prominent throughout the discussion. Symmetry is the aspect of shape which shows most clearly the main thesis of this book: that shape is a manifestation of motion.

6

Matters needing mathematics: an introduction

6.1 The language of rates, and the need for it

The Scientific Revolution of the sixteenth and seventeenth centuries featured above all other concepts that of accelerated motion. At that time, the infinitesimal calculus was developed by Leibniz and, as the "method of fluxions," by Newton. In that language, acceleration is, of course, a second derivative, the rate of change of the rate of change in position. Simultaneous development of the concept and the mathematical language was no accident. When one has to deal with second and higher derivatives, the compact language of the calculus is essential. Not so, I think, if one has to deal only with first derivatives. Aristotle's law of motion, that velocity is proportional to applied force, is correct for motion in a viscous medium (the world as experienced by bacteria). It had been current for many centuries without provoking the invention of such a language as the calculus. If one wishes, for instance, to discuss the law of exponential growth for some quantity C, nowadays one will usually start by writing

$$dC/dt = C; \qquad (6.1)$$

but one need not. It is quite possible to write the compound-interest law (100%)

$$C = C_0[1 + (1/n)]^{nt} \qquad (6.2)$$

and to go to the limit of continuous addition of interest by taking n to infinity and so inventing the number e and writing

$$C = C_0 e^t \qquad (6.3)$$

without using the notation d/dt or the notion of a differential equation at all.

The accumulation or disappearance of material in diffusion involves a second derivative and, like accelerated motion, demands the language of the calculus. Surely this should be, therefore, in commoner use among experimental developmental biologists than it is at present. But one cannot readily rekindle enthusiasm for something three hundred years old as if it were the latest fashion. Nevertheless, that is what is needed for theories of development: a lively enthusiasm for fairly elementary calculus. It is the old stuff in the expertise of mathematicians and physicists which is relevant to biology, because most of it has not yet been used there.

Calculus is a means of discussing continuous variables, and we are here particularly concerned with concentrations of substances as such variables. This does not arise very obviously from the model in Chapter 5, which used small-number statistics with an all-or-nothing character to the contents of every box: either an *A* or a *B*. That simulation is not readily extendable, without the use of much larger numbers, to models for forming stable pattern. Consider the stage at which a two-part pattern has developed, with some *B*'s on the left and some *A*'s on the right. We would like to devise a model, for instance, in which the boundary between these two regions would not go into a random walk but would stay unmoving if it were exactly in the middle of the system, with as many *B*'s on the left as there were *A*'s on the right. How is the boundary to know the length of the region on each side of it? Clearly it does not, and it moves randomly with respect to the ends of the system because its lack of information about them is total.

That is not so if the concentrations of *A* and *B* are continuous variables. Diffusion, and its consequences for buildup or dispersal of material, is quite sensitive to the spatial scale of the pattern of nonuniform distribution of the diffusing material. This, we shall see later (Chapter 7), leads ultimately to the conclusion that reaction-diffusion can set up a quantitative measure of distance – something surely of importance in biological development, but not much emphasized in most accounts. In the following section, the diffusion equation is set up and applied to a sine-wave distribution of material which at once gets rid of the second derivative and leaves us with only first derivatives, time rates of change. The same strategy is used in Chapter 7 to handle Turing's model. This treatment is designed to be intelligible to readers who at some time have taken a first course in calculus but who do not use it every day and therefore are not fluent in the language. Section 6.5 poses some problems intended to assist at this level, and there are brief indications of their solutions in Section 6.6.

There is no purpose in reading on beyond this point unless the reader intends to spend a significant amount of time on all the mathematics in Chapters 6 and 7. Both chapters are needed to reach the point at which one can understand the nature of a pattern-forming reaction-diffusion mechanism. Many biologists may still have doubts that such effort will be worth their while. Indeed, in most published work I have read by experimental biologists who mention reaction-diffusion with any favour at all, such mention usually ends exactly at the point at which this mathematical account is about to begin. To take up again a matter addressed in Chapter 1, I do not believe that this happens merely from an aversion to mathematical language and procedures.

The biologist's customary answer to the problem of stabilizing the position of an A/B boundary (usually with A and B representing differentiation states of cells, rather than molecular species) is the concept of positional information provided by a fixed gradient of another substance. In terms of the number of substances required, this is no less complicated than a reaction-diffusion mechanism. Indeed, the gradient itself needs a simple form of reaction-diffusion to account for its quantitative stability. The contrast between this kind of model and Turing two-morphogen reaction-diffusion is in the kinds of processes envisaged. The usual biologists' explanations are fundamentally hierarchical. Each new step is dependent upon something that is higher in the hierarchy and is not alterable by the subsequent process. You don't try to get back to the boss and change his mind. The essence of a Turing model is the existence of feedback loops in which everything interacts back on everything else and there is no boss. It does not preclude the process being part of a hierarchy of processes on a grander scale. But for the formation of a single pattern, processes with feedback control are far more powerful than are processes with a unidirectional control sequence, and to my mind therefore much more likely to have been found by living things.

6.2   Differential equations, diffusion, and a Cheshire Cat

6.2.1   A Cheshire Cat

Diffusion is flow of material down a gradient of concentration. Fick's law, which gives the simplest mathematical expressions for diffusion and which well represents the behaviour of many diffusing materials, is that the rate of flow of material is proportional to the concentration gradient. The process of diffusion tends to destroy nonuniformities in concentration distribution. One often thinks of this as equivalent to spreading out, in the sense, for example, that a spot of coloured solution placed somewhere in a dish of water will become gradually larger and fainter in colour and have fuzzier boundaries, or that an originally sharp boundary between a coloured solution and water in two sections of a tube will gradually blur (Figure 6.1a,b).

There are circumstances in which the approach to uniformity does not involve spreading in the sense of change of shape of the original distribution. Suppose we have a rectangular dish of water in which some coloured solution is somehow arranged initially in the form of a set of parallel, equidistant, fuzzy-edged stripes (Figure 6.1c). More precisely, the initial distribution is to be (Figure 6.1d)

$$C = A \cos(\omega s) + C_0, \tag{6.4}$$

where C is the concentration of the material, s is distance along the dish perpendicular to the stripes, and ω and C_0 are constants. Because if we start from the crest of the wavelike distribution at one end of the trough ($s = 0$) the

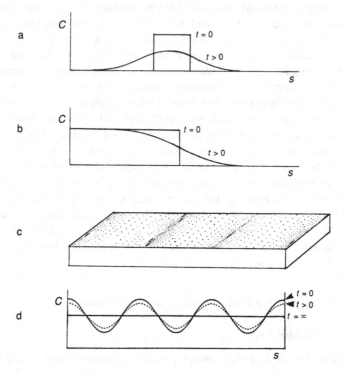

Figure 6.1. Spreading by diffusion. (*a* and *b*) Two cases in which spreading changes the shape of the distribution: (*a*) a spot of solution, containing a diffusible coloured solute, put in a dish of solvent; (*b*) a tube with an initially sharp boundary between solution on the left and solvent on the right. (*c*) A case in which spreading alters the intensity but not the shape of the distribution: Coloured solution is distributed in "fuzzy-edged" stripes along a trough full of solvent. (*d*) The precise indication of what is meant by "fuzzy-edged" for the pattern to fade undistorted: a sinusoidal distribution of concentration along the left-to-right direction. The broken line (*t* > 0) indicates the fading of the pattern, i.e., decrease in amplitude, without change in its sinusoidal morphology. The peak-to-peak distance is the wavelength λ. Throughout this book, I use $\omega = 2\pi/\lambda$. Physicists are more accustomed to the use of ω as an angular frequency in time-periodicity, and $k = 2\pi/\lambda$. I have switched the terminology because, as a chemical kineticist, I wish to use k's extensively for chemical rate constants.

next crest must be at $\cos(2\pi)$ and also at $s = \lambda$ (the wavelength), then it follows that

$$\omega = 2\pi/\lambda. \tag{6.5}$$

I assume that the trough is an integral number of wavelengths long, and so has a crest at both ends.

    What will be seen to happen if we observe this system without disturbing it? The answer is not entirely obvious. Clearly, the stripes will finally disappear, and the whole solution will have the depth of colour corresponding to C_0. What is not obvious is that as the stripes fade, their shapes do not change at all. The amplitude A of the sinusoidal waveform decreases as time goes on, and nothing else changes. The pattern fades like the Cheshire Cat, without change in its features. Thus in the sense of the *intensity* of the pattern, spreading occurs, but in the sense of its *morphology*, there is no spreading. It is clearly important in relation to pattern formation that there is such a thing as a pattern that diffusion cannot change, morphologically.

Anyone who can prove that the pattern fades in this way knows enough mathematics to cope with the rest of this book. The example will serve as an introduction to differential equations for those not yet acquainted with them.

6.2.2 Differential equations

Let us consider a concentration C which varies in some way with distance s, so that its distribution constitutes a pattern or form (or one may prefer the term "prepattern"). Let us suppose that all we are told about it is that it obeys the relationship

$$C = C_0 - (1/\omega^2)d^2C/ds^2. \tag{6.6}$$

This is, obviously, an equation. It looks as if it might possibly contain the information on what function C is of s, that is, what the pattern is. That relationship is called a solution of the equation. Thus we have an equation involving differential coefficients, the solution of which is not a numerical value of C, but an algebraic expression relating C to s. This is a differential equation. When it involves only one independent variable (s, in this case), it is called an ordinary differential equation.

How to go about solving a differential equation is often not at all obvious, and there is no standard method that always works. Guesswork is sometimes as good as anything. (The reader with only an introductory calculus course should have met this same problem in integration, which is closely related to solving differential equations.) In the present case, a correct guess is easy to make, because one has just been given as equation (6.4). Check by differentiation:

$$dC/ds = -A\omega \, \sin(\omega s); \tag{6.7a}$$

$$d^2C/ds^2 = -A\omega^2\cos(\omega s); \tag{6.7b}$$

and on substitution in (6.6),

$$C = C_0 - (1/\omega^2)[-A\omega^2\cos(\omega s)] \tag{6.8a}$$

$$= C_0 + A \, \cos(\omega s). \tag{6.8b}$$

So we have a solution of the differential equation, but not the only solution. If, instead of the cosine function, we used sine, that is,

$$C = C_0 + A \sin(\omega s), \qquad (6.9)$$

it is easy to check by differentiation that we again have a solution. The difference between the two solutions we now have is the condition at one end of the system, $s = 0$. For the first solution, a crest of the waveform was at this position; for the second, a node. To find not *a particular* solution, but *the general* solution, it remains only to see that any intermediate point on the waveform could be its start at $s = 0$. This can be expressed as

$$C = C_0 + A \sin(\omega s - \delta). \qquad (6.10)$$

This expression contains two undetermined constants: the amplitude A of the sinusoidal waveform, and its phase δ. The latter can be determined if we have sufficient information on what is happening at the boundary of the system $s = 0$. There are two undetermined constants because the equation contained a *second* derivative. The order of the highest derivative is the order of the equation and gives the number of arbitrary constants in its general solution.

6.2.3 Diffusion

The example just treated was of a frozen distribution of material, a pattern unchanging in time in any of its characteristics, including amplitude. If the material so distributed is diffusible, C changes with time at every point in the pattern (Figure 6.1d). The complete specification of C is therefore as a function of two variables, position and time:

$$C = C(s, t). \qquad (6.11)$$

Information which would let us determine this function might also be given as a differential equation, but it could now contain two kinds of derivatives, those with respect to position and those with respect to time. Such an equation is called a *partial* differential equation, distinguished from the previous case of an *ordinary* equation. At first encounter, the word "partial" may seem obscure in this context; its rather straightforward meaning should emerge in what follows.

[In his introductory remarks at a discussion hosted by the Royal Society of London on theories of biological pattern formation, J. D. Murray (1981b) said: "At several 'interdisciplinary' meetings it was clear that communication between the various groups was non-existent and, after some of the answers to questions, perhaps it was not even wanted. A reply such as, 'It's probably a secondary Hopf bifurcation in the p.d.e. parameter space' does not have biologists on the edge of their seats – unless to leave." In those remarks, p.d.e. stands for partial differential equation.]

As the reader knows, a derivative such as dC/ds is the slope of a graph of C against s. The problem in the more complicated two-variable situation is this: While s is changing, what is happening to t? The most obvious answer, to give a simple significance to the slopes, is as follows: While measuring dC/ds, keep t constant, and while measuring the time rate of change of C (i.e., dC/dt), keep s constant (i.e., make the measurement at a fixed position) (Figure 6.2a). Each slope so measured gives only a part of the information about how C is varying, and is called a partial derivative. There is a special notation for it. The two just mentioned are $(\partial C/\partial s)_t$ and $(\partial C/\partial t)_s$. This notation is devised to cope with an unlimited number of variables. When we have only two, it is not necessary to use the subscript indicating what is held constant while we are measuring one variation. We may write $\partial C/\partial s$ without specifying that t is to be held constant. The use of ∂ in place of d signals a partial derivative, and there is only one possibility for what the other variable is.

The beginner may have been puzzled since reading in the second paragraph of this chapter that diffusion involves a second derivative, because Fick's law mentions only a first derivative:

$$\text{rate of flow of material } \dot{M} = -\mathcal{D}A_c(\partial C/\partial s), \qquad (6.12)$$

where $\mathcal{D}$ is the diffusivity of the moving material, and A_c is cross-sectional area perpendicular to s. Consider, however, a system with a linear distribution of concentrations (Figure 6.2b); $\partial C/\partial s$ is the same everywhere. Thus, at the point B, the rate of delivery of material from the right is the same as its rate of removal on the left. There is no net gain or loss of material at B (which could be any point along the linear distribution); $\partial C/\partial t$ is everywhere zero. For material to accumulate at B, the slope on the left must be less than the slope on the right, so that less material is removed than is delivered in any time interval. This is illustrated in a finite-difference approximation, such as one would use in programming a computer to calculate diffusion, in Figure 6.2c. For a continuous variation (Figure 6.2d), what we see is that if, as we go past the point B from left to right, $\partial C/\partial s$ is increasing with s, then there is a buildup of material at B (i.e., $\partial C/\partial t$ is positive if the rate of change of $\partial C/\partial s$ with s is positive, i.e., if $\partial^2 C/\partial s^2$ is positive). The reader may now be prepared to swallow, without further ado, the expression of this result as an equation:

$$\partial C/\partial t = \mathcal{D}(\partial^2 C/\partial s^2). \qquad (6.13)$$

The derivative of (6.13) is given as Problem 6.5.5. It is sometimes called Fick's second law, but there is really only one law; equation (6.13) is an inevitable mathematical consequence of equation (6.12).

A curve for which $\partial^2 C/\partial s^2$ is everywhere positive is concave upward, and $\partial C/\partial t$ is everywhere positive; that is, a short while later, the whole curve will be displaced upward, to higher concentrations. By the same token, a curve which is concave downward is moving downward as time goes on. If we apply this to

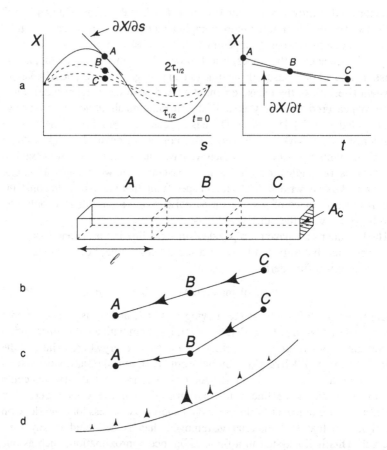

Figure 6.2. (*a*) The meaning of partial derivatives, for a concentration (*X*) dependent upon both position and time. (*b* and *c*) Illustration of why the diffusion equation involves the *second* derivative of concentration with respect to position, shown in the kind of finite-difference approximation one might use in a computation; a tube (shown square-sectioned) of diffusing material is divided into segments of equal length, the average concentration of each is put at its midpoint, and the concentration gradients are approximated by straight lines joining the midpoints; (*b*) When the concentration distribution is linear, there is no buildup of material at *B*; rate of delivery from the right equals rate of removal on the left. (*c*) When the concentration distribution is nonlinear, the concentration at *B* is changing. Here, the rate of delivery exceeds the rate of removal, and concentration at *B* is rising. (*d*) The same as (*c*), but shown with a continuous variation of concentration with position. The rate of rise of concentration is fastest where the curve departs most from a straight line.

the sinusoidal distribution, we get the result shown in Figures 6.1d and 6.2a: The entire top half of a sinusoidal wave is concave downward and the entire bottom half is concave upward, and each is moving toward the direction to which it is concave. The nodes, where $C = C_0$, are at the transition from concave up to concave down. There, $\partial^2 C / \partial s^2 = 0$; that is, the curve has, in fact, no curvature. The concentration at these points therefore remains at C_0: The nodes of the wave do not move. Without further algebra, we are close to showing that the pattern is changing only in amplitude, but we have not yet proved precisely that it remains sinusoidal as it decays.

6.2.4 The Cheshire Cat, with mathematics, and three entities

The step that remains is to put together equations (6.6) and (6.13), and so get rid of the second derivative. From equation (6.6), applied to the shape of the concentration distribution at any fixed time, and therefore with the ordinary second derivative replaced by a partial,

$$\partial^2 C/\partial s^2 = -\omega^2(C - C_0). \tag{6.14}$$

Substituting in (6.13),

$$\partial C/\partial t = -\omega^2 \mathcal{D}(C - C_0). \tag{6.15}$$

This is an ordinary differential equation. Its solution, for C as a function of t, will show how concentration changes with time at any point along the pattern. It is already obvious to us that C is everywhere heading for the value C_0 at infinite time. To represent this progress as concisely as possible, it is useful to introduce the measure of displacement from C_0:

$$U = C - C_0. \tag{6.16}$$

Because C_0 is constant, differentiation yields

$$dU/dt = dC/dt, \tag{6.17}$$

and recognizing that because s has disappeared from the scene the derivative in (6.15) can be written as an ordinary derivative, that equation becomes

$$dU/dt = -\omega^2 \mathcal{D}U. \tag{6.18}$$

This equation probably is already familiar to the reader who has not previously encountered differential equations in general, but who has learned something about chemical kinetics, or radioactive decay. It is the equation for exponential decay (often called a relaxation process). The solution is

$$U = U_0 e^{-\omega^2 \mathcal{D}t}. \tag{6.19}$$

This decay has the well-known property of first-order chemical reactions. The time required for U to fall to some given fraction of U_0 is independent of U_0. Thus, for our sinusoidal pattern, if the crest, where $U_0 = A_0$, drops to $A_0/2$ in some given time, every other point along the curve will have halved its initial displacement from C_0 in the same time. The curve will have retained its sinusoidal shape, but its amplitude will have been halved. This establishes that the Cheshire Cat behaviour described in Section 6.2.1 actually occurs (Figure 6.3, from 0 to $\tau_{1/2}$).

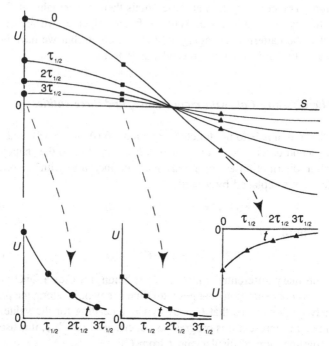

Figure 6.3. Exponential decay of a sinusoidal pattern, as a result of diffusion, with the same half-life ($\tau_{1/2}$) for all positions in the pattern. $U = X - X_0$ is displacement of concentration X from its average value for the whole pattern, which is the same as the value of X at all points at infinite time.

Equation (6.19) is the general first-order decay equation

$$U = U_0 e^{-kt}, \tag{6.20}$$

with the decay constant k given by

$$k = \omega^2 \mathscr{D} = 4\pi^2 \mathscr{D}/\lambda^2; \tag{6.21}$$

that is, the shorter the wavelength, the faster the decay. For a given amplitude, concentration gradients are greater for shorter waves, and therefore diffusion is faster for shorter waves, so we should intuitively expect this sort of variation; but without the mathematics, it would have been difficult to see that the dependence is on $1/\lambda^2$ rather than on $1/\lambda$.

This analysis has shown us that the sinusoidal pattern behaves, because of the occurrence of diffusion of its constituent molecules, as a whole, a single organized entity which fades away as if it were one object, the Cheshire Cat. Hence we recognize two kinds of entities in the system: the whole pattern, and its constituent molecules. A third entity may be useful to think of from time to time: the repeating unit of the pattern, which is one cycle of the sinusoidal wave, occupying length λ of the system. It is informative, instead of thinking of the length of each unit, or the spacing between units (which is the same thing for the sinusoidal pattern), to take its reciprocal and think of how many pattern units are present in unit length, that is, the concentration of pattern units, $1/\lambda$. From this viewpoint, our kinetic law is that pattern units do not like to be crowded, and each one of them fades away in intensity at a rate proportional to the square of its concentration.

6.3   Reaction-diffusion and growth of pattern: departure from uniformity, both ways

To study the mathematically simplest reaction-diffusion equation, we need only to suppose that the diffusing substance of concentration C, or $U = C - C_0$, is also being formed autocatalytically. The kind of autocatalysis required has the peculiarity introduced in Section 2.2: The autocatalytic function is to involve U, not C, so that the catalysis will pull the concentration away from C_0 in either direction, upward or downward:

$$\partial U / \partial t = kU. \tag{6.22}$$

Before we combine this with the diffusion equation, perhaps some justification is needed for its form. It is obviously reasonable to suppose that there may be molecules which catalyze their own formation in a unidirectional sense,

$$\partial C / \partial t = kC, \tag{6.23}$$

but how the bidirectional character of equation (6.22), in which U can have a positive or negative value, arises is not so obvious. In Chapter 9, the algebra of several reaction-diffusion models is presented and shown to give the kind of variation indicated by equation (6.22) for the displacements $U = X - X_0$ and $V = Y - Y_0$ from the spatially uniform (patternless) steady state (X_0, Y_0). What is going on in principle can easily get lost in the algebraic details. Morphogenetic reaction-diffusion mechanisms usually contain a term in X^2 (or the equivalent, A^2, in the Gierer–Meinhardt mechanism) in the growth rate of X. Just why this works to give morphogenetic behaviour is not easily generalized in words. An understanding of it requires discussion of several examples, some of which can be handled quite easily mathematically, but in which the essence of what is going on does not lend itself to verbal description. It is necessary to appreciate that one equation can be worth a thousand words.

One such example, that of bimolecular activation of a site for stereospecific autocatalysis, was discussed in Chapter 5 as far as could reasonably be done without the use of equations. It is taken up again in Sections 6.3.3 and 6.4. An advantage of that system as a first example is that the movement away from equilibrium in two opposite directions can be appreciated as movement toward optical resolution in the two opposite senses. But this same bidirectional movement can occur where there is no such symmetric relationship between two substances, as the following section shows.

6.3.1 Linearization about the spatially uniform steady state: a simple example

Suppose that a substance X decays in first-order manner and also self-catalyzes its own formation in a second-order manner, and is diffusible:

$$\partial X/\partial t = -k_d X + k_f X^2 + \text{diffusion}. \tag{6.24}$$

Let us determine if there is a steady-state value, $X = X_0$, for which X is not changing with time throughout the system. In spatial uniformity, the diffusion term vanishes, because there are no concentration gradients in the system. Then if we write $\partial X/\partial t = 0$, for $X = X_0$,

$$0 = -k_d X_0 + k_f X_0^2, \tag{6.25}$$

whence

$$X_0 = k_d/k_f \tag{6.26}$$

(except for the trivial value $X_0 = 0$; because morphogenetic mechanisms contain squared terms, one often has to solve quadratic equations, and one is then faced with the usual question of which of the two solutions is the physically significant one).

Being assured that such a quantity as X_0 exists, we may now write, as usual, $U = X - X_0$ and rewrite equation (6.24) in terms of U instead of X, in order to enquire about the rate of departure from the spatially uniform steady state. Because X_0 is constant,

$$\partial U/\partial t = \partial X/\partial t, \tag{6.27}$$

and hence, for the moment, with the diffusion term left out, and with $X = X_0 + U$,

$$\partial U/\partial t = -k_d(X_0 + U) + k_f(X_0 + U)^2,$$

or, expanded,

$$\partial U/\partial t = -k_d(X_0 + U) + k_f(X_0^2 + 2X_0 U + U^2). \tag{6.28}$$

Subtract from this

$$0 = -k_d X_0 + k_f X_0^2. \tag{6.25}$$

Hence

$$\partial U/\partial t = (-k_d + 2X_0 k_f)U + k_f U^2, \tag{6.29}$$

which, upon inserting the value of X_0, becomes

$$\partial U/\partial t = k_d U + k_f U^2. \tag{6.30}$$

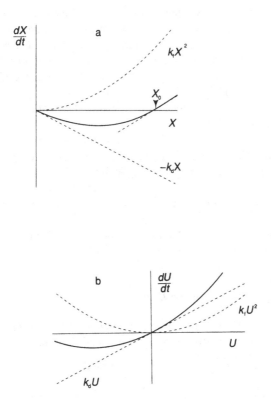

Figure 6.4. Rates of change of (*a*) concentration X and (*b*) displacement from steady state $U = X - X_0$ plotted against X and U, to illustrate equations (6.24) (without diffusion term) and (6.30). In each case, the solid curve is the overall rate of change, and the broken lines are the separate linear and quadratic contributions to it.

This is very straightforward algebra, but it has led to a result with some remarkable features (Figure 6.4), which are typical of the more complicated cases in Chapter 9. When the system starts to move away from the steady state $U = 0$, the term in U in equation (6.30) will at first be much larger than the term in U^2; so, for small displacements, we may neglect the latter. This gives us the equation we wanted, equation (6.22), and indicates that movement away from $U = 0$ will indeed occur, and either upward or downward, depending on what sort of disturbance starts the process on its way. Whenever the U^2 term can be safely ignored, equation (6.30) becomes a linear equation; and its important feature for our purpose is that the coefficient of U is positive. If we trace back through the algebra where that value of $+k_d$ came from, we find $-k_d$ from the original linear term in equation (6.25) or (6.28), and $2X_0 k_f = 2k_d$ from the nonlinear term. *Thus the behaviour that we want, although it can be expressed as a linear term, arose from the nonlinear part of the original expression.*

Evidently the second-order nature of the autocatalysis is very important; therefore, one might imagine that the rate constant for that catalysis, k_f, would dominate the movement away from the steady state. Not so: The coefficient of our linear term is k_d, the decay constant of X, not k_f. Of course, when the pattern amplitude becomes so large that we cannot neglect the U^2 term, the *further* development of the pattern will be governed by k_f. But what pattern develops may be strongly dependent on the *initial* stages of that development, in the linear region.

Meinhardt (1984) has stressed that reaction-diffusion models have "counter-intuitive properties." The difficulties in recognizing that the expression for $\partial U / \partial t$ would have a linear term in U (applicable to both positive and negative values of that variable), that the coefficient of that term would be positive, and that it would arise from the decay rate of X are three such counterintuitive matters.

The diffusion term will be added in Section 6.3.4, following brief notes on two other linearizations.

6.3.2 A brief comment on the Brusselator

The Brusselator is discussed in more detail in Chapter 9. It is formulated as a hypothetical chemical reaction mechanism with four steps:

$$A \rightarrow X \qquad \text{(rate constant} = a), \qquad (6.31a)$$

$$B + X \rightarrow Y + D \qquad \text{(rate constant} = b), \qquad (6.31b)$$

$$2X + Y \rightarrow 3X \qquad \text{(rate constant} = c), \qquad (6.31c)$$

$$X \rightarrow E \qquad \text{(rate constant} = d). \qquad (6.31d)$$

The overall reaction is (as an unbalanced equation)

$$A + B \rightarrow D + E, \qquad (6.32)$$

and the intermediates X and Y are the morphogens.

When the kinetic equations are written down for this mechanism and linearized about the spatially uniform steady state in a more complicated version of the procedure described in the preceding section, it turns out that the rate of displacement of X from its steady state has terms linear in both $U = X - X_0$ and $V = Y - Y_0$:

$$\partial U / \partial t = k_1 U + k_2 V + \text{nonlinear terms} + \text{diffusion}. \qquad (6.33)$$

Without doing the algebra, can the reader guess what the expression for k_1 is likely to be, in terms of a, b, c, and d? Most probably, the previously uninitiated reader would have gone for c, the rate constant of the all-important autocatalysis, as a dominant part of k_1, but will, after reading the preceding section, have realized that c is not going to be in k_1. Anyone who further guesses that either b or d, the rate constants for the two reactions consuming X, is going to be there is along the right lines. The expression is

$$k_1 = bB - d. \qquad (6.34)$$

Note that b appears with a positive sign, but d with a negative sign; and for the full treatment, see Chapter 9.

6.3.3 The mechanism for optical resolution

In the previous two examples, symmetrical departure from the steady state arose despite the absence of any such symmetry in the chemical makeup of the system. Spontaneous optical resolution is the only case in which the symmetry of the kinetics corresponds to the symmetry of chemical composition. Hence the behaviour of this system is not nearly so markedly counterintuitive as that of most other systems. That is why this system is a good lead-in to the topic for biologists. The structural symmetry, which is probably the first thing the biologist perceives, corresponds to the symmetry of the algebra.

The model described in Chapter 5, following Mills, assumes random association of the two enantiomers in pairs on potentially active catalytic sites. If we use A and B to represent the concentrations of those substances in the solution in contact with the catalyst, and suppose that the sum of these concentrations is always high enough to keep the catalytic sites saturated with attached A or B, then for a fixed number of catalytic sites, the number activated by two attached A's will be proportional to the square of the fraction $A/(A + B)$; and similarly for activation by two B's. This gives, for the rates of the reactions,

$$S \overset{2A}{\rightleftharpoons} A \qquad (6.35a)$$

and

$$S \overset{2B}{\rightleftharpoons} B, \qquad (6.35b)$$

with diffusion still omitted from the equations,

$$\partial A/\partial t = k_f SA^2/(A + B)^2; \qquad (6.36a)$$

$$\partial B/\partial t = k_f SB^2/(A + B)^2. \qquad (6.36b)$$

Now the optical asymmetry of the solution (being the quantity one would measure as being proportional to the rotation found in a polarimeter) is proportional to $A - B$. The rate of increase of this asymmetry is found by subtracting (6.36b) from (6.36a) and simplifying the result by using

$$A^2 - B^2 = (A - B)(A + B). \qquad (6.37)$$

Thus,

$$\partial (A - B)/\partial t = k_f S(A - B)/(A + B). \qquad (6.38)$$

Because equilibrium between the two enantiomers is the racemic state $A = B$, we may use $A - B$ as a measure of departure from true thermodynamic equilibrium (not just a spatially uniform steady state, as in the previous examples). If we write

$$\text{asymmetry } U = A - B \tag{6.39a}$$

and

$$\text{total product } P = A + B, \tag{6.39b}$$

then (6.38) becomes

$$\partial U / \partial t = (k_f S / P) U. \tag{6.40}$$

Thus, once again, we have an equation indicating that the kinetics of the system will take it away from equilibrium (in this case, the racemic equilibrium of the enantiomers) in either direction.

Linear and nonlinear terms are not clearly separated in equation (6.40). If P were constant, and the supply of reactant S to the system were externally managed so that S remained constant, the system would behave exactly linearly. It is quite straightforward to envisage the supply of S being so managed, though while doing so we should not neglect to observe that this brings in the whole question of the interaction of system and surroundings, and hence the compatibility of pattern formation with the second law of thermodynamics, as discussed in Section 6.4. But from the kinetic equations we have used, P cannot be constant, as we may see simply by adding together equations (6.36a) and (6.36b) and obtaining a nonzero sum for the rate of increase of $A + B$.

P can actually be held constant by the washout process, which was in the model in Chapter 5 but has not been included in the foregoing equations. This is done in Chapter 9, and equation (6.40) is thereby resolved into its linear and nonlinear terms. For the present purpose, it is sufficient to recognize that circumstances can readily be envisaged in which the asymmetry U is changing rapidly while the total product P is changing only slowly. The linear form of equation (6.40) is then a good approximation. The treatment of the washout model in Chapter 9 shows that when the system is 10% of the way from the 50/50 racemic mixture to resolution (i.e., it is at 55/45 or 45/55), the correction to the rate for nonlinearity is only 1%.

6.3.4 Reaction-diffusion: rate versus wavelength for a single morphogen

For a displacement $U = X - X_0$, because X_0 is constant it disappears from the derivative in the diffusion equation as it did from the kinetic derivatives, that is,

$$\partial^2 U / \partial s^2 = \partial^2 X / \partial s^2. \tag{6.41}$$

This makes it appear, formally, in a diffusion equation, as if U were the concentration of a diffusing substance, whereas it is actually a measure of displacement from a steady state. Combining this with the linear self-enhancement of U, as in the previous three examples, we have the reaction-diffusion equation

$$\partial U/\partial t = kU + \mathcal{D}\, \partial^2 U/\partial s^2. \tag{6.42}$$

Let us now suppose that to a system obeying this dynamic equation, and in the state $U = 0$ everywhere, there is applied a small disturbance in the form of a sine wave of wavelength λ, alternatively specified by $\omega = 2\pi/\lambda$. For such a disturbance,

$$\partial^2 U/\partial s^2 = -\omega^2 U. \tag{6.43}$$

On substituting this in equation (6.42), we have

$$\partial U/\partial t = kU - \omega^2 \mathcal{D} U,$$

or

$$\partial U/\partial t = (k - \omega^2 \mathcal{D})U. \tag{6.44}$$

This is the differential equation for an exponential growth or decay, according as the quantity in parentheses is positive or negative. The solution for U as a function of t is

$$U = U_0 e^{(k - \omega^2 \mathcal{D})t}. \tag{6.45a}$$

The exponential growth or decay constant k_g is

$$k_g = (k - \omega^2 \mathcal{D}) = (k - 4\pi^2 \mathcal{D}/\lambda^2). \tag{6.45b}$$

Clearly, at short wavelengths the negative term is dominant, and the initial disturbance will decay away. At a certain threshold wavelength λ_0, the growth constant will reach zero, and the disturbance will neither decay nor grow; and at longer wavelengths, k_g will become positive, and the disturbance will grow exponentially. Its growth rate will become greater with increasing wavelength, reaching $k_g = k$ at infinite wavelength.

These equations formalize mathematically the model for spontaneous optical resolution. In a system with patches of partially resolved material, which may be regarded as fragments of patterns of various wavelengths, the ones of longer measure are growing faster than the shorter ones. It follows that, in the eventual outcome, one patch must take over the whole system. This wavelength dependence of k_g for unstable pattern (Figure 6.5) also establishes the behaviour of a single self-enhancing morphogen, so that we may see what more is required to form stable pattern. This has already been shown in Figure 3.2: k_g must not go on increasing with λ, but must pass through a maximum and start to decrease as λ rises. This needs a second morphogen.

Anyone wishing to continue pursuing this aspect should jump to Chapter 7;

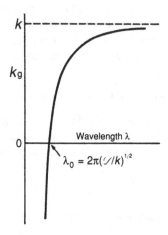

Figure 6.5. Exponential growth rate constant k_g of pattern amplitude versus wavelength λ for a one-morphogen model, equation (6.45). Patterns with repeat unit shorter than λ_0 decay (k_g is negative), and patterns with repeat unit longer than λ_0 grow. Their growth rate increases monotonically with wavelength, toward a maximum which is k, the autocatalytic rate constant in equation (6.42). As in Figure 6.4, this k may not be the rate constant of the autocatalytic step in a reaction mechanism; in Figure 6.4, $k = k_d$, the rate constant of a decay step.

but here we have some unfinished business to do with optical resolution and the second law of thermodynamics (Section 6.4).

6.3.5   Must the self-enhancement involve a squared concentration?

Second-order autocatalysis, as postulated in equation (6.24) and in most nonlinear models (Chapter 9), often needlessly worries biochemists who want to identify it at once with a known mechanistic step. (Curiously, when biochemists find that a protein works only dimerically or tetramerically as an enzyme or a transcriptional activator, they do not usually identify this potential dynamic significance. For my response when I first learnt of protein tetramerization, see Section 9.2).

One should keep in mind the perspective of Mills (p. 157). The dynamics may involve any power, integral or fractional, above the first, and all the mechanistic complexity of biological growth. I define the dynamics by rate equations which could arise from a few steps in a chemical mechanism, but which could also be metaphors for larger-scale feedback controls. Even when there is a simple mechanism, the right nonlinearities can appear without a specific autocatalytic step (e.g. equations (9.41)). An inhibition of an inhibition is equivalent to an autocatalysis.

If, however, one takes the chemical dynamics quite literally, equation (6.24) shows the two commonest concentration dependences in chemical kinetics: first power and second power of concentration. The latter, however, stems not from an overall bimolecular reaction, but from an autocatalysis which is bimolecular in the catalyst and at least unimolecular in a substrate; for example,

$$S + 2A \rightarrow 3A, \qquad (6.46)$$

which is another way of writing equation (6.35a); or, from the Brusselator mechanism,

$$Y + 2X \rightarrow 3X. \qquad (6.31c)$$

In elementary accounts of chemical kinetics, which refer mainly to gas-phase reactions and often fail to stress that many aspects are different in liquid solution, a termolecular step is usually indicated as improbable in a reaction mechanism, because three-body collisions are very infrequent. Indeed, in some of their earlier writings, the devisers of the Brusselator were clearly worried about the termolecular step; for example, Prigogine (1967) wrote that "this reaction scheme is physically unrealistic because it involves the tri-molecular step [(6.31c)]."

This is a problem only if one takes the equation literally as a *single step* in a reaction mechanism. In condensed phases, the same kinetic expressions that one would write for (6.31c) may very readily arise through the quasi-equilibrium formation of a catalytically active complex followed by the kinetic exercise of its activity. This is just what I proposed in the optical-resolution model of Chapter 5, and it is in the spirit of Mills' (1932) suggestion, which envisaged something like a quadratic concentration dependence as a very generalized feature of complex reaction mechanisms. The optical-resolution mechanism may be written, with M as the site which becomes active on attachment of product, as

$$M + 2A \rightleftharpoons MA_2 \qquad (6.47)$$

$$S + MA_2 \rightarrow SMA_2 \rightarrow MA_2 + A. \qquad (6.48)$$

The first step still looks termolecular, but in fact makes no claim to be a single step. It is written as an equilibrium, from which an equilibrium concentration of the complex can be calculated no matter how many steps it takes to achieve the formation of MA_2. One would usually expect a sequence of bimolecular processes such as

$$M + A \rightleftharpoons MA \qquad (6.49a)$$

and

$$MA + A \rightleftharpoons MA_2. \qquad (6.49b)$$

Such a sequence might, however, stop at the first step or go on to the formation of higher complexes MA_n, with $n > 2$. This would lead to kinetic expressions similar to equation (6.24) but with powers other than 2 in the autocatalytic term:

$$\partial X/\partial t = -k_d X + k_f X^n. \tag{6.50}$$

It is left to the reader (Problem 6.5.4) to repeat the algebra of equations (6.24) to (6.30) in this more generalized form, and hence to show that the required expression for departure from equilibrium,

$$\partial U/\partial t = kU + \text{nonlinear terms, with } k > 0, \tag{6.51}$$

is obtained if $n > 1$, whereas if $n < 1$, k turns out to be negative, and a displacement from equilibrium would decay away. In this exercise, one need not think of n as being capable only of assuming integral values. Very simple mechanisms can give $\frac{3}{2}$-order rate equations, and if one is thinking of these simple mechanistic schemes as being abbreviations for much more complicated ones, then one should think of n as a continuous variable.

The threshold $n = 1$ for production of out-of-equilibrium structure arises, of course, because the X^n term stands compared with a first-order decay. One may question whether the latter is an arbitrary choice (although biologists, whatever they have doubts about in proposed kinetic expressions, usually seem prepared to admit a probable first-order decay of almost anything). The optical-resolution model of Chapter 5 calls attention to the importance of envisaging biological systems as flow systems. Thus, the disappearance of X (corresponding to A or B in that model) in proportion to its concentration is nothing more than the washout in the continuous flow. The parameter k_d is then a measure of the rate at which the developing system interacts with the rest of the universe, which is somehow maintaining the flow and reprocessing the removed products into fresh reactants. The term $-k_d X$ is then far from arbitrary in form; its linearity corresponds to a linear measure of flow rate and of system–surroundings interaction. Because by the second law of thermodynamics the production of order in the system (optical resolution or pattern formation) requires a system–surroundings interaction in which the latter is increasing in entropy, is it any longer surprising that the rate parameter for departure from equilibrium turned out to be k_d? In the more extensive analysis of this aspect in the next section, a corresponding constant is called k_{ext}, to stress its relation to external influences.

6.4   Thermodynamics, thresholds, bifurcations, and catastrophes

Thermodynamics, as the name implies, arose from the study of movement of heat. But the chemical thermodynamics of an open system can often be discussed primarily in terms of flow of chemical substances. The optical-resolution model of Chapter 5 is formulated in a way which facilitates such a discussion. There is no specific requirement that any of the reactions be endothermic or exothermic, and no need for any flow of energy per se between fixed masses of material. All exchange between system and surroundings is represented by the flow of S into the system and A and B out of it. These are analogues of the supply of food and oxygen or carbon dioxide to a living organism, or a population (for the model may be read as developmental or ecological), and the removal from it of waste products to be recycled into food.

The following equations do not represent exactly what is shown in Figure 5.5. There, the enduring analogue of the living system is a fixed amount of catalyst with a fixed total of A and B attached. To achieve this constancy, all A and B produced in solution, if not exchanged back onto the catalyst, is removed in the flow, which implies that the rate of removal is increasing as the catalyst increases in total activity by loss of inactive AB pairs. In the present account, the system is the catalyst with a fixed volume of solution, catalyst bed in Figure 6.6a. The flow rate in and out of the bed is constant; hence A and B are removed at rates proportional to their concentrations. The catalyst is again constant in amount and is always saturated with A and B attached at random, in pairs, to each site. The total amounts of A and B in the system are thus not held arbitrarily constant, but are subject to changes which can be calculated from the kinetic equations.

Those same rate equations can, however, be set up if the system is envisaged as closed, but with a flow of energy into and out of it. One may suppose, for instance, that A and B are destroyed photochemically and thereby converted into something from which S can spontaneously regenerate. The energy flowing into the system most probably would become thermalized and flow out of the system as heat (Figure 6.6b).

6.4.1 *Threshold flow rate: kinetic analysis*
of the optical-resolution model

For either the open system with material flow or the closed system with energy flow, the rate of removal of products can be written in the rate equations as

$$\text{rate of removal of } A = k_{ext}A, \tag{6.52a}$$

$$\text{rate of removal of } B = k_{ext}B. \tag{6.52b}$$

In the case of the open system, occupying a length L of the flow path in which linear velocity of flow is V, the equivalent of k_{ext} is simply the reciprocal of the time taken for the flow to traverse the system,

$$k_{ext} = V/L \tag{6.53}$$

The flow of matter or energy which appears in the equations only as represented by the parameter k_{ext} drives the nonspontaneous order-creating process in the system. But up to this point, throughout the discussion in Chapters 5 and 6, it has not been especially obvious that the process needs

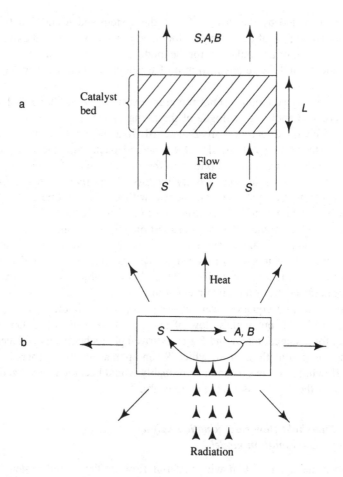

Figure 6.6. (*a*) Open system, with substrate *S* flowing in at a concentration above equilibrium with *A* and *B*, and flowing out at equilibrium with them. (*A* and *B* may or may not be at the racemic equilibrium with each other.) (*b*) Closed system with radiation entering not at the temperature of the system (e.g., solar radiation, which has an effective thermodynamic temperature of about 5,000K, in equilibrium with the surface of the sun) and photochemically converting *A* and *B* back to *S* (above equilibrium). Energy leaves as heat, at the temperature of the system.

driving. This is because each reaction has been written with an arrow pointing only to the right. Such an assertion of total irreversibility represents an infinite thermodynamic driving force, and is always unrealistic. In rigorous thermodynamic discussion, every chemical reaction must be shown as reversible, and the rate of its reverse taken into account. In treating the development of the system in this way, one is replacing thermodynamic by kinetic formalism. This is often indicated as dangerous, but is always legitimate if one can cope with the problem of writing, for every forward reaction shown, the *precise* reverse reaction. The present case is not difficult. For the reaction

$$S \underset{MA_2 \text{ catalyst, } k_r}{\overset{MA_2 \text{ catalyst, } k_f}{\rightleftharpoons}} A, \qquad (6.54)$$

the reverse reaction involves the same catalytic function as the forward, and is therefore stereospecific, in contrast to the destructions of A and B by the k_{ext} process. This contrast confers on the removal process of A and B by matter flow or energy flow the ability to drive the order-generating process. A fraction $A^2/(A + B)^2$ of the catalyst has $2A$ molecules attached. Hence, the rate of removal of A by reverse reaction is

$$k_r AA^2/(A + B)^2 = k_r A^3/(A + B)^2, \qquad (6.55)$$

where the constant k_r is proportional to the total amount of catalyst in the system.

The gist of the role of flow rate in permitting resolution to occur is that the forward reaction has a rate proportional to A^2, whereas its reverse has a rate proportional to A^3. This higher power prevents the system from leaving the racemic state, unless the dominant reverse reaction is the one involving external interference, which has a rate proportional to A. The algebraic proof of this assertion is as follows (Harrison, 1974). The overall rates of change of A and B in the system are (Figure 6.7)

$$dA/dt = k_f SA^2/(A + B)^2 - k_r A^3/(A + B)^2 - k_{ext}A, \qquad (6.56a)$$

$$dB/dt = k_f SB^2/(A + B)^2 - k_r B^3/(A + B)^2 - k_{ext}B. \qquad (6.56b)$$

This treatment will consider only the possibility of resolution in the well-stirred solution in which concentrations are spatially uniform. Hence there are no diffusion terms, and the time derivatives can be written as ordinary derivatives. From these two equations, we may determine the steady states of the system by setting both rates equal to zero. The one we need, because it is the starting point for resolution, is the racemic state:

$$A = B = Sk_f/[k_r + 4k_{ext}]. \qquad (6.57)$$

[There are also resolved states:

$$A = 0, B = SK_f/(k_r + k_{ext}), \qquad (6.58)$$

and the same with A and B reversed.]

From equations (6.56) we may also calculate the rate of change of the fraction of A in the system, $d[A/(A + B)]/dt$. If one would prefer to look at the rate of change of (optical asymmetry)/(total product) $= d[(A - B)/(A + B)]/dt$, one has only to double the previous quantity. After some algebra, the result is

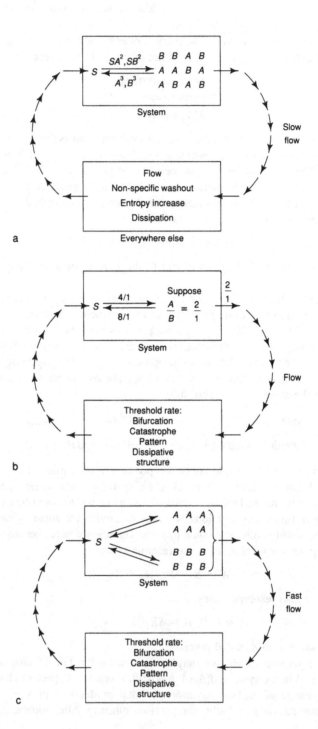

$$d[A/(A + B)]/dt = [AB/(A + B)^4]k_f(A - B)[S - (k_r/k_f)(A + B)].$$

(6.59)

This gives the rate of optical resolution for any values of A and B. Because the first square bracket is necessarily positive, if $A > B$ the rate of change of $A/(A + B)$ is positive or negative according as the second square bracket is positive or negative. We wish to consider starting from the racemic state, equation (6.57), and applying a small disturbance which will make $A - B$ slightly positive without much change in the second square bracket. The condition for the latter to be positive is

$$S > (k_r/k_f)(A + B) = 2(k_r/k_f)Sk_f/[k_r + 4k_{ext}]$$
$$= 2Sk_r/[k_r + 4k_{ext}].$$

(6.60)

Clearly, this inequality is satisfied if and only if

$$4k_{ext} > k_r.$$

(6.61)

This puts on a quantitative basis the earlier statement that for resolution to occur, the dominant reverse reaction is the one involving external interference. Expression (6.61), written as an equation rather than an inequality, is a threshold condition on the rate of flow of matter or energy through the system for optical resolution to occur. Similar conditions may usually be expected to govern the occurrence of any order-generating or pattern-forming processes in living systems.

The parameter k_{ext} appears, most obviously, as something concerned with the rate of removal of material from the system. If we think of the model in an ecological sense, with each A or B particle as an organism, then the terms $-k_{ext}A$ and $-k_{ext}B$ in fact represent death. One of my thermodynamically minded colleagues used to describe the foregoing argument to me as "your 'death dominates' thing." One may also go right back to the origins of thermodynamics in practical engineering and note that James Watt's success in getting steam engines to work well arose largely because he worried about the condenser, the remover of heat from the system.

It is, however, important to appreciate that the flow of the significant reactants and products in this process, through system and surroundings, can

Figure 6.7. *(opposite)* Aspects of the optical-resolution model in an open system with material flow. (*a*) Slow flow produces the racemic mixture, with no resolution or pattern formation. (*b*) Numerical illustration of how the internal and external reverse processes compete to destroy and assist, respectively, the asymmetrization. (*c*) Fast flow produces the kind of sequence of events shown in Figure 5.7, pattern formation followed by resolution, shown here at the intermediate stage. The rest of the universe contains a chaotic mixture of thermodynamic and mathematical jargon, which is getting worse all the time. From Harrison (1982), © Alan R. Liss with permission from Wiley-Liss, a division of John Wiley and Sons, Inc.

be formulated cyclically. In that sense, k_{ext} is not related to any particular point in the cycle, but to the rate of rotation of material around the whole of it. The important aspect may be seen not as the destruction of A and B, but as the building up of an adequate supply of the reactant S. Thresholds such as expression (6.61) often may be recast as lower limits on reactant concentrations for order-generation. The Brusselator is interesting in this respect. Of the two reactants A and B (which are analogous to the S of the present discussion), in some important regions of pattern-forming behaviour, B must exceed a threshold value, but A must be *less than* a threshold value for the order-generating behaviour to occur. Such thresholds are potentially important in connection with the establishment of the boundaries of a morphogenetic field (Chapter 9, especially Section 9.1.3).

The optical-resolution model may be thought of as more evolutionary than developmental. For the stabilization of pattern, it is incomplete. But it certainly takes a system irreversibly away from its starting configuration toward two possible states each of which has more order than the starting configuration. This is variation, or even the most primitive example of speciation. Here, a partial failure of one of my favourite analogies may be very significant. The autocatalytic terms, bimolecular in A and B, are seen in the ecological aspect of the model as representing sexual reproduction. But once a system has reached the complexity of a unicellular organism, and the chemical reaction step has been transferred in meaning to something to do with the whole biochemistry of two cells, then the reverse reaction becomes, if I may be excused the double entendre, inconceivable. For sexual reproduction, $k_r = 0$. It therefore appears to be a process having the character of an infinite thermodynamic driving force. I have not been thinking about evolutionary theory for long enough to have arrived at any personal assessment of how important this may be. I offer it for stimulation to evolutionary thermodynamicists (though they seem to need none, because the field is currently one of strong controversy).

6.4.2   The same threshold condition with more entropy and less algebra

Although the word "thermodynamic" is used extensively in the preceding section, the argument presented in detail is kinetic. Because the matter concerns relative rates, any treatment necessarily has kinetic aspects, even if it is made ostensibly fully thermodynamic by disguising all rates as rates of entropy increase rather than rates of production and consumption of matter. Here, I do not attempt a complete recasting of the treatment in such terms. Instead, I introduce entropy criteria, use the racemic steady-state equation (6.57) arising from the kinetic equations (6.56), but avoid asking the reader to perform or take for granted the algebraic chores involved in deriving equation (6.59). With this step (which involved differentiating a quotient) bypassed, we actually need no formalisms of the calculus. Instead of dA/dt, we can write "rate of formation of A" and conduct the argument just as well. The less mathematically prepared reader should try this approach wherever in this book expressions are used containing only *first* derivatives.

We may take a step toward analysis of the requirements of pattern formation by abandoning the well-stirred system and the departure of the overall A/B ratio from unity and returning to the intermediate stage of sorting out a mixture of A and B into a pattern of two parts, that is, the first 23 time steps of the computation shown in Figure 5.7, but idealized to the case in which the lengths of the two parts are exactly equal. Let us assume that solutions of S, A, and B behave ideally and that there are no energy changes in the chemical reactions. The pattern formation can then be described thermodynamically by an entropy of unmixing. For one mole total of A and B, this is $-R \ln 2$, where R is the gas constant. Its negative sign is the problem, for occurrence of the process. Actually, we do not need to use this expression, but only to know that it is the entropy decrease of one mole upon compression into half its original volume. (In an ideal mixture, each component behaves as if it occupied the whole volume entirely independently of the other components. Here, upon unmixing, $\frac{1}{2}$ mole of A has been compressed from the complete volume to half of it, and likewise for B, a total compression of one mole by a factor of 2.) If we look at the flow system batchwise, and suppose that a batch of S has been delivered to a system with randomly distributed active catalyst for A and B production, we may then treat the system as isolated while the reaction occurs. For the reaction to include spatial unmixing, there must somewhere be a compensatory entropy increase. This is present if S is supplied at a concentration at least twice its equilibrium concentration with A and B. Then the process involves, effectively, a dilution of one mole of S by a factor of 2 for the production of one mole total of A and B. The entropy increase associated with this dilution compensates for the entropy of unmixing of A and B (Figure 6.8).

Now the equilibrium relationship among A, B, and S can be obtained by setting $k_{ext} = 0$ in equation (6.57). Then, only the forward reaction and its precise thermodynamic reverse on the same catalyst are being considered, and

$$A = B = S_{eq} k_f / k_r. \tag{6.62}$$

This is the familiar representation of an equilibrium constant as the ratio of forward and reverse rate constants. We require that this concentration of S be no more than one-half of the "forced" steady-state concentration S_{fo} which is the S in equation (6.57):

$$A = B = S_{fo} k_f / (k_r + 4k_{ext}). \tag{6.57}$$

For unmixing to occur at some finite rate, then, we require $S_{fo} > 2S_{eq}$, or

$$k_r + 4k_{ext} > 2k_r. \tag{6.63}$$

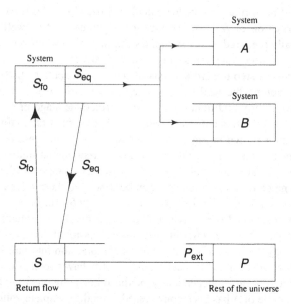

Figure 6.8. A hypothetical "optical-resolution machine" illustrating the essential entropy changes in the irreversible cycle. The three locations marked "System" would all be the same location in practice. Thus the two compressions of A and B shown separately represent the unmixing of them. These two pumps, as well as the two concerned with S, are active-transport pumps with semipermeable pistons (stereospecific, for A and B). The A and B compressions are each by a factor of 2 and can be driven by S if $S_{fo} > 2S_{eq}$. The P pump is envisaged as a compressed perfect gas. Its noncyclic expansion is the dissipation in the surroundings. P/P_{ext} must exceed S_{fo}/S_{eq}. This machine contains an antecedent, on the macroscopic scale, for the pattern formation: The two pistons pumping A and B are, in effect, separated enantiomers. This does nothing to relieve the thermodynamic requirements of separating the continuing production of A and B as a mixture. Any attempt to locate the two pistons in the actual geometry of the system would confer a somewhat supernatural character upon them: They would have to pass through each other.

Subtracting k_r from each side gives us again inequality (6.61), which we now see also as a threshold of reactant concentration, being the requirement that reactant be delivered in twofold excess over equilibrium.

One is obliged to worry about the entropy of the universe, and hence about the reprocessing of S outside the system for redelivery as new input. Because S leaves the system at equilibrium with A and B, compression by a factor of at least 2 is needed. Every student of biology learns that such compression of a solute in solution is the proper business of an active-transport pump. This can be formally envisaged as a piston made of a semipermeable membrane which passes the solvent but not the solute. The minimum compression needed is equivalent to doubling the pressure and halving the volume of a mole of perfect gas, and again has an entropy decrease of $-R \ln 2$. To complete the set of changes in the whole universe, we may now suppose that, to achieve a finite rate of compression, the semipermeable piston is connected to another on a cylinder of perfect gas at somewhat more than the osmotic pressure of the higher concentration S_{fo}. If any restraint on the pistons is released, the required process will take place with an entropy increase exceeding $R \ln 2$ in the perfect gas. We have thus taken into account four entropy changes: $-R \ln 2$ in the pattern formation; two equal and opposite changes, both exceeding $R \ln 2$, in the effective dilution of S within the system and its reconcentration outside; and an increase exceeding $R \ln 2$ in the external driver (the perfect gas, or whatever). The difference of the first and fourth of these terms is the entropy increase of the universe.

6.4.3 An assortment of jargon: bifurcation, instability, catastrophe

The biologist is accustomed to use terms descriptive of shape for the real morphologies of living things. The mathematician, however, often uses similar terms in an abstract, figurative sense which has proceeded beyond the need for pictorial representation. Most people describe x^2 as a square and x^3 as a cube, without trying to visualize any such actual shape. The mathematician behaves similarly in many instances much less familiar to other people. There is sometimes danger of confusion between concrete and metaphorical usages.

Bifurcation means division into two branches. To the biologist, it is a synonym for dichotomous branching. To the mathematician, it means, for a set of differential equations, the point at which a new solution arises as one continuously changes the value of some adjustable parameter in the equations. This mathematical concept can be involved in theoretical explanations of real dichotomous branching. But it could equally well arise in explanations of the initiation of any change of shape in a living organism, regardless of whether or not that change is morphologically a bifurcation. Because of this possible confusion, and the wide prevalence of the mathematical usage of "bifurcation," I tend to avoid the term when describing the real phenomenon.

As an example of the mathematical usage, consider the system of two enantiomers A and B initially at racemic equilibrium, and with an autocatalytic asymmetrizing mechanism, as discussed in Sections 6.3 to 6.4.2. If $U = A - B$ and $P = A + B$, then we may write a measure of asymmetry, indifferent to whether A or B is in excess, as

$$Z = |A - B|/(A + B). \tag{6.64}$$

If we use equation (6.59) for the growth rate of asymmetry, with equation (6.57) for the racemic starting concentrations, then the exponential growth rate of asymmetry at the beginning of departure from the racemic state,

$$k_g = (1/Z)dZ/dt, \tag{6.65}$$

becomes the rather simple expression

$$k_g = k_{ext} - k_r/4. \tag{6.66}$$

Suppose that, as an instance of parameter variation in differential equations, we increase the flow rate around the cycle of Figure 6.6a or 6.7. This amounts to increasing k_{ext} at constant k_r. At slow flow rates, k_g is negative, and any small displacement from racemic proportions will decay back to zero. At the critical value already indicated in equation (6.61), and evident again from equation (6.66), k_g switches sign, and a displacement will grow exponentially instead of decaying. This growth will continue, but must eventually slow down because the system has limited capacity to hold material, so that U is limited to some maximum value $U_{max} = P_{max}$. We do not need the details of the growth law to draw a bifurcation diagram representing the destiny of the system at infinite time (Figure 6.9).

A *threshold value* of a parameter for the onset of a particular kind of dynamical behaviour arose in the last example. To my mind, the concept of thresholds is a better one for the biologist to keep in mind than is the concept of bifurcations, though frequently the two words draw attention to the same changeover point in behaviour.

In this book, such thresholds are treated throughout in terms of the switch of an exponential growth constant k_g from negative to positive (i.e., from decay to growth), or occasionally in terms of competitive growth of two patterns according to which has the larger positive k_g. This kind of treatment arises because this book is about kinetic theory, and kinetics often simplifies to the matter of exponential growth or decay, at least at the onset of any new process. In other respects, however, the thresholds mentioned can be distinguished into three kinds:

1. *Threshold wavelengths* for growth of pattern, as exemplified by the λ_0 of Figure 6.5 for the "one-morphogen" case. This kind of threshold has to do with possible variable input to a system with a definite mechanism for pattern formation, with fixed parameters. This kind of threshold is very useful to think about in trying to arrive at an understanding of how a pattern-forming mechanism manages to produce order out of chaos. For the matter of what will be observed in practice in undisturbed natural development, however, the disorderly input which one can expect to be present always will contain wavelengths on both sides of the threshold. The system therefore will always show the behaviour corresponding to the k_g-positive side of the threshold.
2. *Kinetic-parameter variations corresponding to the second law of thermodynamics*. This kind of threshold is illustrated by equation (6.64) and the discussion in Sections 6.4.1–6.4.3. Without that discussion, it is not immediately obvious that the threshold condition, which can so easily be derived from simple consideration of relative rates of a few reactions, has to do with ensuring that the entropy of the universe increases as it must.

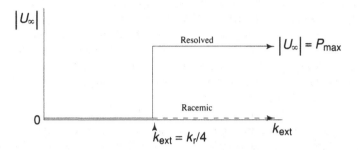

Figure 6.9. Bifurcation diagram for the flow system of Figures 6.6a and 6.7. The diagram is drawn for the absolute value of asymmetry, so that resolutions as A or B appear as the same line, although U is positive for resolution as A and negative for resolution as B. The absolute value is used to stress that the bifurcation is not so called because two new states have arisen at the bifurcation point (resolution as A or B), but because *one or more* new states have branched away from the possible continuation of the racemic state, shown as a broken line. The system would remain in this state if no displacement at all ever arose to start the movement toward resolution; but the racemic state is kinetically unstable beyond the bifurcation point. The line representing equilibrium behaviour (in this case, the racemic case) is called in the writings of the Prigogine school "the thermodynamic branch."

3. *Kinetic-parameter variations not corresponding to the second law.* Proper discussion of the thermodynamic criterion required the introduction of reverse reactions into the discussion, and the resulting condition involved only the reverse rate constants. Very often, kinetic mechanisms for pattern formation are analyzed with the reverse reactions omitted. Then, it is being assumed (usually, tacitly) that the flow rate, or excess of reactant concentration over the thermodynamic threshold (where reactant concentration determines flow rate and hence is hidden in k_{ext}), is very large, so that the system is far from the thermodynamic threshold.

In this book, this is done for the Turing model (Chapter 7) and the Brusselator (Chapter 9). But in those accounts, some more threshold conditions appear in the form of restrictions on the relative values of various rate parameters. For instance, Figure 7.5 is a threshold (or bifurcation) diagram showing how different kinds of behaviour of the two-morphogen Turing model depend on the autocatalysis (or self-inhibition) constants for the two morphogen displacements U and V. These relative values determine, for instance, whether stable pattern will form or whether the system will go into oscillations or whether all disturbances will decay into uniformity. But these distinctions are not concerned with satisfaction of the second law.

"Structure," to many people, would seem to be a word exclusively specifying concrete, physical realities. But again, mathematicians use it in a figurative sense. To a mathematician, a set of differential equations is a *thing,* and changes to its structure are new terms in the equations. Thus, for instance, of the three categories of thresholds just listed, the first is concerned with variable input to a system with fixed dynamic capabilities, whereas the other two are concerned with changing those capabilities by changing parameters in the equations. The mathematician sees the equations as an engine, and the distinction as being that between the driver trying to start the engine and a mechanic tuning it up (changing parameter values) or more drastically changing its *structure* (adding new terms).

It is in this mathematical sense, quite different from the sense of my distinction among structure, equilibrium, and kinetics, that Andronov introduced the term "structural stability" into the theory of oscillators (Andronov and Pontryagin, 1937; Andronov, Vitt, and Khaikin, 1937). An example often given to explain the meaning of structural stability is the simple harmonic oscillator, in the form of either a swinging pendulum or a weight going up and down on the end of a spring. In an elementary account, such situations are described in terms of the ordinary differential equation

$$d^2x/dt^2 = -kx, \qquad (6.67)$$

where x is displacement of the weight or the pendulum bob from its rest position. (For the pendulum, one should have $\sin x$ on the right-hand side, but it approximates to x for small amplitudes.) Solutions of this equation indicate a sinusoidal oscillation going on forever; for example,

$$x = A \sin(k^{1/2}t). \qquad (6.68)$$

In practice, if we let an oscillating spring or a pendulum go on without external interference, it eventually will come to rest because of the existence of something called friction in the real system which was not represented in the differential equation. It can be represented by adding a term $(-k_f\, dx/dt)$ on the right-hand side to show a drag proportional to velocity. The solutions of the equation will then remain periodic in time, except that the amplitude A will acquire a time dependence, decaying exponentially to zero, and the frequency of oscillation will be somewhat altered. Equation (6.67) is said, in Andronov's terminology, to be *structurally unstable* to the addition of a friction term.

Suppose now that we add to the system some mechanism to keep on supplying the energy which friction removes. For a pendulum clock, this will be the mainspring, plus the escapement mechanism by which it and the pendulum communicate with each other. The oscillation can now go on forever, if we assume the presence of someone to wind up the mainspring. An equation with an extra term for the transfer of energy through the escapement will have solutions rather similar to those of equation (6.67): a slightly different frequency of oscillation, quantitatively, but qualitatively an approximately sinusoidal oscillation which will go on forever. The equation with three terms on its right-hand side has become structurally stable with respect to any additional minor amendments we might think of to fine-tune the model to the physical reality, and the upshot is that the one-term equation (6.67) can be used as a good first representation of what is going on.

The close analogy between this example and the one-morphogen model for optical resolution should now be clear. Mills' second-order autocatalysis mechanism seems to be destroyed by putting in a realistic reverse reaction, just as friction destroys simple harmonic motion as a long-term phenomenon. But an external drive will keep the move toward optical resolution running, just as it keeps a clock pendulum running. Once we are assured of this, often we may drop both the additional terms, as is done throughout the discussion of two-morphogen models in the next three chapters.

For systems which may have behaviour periodic in time or space or both, the theory of stability is complicated and leads to various categories: Liapunov stability, marginal stability, orbital stability, structural stability, and so forth. For those who want to know whether a mathematical idealization is irrelevant to real systems because small changes in the model totally change its predictions, or whether it is "coarse" enough (another of Andronov's words) that its properties resist small perturbations, structural stability is the concept that matters.

"Catastrophe" has, both in French and in English, a connotation corresponding fairly precisely to its Greek etymology: a downturn. It is therefore a somewhat strange word to characterize events which we usually view as upturns, or progress to higher forms, in the development of a system. Yet in 1972 R. Thom (1975) introduced the term for just such changes. (Theoreticians, who should be sensitive to the need to persuade an unreceptive audience of experimentalists, seem to have had little regard for rhetoric: "catastrophe," "instability," and "dissipative" are all pejorative terms.)

Thom's strategy in developing catastrophe theory is concordant with the approach of physicists and applied mathematicians, as discussed here in Section 1.2.1, especially Figure 1.1: He wishes to proceed from observed morphological change to analysis of the type of process which could account for it. Thus he wrote (1975, p. 60): "This algebraic behaviour has fundamental importance in our model, *as it shows that the type and dynamical origin of a catastrophe can be described even when all the internal parameters describing the system are not explicitly known*" (Thom's italics).

Despite this philosophical resemblance of Thom's basic attitude to my own, I have not yet found catastrophe theory to be particularly useful as a practical formalism for trying to relate theory to experiment in pattern formation. This may signify only that I have not had time to work with the theory so as to get a feeling for it. But I suspect the cause is something less personal: that catastrophe theory is at present, in its level of abstraction, one big step further from experiment than is, for instance, reaction-diffusion theory. Thus, to the biologist the living organism is the engine, and a set of differential rate equations is as far as a biologist would want to go in level of abstraction to describe how the organism works. But, to repeat my earlier statement, to the mathematician the set of differential equations is the engine. The theory of structural stability, which is catastrophe theory, is the description of how that abstract engine works. It is therefore not surprising that the language used in that theory, even when specifically biological problems are being addressed, is for mathematicians only, and quite inaccessible to most experimental biologists and also to people like myself with some pretensions to a moderate degree of mathematical literacy. As a possible participant in the scientific enterprise of strong interaction between experiment and theory to the benefit of both, catastrophe theory may be some decades ahead of its time.

I am, of course, expressing a bias toward my own discipline in suggesting that the right level of abstraction just now is that biological systems are chemical systems and their dynamics are in the realm of the physical chemistry of the 1920s to 1940s, which has not yet been fully exploited to this end, though Turing pointed the way in 1952. The physical chemist tends to be quite dogmatic in distinguishing between equilibrium and kinetics, which is an appropriate and important attitude *at that level of abstraction*. Both field theory (Section 4.1.5) and catastrophe theory tend to blur that distinction. To see how this happens in terms of level of abstraction, consider Figure 6.10. Figure 6.10a represents the rate of formation of a substance, concentration C, obeying the kinetic law:

$$dC/dt = -k_r C^3 + k_f S C^2 - k_{ext} C. \tag{6.69}$$

This is a bimolecular autocatalysis model with formation and destruction terms for C similar to those discussed for the optical-resolution model (though it is not the same as that model). It corresponds more closely to the switching behaviour in Oster's mechanochemical model (Section 4.4.4).

Figure 6.10b shows the pressure–volume curves for a fluid which obeys the van der Waals equation. This can be written

$$V^3 - (b + RT/P)V^2 + (a/P)V - ab/P = 0, \tag{6.70}$$

where V is the molar volume. Here again there is a switching phenomenon between two different concentration states of a substance: the evaporation–condensation switch between liquid and vapour.

The chemist distinguishes these two phenomena, the one as belonging to kinetics, the other to equilibrium behaviour. The mathematician notices that both expressions are cubic (one in C, the other in V, which is $1/C$ for the substance concerned). Because a cubic equation can have three real roots, both equations indicate a triple-valued property for some numerical ranges of the parameters. In relation to the kinetic system, if we seek steady states by setting $dC/dt = 0$, there are three values of C for any $k_{ext} > 0$. In the equilibrium system, there are three values of V for a constant pressure at any temperature below the critical temperature. In both cases, the middle of the three values is unstable, but for two very different reasons.

Just as the foregoing example places disparate phenomena within the math-a   Kinetics   b   Equilibrium

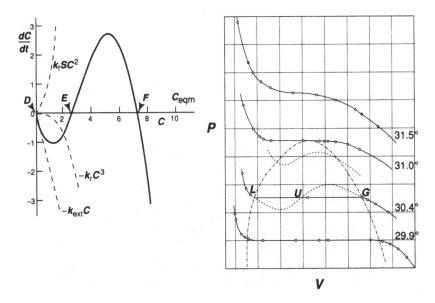

Figure 6.10. Multivalued properties in kinetics and equilibrium, both in the guise of solutions of cubic equations, to illustrate unity versus diversity at different levels of theoretical abstraction. (*a*) Rate of formation of a substance *C* versus concentration, equation (6.69), with $k_r = 0.1$, $k_f S = 1$, and $k_{ext} = 2$. $C_{eqm} = k_f S / k_r$ is the thermodynamic equilibrium concentration. Of the three steady states *D*, *E*, and *F*, the middle one (*E*) is unstable, because if *C* increases slightly, dC/dt becomes positive, leading to further increase, and likewise negative dC/dt arises if *C* falls. Compare the point $U = 0$ in Figure 6.4b. *D* and *F* are stable. (*b*) Pressure–volume relationships in carbon dioxide near to its critical temperature (31.013°C). The sigmoid curve shown as a broken line from *G* to *L* at 30.409°C represents solutions of the van der Waals equation (6.70) which are not physically realized. *G* is the gaseous state, *L* the liquid, and *U* an unstable state, for which greater pressure would lead to greater volume. From *Philos. Trans. R. Soc. London*, 1937, with permission.

ematical discussion of the solutions of cubic equations, so Thom's catastrophe theory seeks to unify bifurcations, or changes in structural stability, into a few shapes of surfaces with interesting folds and pleats in them. Some of these have been given picturesque names (the butterfly, the swallow's tail, etc.). All are plots of polynomials with two variable parameters, just as one might convert all the curves of Figure 6.10b into a surface by adding a temperature axis as the third dimension. Both this surface and the related but different one that could be generated by plotting Figure 6.10a for different values of k_{ext}, and then turning the set of curves into a surface by adding a k_{ext} axis to the

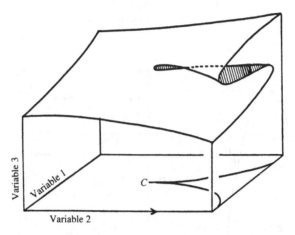

Figure 6.11. The simplest example of the kind of diagram used in Thom's catastrophe theory: the cusp catastrophe. The cusp, at C in the projection of the surface onto the plane for which variable 3 = 0, corresponds to a value of variable 2 which must be exceeded before there can be any "catastrophic" behaviour. This is similar to what happens in the systems of Figure 6.10a,b when, respectively, k_{ext} exceeds zero or the temperature falls below the critical value. Variable 3 becomes triple-valued for certain ranges of the other two variables. But this diagram does not correspond in any more precise way to those obtained by the three-dimensional representations of Figure 6.10.

diagram, bear some relationship to Thom's cusp catastrophe surface (Figure 6.11). But neither example is exactly the same as the cusp catastrophe, which one does not need to know about in order to construct and use diagrams like Figure 6.10.

Catastrophe theory is capable of covering both equilibrium and kinetics. But applications of it to biology have been concerned largely with the latter. Thus, Zeeman (1974) elaborated Thom's earlier work on embryology into an extensive account of gastrulation, neurulation, and somite formation. The entire account envisages differentiation as switches between different states of homeostatic cellular metabolism. Points D and F in Figure 6.10a are two homeostases for substance C. In catastrophe theory, an entire surface like that of Figure 6.11 can consist of possible states of homeostasis.

The reader may detect some inconsistency between my desire to draw physical distinctions here and the unifying philosophy advocated in Chapter 1, especially the paragraph in Section 1.1.1 in which a biological referee's views on mathematicians are mentioned. The resolution of this paradox is that I am primarily concerned with seeking physicochemical explanations of biological phenomena, with mathematics coming into use as a language where necessary, rather than as the ultimate unifier. We are still a long journey from the One Equation.

6.5   Problems illustrating principles

This book is not a primer in calculus. It is intended for readers who have at some time taken at least a first course in that branch of mathematics. The calculus is a language, possessing the powers of description, explanation, and enlightenment of the understanding that any sophisticated language should. Those who need reminders on the basic vocabulary and grammar of the language of calculus should look first at mathematical texts of the appropriate level. Anyone who does not immediately understand the statement that the fundamental basis of calculus is the notion of a limit needs a course in the subject, taken at that leisurely pace which is essential to assimilation of new concepts. Because it is based on the idea that changes can be taken down to infinitesimal limits, calculus is primarily concerned with continuous variables. Molecular systems approximate these when the numbers of molecules become very large. Their individual properties then blur into continuum properties. In place of the random walks in the small-number illustrations of Chapter 5, we have movement along a concentration gradient dC/dx, which can be assigned a value at any *point*.

In the analogy of language, then, what follows next is a set of easy reading pieces, in which the language is used in the simplest possible way to comment on the physicochemical principles of biological morphogenesis. They should be approached in that spirit, rather than as instruction in mathematics.

Problem 6.5.1   Exponential growth

This problem essentially defines the point at which verbal reasoning without mathematics starts to lead one astray. I am indebted to Paul Green for pointing out to me the existence of the trap set in this question, and for telling me that many biology students of high calibre have fallen into it.

Two plants, A and B, elongate in a 24-hour period, A to twice its original length, and B to three times its original length. The growth is exponential, because every part of the plant stem is growing continuously in proportion to its length at any time. Three substances X, Y, and Z are found in the plants in the following concentration (C) ratios:

Substance	$(C$ in $B)/(C$ in $A)$
X	3.0
Y	2.0
Z	1.6

Which substance is most likely to be the growth controller?

Problem 6.5.2   Equilibrium shape and Wulff's theorem

An aspect of development which overrides the distinction between kinetics and equilibrium is that when nature designs a shape, just as much as when an engineer designs one, some kind of optimization is being done. This usually requires the finding of a maximum or minimum, and is an instance in which it is difficult to get along without calculus, even though only first derivatives are used.

(a) A two-dimensional crystal is rectangular. It contains enough material to form an area A, which must therefore be taken as constant. The sides of the rectangle are of lengths a and b. The edge excess free energies per unit length are G_a for the two sides of length a and G_b for the two sides of length b. Write the total edge free energy G in terms of G_a, G_b, a, and A (i.e., eliminate b). Find the minimum by differentiating G with respect to a. Hence, show that this minimum gives the shape $b/a = G_a/G_b$. This proves Wulff's theorem for the special case of this very simple shape.

(b) Refer to Section 4.2.2, especially Figure 4.8. Again area A is constant. Write an expression for the total edge free energy in terms of l, a, n, and G_{10}, and eliminate l by using A. Minimize and show that at the minimum, $l_{11}/l_{10} = n$.

Problem 6.5.3   Kinetics of dendritic growth

Refer to Section 4.2.3, especially the hexagonal two-dimensional crystal in Figure 4.9. This problem illustrates the structure-kinetics relationship. Faces of different structure develop differently, but only because different structures lead to different rates of processes. The problem asks for the minimal algebra needed to cope with this. The general algebraic strategy will be familiar to readers who have taken a course in chemical kinetics including the use of the steady-state approximation to get from mechanism to rate law. The additional feature in kinetic determination of shape is that the steady states of reaction intermediates are spatially nonuniform. Here, the nonuniformity is the simplest possible: two different concentrations on two kinds of crystal faces. In Figure 6.12, the intermediate between gas and solid is an adsorbed layer on the surfaces, concentration C_1 on (11) faces, and C_2 on (12) faces. The rates of adsorption, transfer by flow along the surfaces, and incorporation into the crystal are shown in the diagram. Adsorption and incorporation rates are per unit length, but the "round-the-corner" transfer is not. Therefore, contributions to the rate of change of concentration by transfer are $k_2 C_2/l_1$ and $k_1 C_1/l_2$. Write expressions for dC_1/dt and dC_2/dt. For the steady-state approximation, set both rates of change equal to zero. Hence, express C_2/C_1 in terms of the rate constants and lengths. If transfer between faces is very rapid compared with incorporation into the crystal, show that C_2/C_1 is approx-

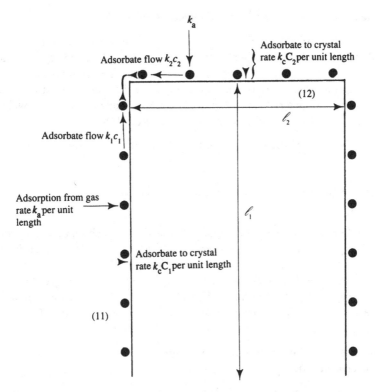

Figure 6.12.

imately k_1/k_2, which therefore gives the relative rates of linear advance of (12) and (11) faces.

Problem 6.5.4   *Kinetic instability of the spatially uniform steady state*

The square of a concentration in a rate expression, or second-order kinetics often related to a bimolecular mechanism, is used so extensively as the self-enhancement term in pattern-forming mechanisms that one can easily overlook the fact that the authors very often will have stated at the outset that the power does not need to be 2; it just needs to be anything greater than unity. This problem is stated following equation (6.50). Use an autocatalysis of arbitrary order n, and show that the k in equation (6.51) is positive if $n > 1$, by repeating the algebra of equations (6.24) to (6.30) with n replacing 2 as exponent. All terms in U^2 and higher powers of U can be dropped from the equations. Only the coefficient of U is wanted.

Problem 6.5.5   Diffusion and the second derivative

The problem is to derive "Fick's second law," equation (6.13), from "Fick's first law," equation (6.12). Consider region B in Figure 6.2, and suppose it to be very thin, with its left-hand boundary at s and its right-hand boundary at $s + \delta s$. The rate of increase of concentration in B is the difference between the rate of arrival of material at the right-hand boundary and the rate of removal at the left, divided by the volume of the region. Recall that

$$d^2y/dx^2 = [(dy/dx)_{x+\delta x} - (dy/dx)_x]/\delta x.$$

Problem 6.5.6   Chemical systems that measure distance: positional signals

Refer to Section 2.1, especially Figures 2.1b and 2.2. A substance, concentration C, is produced at a localized source at position $s = 0$ in a one-dimensional system. The source maintains the concentration $C = C_0$ at $s = 0$ at all times. The substance diffuses with diffusivity $\mathcal{D}$ and is everywhere subject to first-order decay, with rate constant k. Show that for all $s > 0$, C falls off with increasing s so that in the steady state ln C is linear in s with slope $-(k/\mathcal{D})^{1/2}$.

Problem 6.5.7   Concentration distribution with a central minimum

Kinetic theory of morphogenesis is very much concerned with the establishment of regions of maximum or minimum concentration of some substance. For models of localized sources combined with diffusion and decay of the diffusing material elsewhere, the maxima can only be at the positions of the sources. If there is a single source, the only minimum is reached asymptotically at infinite distance from it. But if there are two sources, an intermediate minimum is possible. Such a model, from Nicolis and Prigogine (1977), is used later in discussion of the Brusselator (Section 9.1.3, especially Figure 9.3).

The model is the same as that of the previous problem, but the boundary conditions are $C = C_0$ at both $s = 0$ and $s = 1$. Show that an appropriate solution of the differential equation is

$$C = C_0(e^{\alpha[s - (1/2)]} + e^{-\alpha[s - (1/2)]})/(e^{\alpha/2} + e^{-\alpha/2}),$$

where $\alpha = (k/\mathcal{D})^{1/2}$.

Problem 6.5.8   First-order relaxation to stable equilibrium

A substance X is formed by first-order autocatalysis (rate constant k_f) and destroyed only by the thermodynamic reverse of this formation reaction.

Show that there is a stable steady state X_0 such that if $U = X - X_0$, and terms in U^2 are neglected,

$$dU/dt = -k_f U.$$

This problem is put in just to remind readers what "normal" chemistry is like, in which systems tend to reach equilibrium and return to it if displaced. It is useful to sketch a graph of dX/dt against X and compare it with Figure 6.10a. The comparison raises another problem: Point E is the "point of departure" for kinetics leading to morphogenesis. But in view of its instability, how does a system get near to it in the first place?

6.5.9 Periodic solutions versus growth and decay; phase planes and parameter spaces

It has been stressed already that the behaviour of a system may depend upon the relative values of two rate constants, and there is going to be much more of this sort of thing in Chapters 7 and 9. One particularly common feature of the pairs of differential rate equations for two-morphogen systems is that for some ranges of values of the rate constants they give morphogenetic behaviour, based on exponential growth and decay, whereas for other ranges they give oscillatory behaviour. For our purposes, we want to know about the latter mainly to know how to avoid it. To ecologists studying predator–prey interactions, the time-periodic solutions often represent the reality of their systems and are of the most interest. If two substances X and Y are both varying periodically, with their changes linked, one may ask, If X has some given value, what is the value of Y at the same time? This question can be answered by a graph of Y versus X in which the time variable does not appear. This is called a "phase plane." I do not find much use for this representation in thinking about morphogenesis, but the reader should be aware of its existence. Much more important for the content of the following chapters is the graphing of different kinds of behaviour in a space with two rate constants as its coordinates (k_2 and k_3 in this problem). This is a parameter space. The biochemically minded should think of this sort of representation as being close to their interests and essential to bridging the gap between the macroscopic and the molecular, because the rate constants are related to such things as enzyme activities.

As usual, we have two substances X and Y, and $U = X - X_0$, $V = Y - Y_0$. They interact with each other according to

$$dU/dt = k_2 V, \qquad dV/dt = k_3 U.$$

Here, k_2 and k_3 may each have either positive or negative values; that is, the effect of each substance on the other may be either a catalysis or an inhibition of departure from the steady state (X_0, Y_0).

(a) Show that there are time-periodic solutions if k_2 and k_3 have opposite signs, and solutions showing exponential decay or growth if k_2 and k_3 have the same sign. (Hint: Differentiate dU/dt a second time with respect to t.)

(b) For the periodic solutions, draw a phase-plane representation (X, Y coordinates), mark the state at $t = 0$ as you have written the solutions with any extra conditions needed to make this point unambiguous, and show the direction of circulation around the phase-plane diagram. (It should be a closed loop of some kind. For simplicity, use $k_3 = -k_2$.)

(c) Draw the parameter space (k_2, k_3) and mark in it where the periodic solutions are, and where the growth-or-decay solutions.

Problem 6.5.10   Turing's model without diffusion: parameter space again

The use of parameter space in the previous problem may seem rather trivial. It is, however, a first step along the road to understanding Figure 7.5, which clearly is nontrivial. This problem presents the second step. The differential equations are now

$$dU/dt = k_1 U + k_2 V,$$

$$dV/dt = k_3 U + k_4 V.$$

Again, each of the rate constants may be positive or negative. Consider the special condition $k_4 = -k_1$. For this condition, show that there are periodic solutions if k_1^2 is less than some threshold value, and growth-or-decay solutions if it exceeds that value. Show that the algebra can be set down most concisely by using the complex-number representation of periodic functions. (The solution to this problem may serve as a reminder of this representation, and it is all that one needs to know about the topic for the discussion in the ensuing chapters. Just as for the calculus, the account of complex numbers here is intended as a reminder, not an introduction.) Draw a parameter space in which the coordinates are $k_1' = k_1/|k_2 k_3|^{1/2}$ and $k_4' = k_4/|k_2 k_3|^{1/2}$. (Absolute value of $k_2 k_3$ is specified here because, as emerges in Chapter 7, one of these quantities k_2 and k_3 is usually negative, and the other positive.) In the quadrant ($k_1' > 0$, $k_4' < 0$), draw the conditions you have derived for oscillatory behaviour and for growth-or-decay behaviour.

Problem 6.5.11   Structural stability of dynamic behaviour

Consider equation (6.69) and the plot of it in Figure 6.10a. Remove the k_{ext} term from the equation, leaving only the autocatalytic term and its thermodynamic reverse. Indicate whether or not the dynamics of this system are structurally stable against addition of the k_{ext} term, explaining your answer with diagrams.

6.6 Brief indications of solutions to problems 6.5

Problem 6.5.1

Plant A, length initially L_{0a}, finally L_a; likewise for B. Exponential growth $L_a/L_{0a} = e^{k_a t}$; likewise for B. The growth controller is likely to promote growth in proportion to its concentration, so we want the ratio k_b/k_a, which is NOT 2.0. For the same time t in both cases,

$$k_a/k_b = \ln(L_b/L_{0b})/\ln(L_a/L_{0a}) = \ln 3.0/\ln 2.0 = 1.585.$$

Z is most likely to be the growth controller. This question is just about understanding compound interest. It could be recast as follows: A and B each invest the same amount of capital in different places. After a given time, A's capital has doubled, and B's has tripled. What is the ratio of the interest rates, for *continuous* compounding?

Problem 6.5.2(a)

$$G = 2aG_a + 2bG_b = 2aG_a + 2(A/a)G_b.$$

Differentiating with A constant, $dG/da = 2G_a - 2(A/a^2)G_b = 0$ for a minimum or maximum. At the minimum or maximum, $G_a/G_b = A/a^2 = ab/a^2 = b/a$. We can look at the sign of the second derivative to show that this condition is a minimum, not a maximum.

Problem 6.5.2(b)

There are four (11) edges of length a and four (10) edges of length $l - 2^{1/2}a$.

$$G = 4aG_{11} + 4(l - 2^{1/2}a)G_{10} = G_{10}[4an + 4(l - 2^{1/2}a)],$$

$$A = l^2 - a^2, \text{ whence } l = (A + a^2)^{1/2},$$

$$G = 4G_{10}[na - 2^{1/2}a + (A + a^2)^{1/2}],$$

$dG/da = n - 2^{1/2} + (1/2)2a/(A + a^2)^{1/2} = 0$ at minimum or maximum.

Taking $n - 2^{1/2}$ to the other side and squaring,

$$(2^{1/2} - n)^2 = a^2/(A + a^2) = a^2/l^2.$$

The ratio a/l is one way of specifying the equilibrium shape. To convert it to distances from the centre,

$$l_{11} = 2^{1/2}l_{10} - a/2, \text{ where } l_{10} = l/2,$$

$$l_{11}/l_{10} = 2^{1/2} - a/l = 2^{1/2} - (2^{1/2} - n) = n.$$

In Problems 6.5.2 and 6.5.3, the really nontrivial aspect is recognizing what is constant and what is variable in setting up an expression for the quantity to be optimized. We are looking for the shape which will minimize the free energy of a fixed amount of material, so A must be constant.

Problem 6.5.3

$dC_1/dt = k_a - k_c C_1 - k_1 C_1/l_1 + k_2 C_2/l_1 = 0$ in steady state,

$dC_2/dt = k_a - k_c C_2 - k_2 C_2/l_2 + k_1 C_1/l_2 = 0$ in steady state.

Subtract the second equation from the first, and collect C_2 terms on the left-hand side, and C_1 on the right-hand side:

$$C_2(k_c + k_2/l_2 + k_2/l_1) = C_1(k_c + k_1/l_2 + k_1/l_1).$$

Abbreviate by writing $1/l_1 + 1/l_2 = 1/l_r$.

$$C_2/C_1 = (k_c + k_1/l_r)/(k_c + k_2/l_r).$$

If the transfer between faces is very rapid, the second term in each set of parentheses is much greater than the first, and we can neglect k_c. Then

$$C_2/C_1 = k_1/k_2.$$

Because the linear rate of advance of each face as the crystal grows is proportional to C for the adsorbed layer on that face, the ratio of transfer rates between faces controls the developing shape of the crystal.

Problem 6.5.4

$$dX/dt = -k_d X + k_f X^n,$$

$$0 = -k_d X_0 + k_f X_0^n,$$

$$X_0 = (k_d/k_f)^{1/(n-1)},$$

$$dX/dt = dU/dt = -k_d(X_0 + U) + k_f(X_0 + U)^n,$$

$$dU/dt = -k_d(X_0 + U) + \\ k_f X_0^n + k_f n X_0^{n-1} U$$

$$+ \text{ terms in } U^2 \text{ etc.}$$

Subtract $0 = -k_d X_0 + k_f X_0^n$:

$$dU/dt = -k_d U + k_f n X_0^{n-1} U + \text{ terms in } U^2 \text{ and higher powers.}$$

Substitution of the value found earlier for X_0 yields, for the term in U,

$$dU/dt = -k_d U + n k_d U = k_d(n - 1)U.$$

This is an equation for exponential growth if $n > 1$, and decay if $n < 1$. No calculus is needed for this problem. The notations dX/dt and dU/dt could be replaced by verbal statements: "rate of formation of X (or U)." The rest is elementary algebra.

Problem 6.5.5

The flows suggested by the wording of the problem and shown by arrows in Figure 6.2 are in the reverse sense to the usual meaning of $\dot{M}$ (flow to positive s direction), so the minus sign can be omitted from the expression in equation (6.12):

$$\frac{\text{rate of accumulation in region } B}{\text{volume of region } B}$$

$$= \mathscr{D} \; \frac{A_c[(\partial C/\partial s)_{s+\delta s} - (\partial C/\partial s)_s]}{A_c \delta s}$$

From the definition given in the problem, this can at once be rewritten as equation (6.13).

Problem 6.5.6

$$\partial C/\partial t = -kC + \mathscr{D}(\partial^2 C/\partial s^2),$$

steady state: $0 = -kC + \mathscr{D}(d^2 C/ds^2)$.

(There is now only one variable, s; so the partial derivative sign is no longer needed.) The equation to be solved may also be written

$$d^2 C/ds^2 = (k/\mathscr{D})C.$$

Proportionality between a function and its own second derivative indicates exponential variation with the variable if the constant of proportionality is positive (which $k/\mathscr{D}$ is), or a periodic function (e.g., sin or cos) if the constant is negative. The obvious simplest solution is

$$C = C_0 e^{-(k/\mathscr{D})^{1/2} s}.$$

This satisfies the boundary condition $C = C_0$ at $s = 0$ (the same expression, but with a plus sign in the exponent, would also satisfy the equation and the boundary condition, but would indicate C increasing away from the source, a distribution which could not be generated without other sources):

$$\ln C = \ln C_0 - (k/\mathscr{D})^{1/2} s.$$

Problem 6.5.7

Satisfaction of the boundary conditions is easily checked. As to the satisfaction of the steady-state equation, which is the same as in the previous problem, the two exponential functions of s are both functions which give α^2 times themselves on differentiating twice, so

$$d^2C/ds^2 = \alpha^2 C,$$

which is the steady-state equation.

Problem 6.5.8

$$dX/dt = K_f X - k_r X^2 = X - (\tfrac{1}{3})X^2$$

for Figure 6.13. For any positive values of k_f and k_r, the curve of dX/dt versus X starts out from the origin to positive values, because the X^2 term has zero slope at the origin, and later goes negative as the X^2 term becomes dominant. X_{eqm} is clearly a stable steady state like point F in Figure 6.10a. The latter is displaced from C_{eqm} by the k_{ext} term, which we do not have in this problem:

$$dX/dt = dU/dt = k_f(X_0 + U) - k_r(X_0 + U)^2,$$

$$\text{steady state: } 0 = k_f X_0 - k_r X_0^2.$$

Subtract $$dU/dt = k_f U - 2k_r X_0 U - k_r U^2:$$

$$X_0 = k_f/k_r,$$

$$dU/dt = k_f U - 2k_f U - k_r U^2$$

$$= -k_f U$$

when the term in U^2 is neglected.

For an unstable state such as point E in Figure 6.10a, we have to consider what is happening in the system before the morphogenetic action is "switched on" and what kind of change the switching is. Quite possibly the switch-on is an increase in the metabolic flow which at once supplies S and removes C at rate k_{ext}. Before this increase, S probably had a lower value. Now $C_{eqm} = k_f S/k_r$ and is therefore lower before the switch-on than afterward. It is quite likely that C_{eqm} before the switch-on lies near to point E or at least in the approximately linear region of positive slope in Figure 6.10a, which is what we need to start the morphogenetic process in the region represented by linear equations.

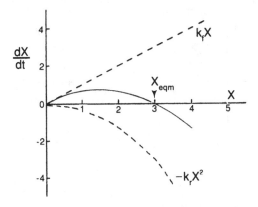

Figure 6.13.

Problem 6.5.9(a)

$$d^2U/dt^2 = k_2\, dV/dt = k_2k_3U.$$

This is the familiar proportionality between a function and its own second derivative, which has exponential growth-or-decay solutions if the constant, here k_2k_3, is positive, and periodic (sin or cos) functions if the constant is negative. For positive k_2k_3,

$$U = U_0 e^{\pm(k_2k_3)^{1/2}t},$$

and similarly for V. For negative k_2k_3, the same solution can be used if one likes to represent periodic functions by complex exponentials, and most mathematicians do prefer that to sine and cosine, for the reason exemplified here that nothing extra now needs to be written. In terms of trigonometric functions,

$$k_2k_3 = -\omega^2,$$

$$U = A\,\sin(\omega t - \delta),$$

$$V = (1/k_2)\, dU/dt = (A\omega/k_2)\cos(\omega t - \delta) = A(-k_3/k_2)^{1/2}\cos(\omega t - \delta).$$

Because $\sin^2 x + \cos^2 x = 1$, we may combine these expressions into an equation for a curve in the (U, V) phase plane, which is just the (X, Y) phase plane with the origin shifted to (X_0, Y_0):

$$U^2 + (-k_2/k_3)V^2 = A^2.$$

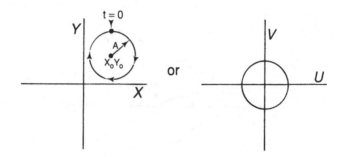

Figure 6.14.

This is the equation for an ellipse; if $k_3 = -k_2$, it is

$$U^2 + V^2 = A^2,$$

which is the equation of a circle of radius A and centre at $U = V = 0$, that is, at (X_0, Y_0). The starting point at $t = 0$ depends on the phase angle δ. If that is zero, then at $t = 0$, V is at its maximum, and $U = 0$; this is the top of the circle. As t becomes positive, U increases and V decreases: clockwise circulation (Figure 6.14).

The (k_2, k_3) parameter-space representation is quite simple. Two quadrants are occupied by growth-or-decay behaviour, and two by oscillatory behaviour (Figure 6.15).

Problem 6.5.10

$$
\begin{aligned}
d^2U/dt^2 &= k_1(dU/dt) + k_2(dV/dt) \\
&= k_1^2 U + k_1 k_2 V + k_2 k_3 U + k_2 k_4 V \\
&= (k_1^2 + k_2 k_3)U + k_2(k_1 + k_4)V.
\end{aligned}
$$

If $k_4 = -k_1$, then

$$d^2U/dt^2 = (k_1^2 + k_2 k_3)U.$$

Once again, we have the relation of something to its own second derivative, which makes the something periodic if the constant of proportionality is negative, and gives it growth-or-decay behaviour if the constant is positive.

Periodic solutions will therefore arise if one of k_2 and k_3 is negative and the other positive, and $k_1^2 < k_2 k_3$ or $k_1' < 1$. In the parameter space (k_1', k_4'), the condition $k_4 = -k_1$ is a straight line through the origin, of unit negative slope, and this condition divides it at the point A in Figure 6.16.

Reminders about complex numbers and periodic variations: Suppose a variable U is indicated as being related to time by

$$U = U_0 e^{ikt},$$

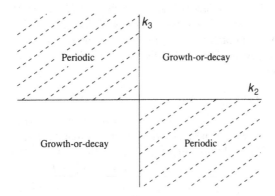

Figure 6.15.

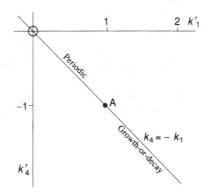

Figure 6.16.

where k is a real constant, and $i^2 = -1$. On differentiating twice,

$$d^2U/dt^2 = i^2k^2U_0e^{ikt} = -k^2U.$$

This is the now-familiar differential equation for which the solutions are periodic functions previously written as sine or cosine. Hence e^{ikt} is such a periodic function. It is shown in elementary accounts of complex numbers that

$$e^{ikt} = \cos kt + i \sin kt.$$

In the ensuing chapters, expressions will turn up of the form

$$U = U_0e^{k_g t},$$

in which the exponential growth constant k_g is obtained as the solution of a quadratic equation. Such equations have solutions which are sometimes real and sometimes complex, according to the numerical values of their coefficients. When k_g is real and positive, U grows. When k_g is real and negative, U decays. When k_g is complex, U oscillates, but the oscillations may grow in amplitude or die away according as the real part of k_g is positive or negative. To see this, write

$$k_g = k_{re} + ik_{im} \quad \text{and} \quad e^{k_g t} = (e^{k_{re} t})e^{ik_{im} t}.$$

The expression in parentheses is a real growing or decaying function; it multiplies a complex exponential, which is an oscillating function. Problems 6.5.9 and 6.5.10 could have been discussed in this format, with, in Problem 6.5.9, $k_g^2 = k_2 k_3$, and, in Problem 6.5.10, $k_g^2 = k_1^2 + k_2 k_3$.

Problem 6.5.11

For the parameter values used in Figure 6.10a, at low concentrations $k_f S C^2$ rises faster than $-k_r C^3$ falls. Thus if the term $-k_{ext} C$ is omitted from equation (6.69), as C increases from zero, dC/dt will rise to positive values instead of falling to negative values as shown for the sum of all three terms. Thus without the k_{ext} term, the curve is generally similar to the solid line in Figure 6.13, showing only one stable steady state, at true equilibrium, and an unstable one at zero concentration. Addition of a k_{ext} term, however small, changes the form of the curve to the three-steady-state shape of Figure 6.10a. Equation (6.69), without the k_{ext} term, is therefore structurally unstable to addition of that term.

7

Kinetic models for stable pattern: an introduction

The optical-resolution mechanism of the preceding two chapters is the basic self-organizing asymmetrizer. But its enthusiasm outruns its discretion. Its final outcome is to leave the system, macroscopically, as symmetric at the end as it was at the start, by the rejection, on the molecular scale, of all molecules of one symmetry. (Even for this simplest model, the interplay of molecular and large-scale symmetries and asymmetries is rather subtle.) The game is played to this extremity because, as explained in Section 6.4, the exponential growth rate constant k_g of pattern amplitude increases monotonically with wavelength. Thus the longest wavelength, spatially infinite uniformity, is ultimately favoured.

The basic difference needed to produce stable nonuniform pattern was described qualitatively in Section 3.1.2 with reference to Figure 3.2. Above some finite wavelength designated λ_m the growth rate k_g must fall. We must now consider what has to be added to the simplest reaction-diffusion model to give it this characteristic. In general, at least one more morphogen is needed. The interplay of two morphogens cannot be analyzed fully without mathematics. An illustration devised by Maynard Smith (1968) is described in Section 7.1. It shows the need for the form of the cross-interactions between morphogens X and Y and for the relative values of their diffusivities. Such a simple illustration cannot, however, reveal all the important properties of such complex and subtle dynamics. It shows that X and Y may settle down into standing waves in phase with each other. It does not show what factors determine the wavelength, nor how the amplitude of the waves will grow or decay as time goes on. It does not show that V should be self-inhibiting (where $V = Y - Y_0$), nor does it show that the standing-wave behaviour arises only for restricted ranges of the rate constants and diffusivities. To use reaction-diffusion models, one must study them mathematically.

7.1 Turing's model without equations

7.1.1 Maynard Smith's illustration

In Chapter 6, the concept of a self-catalyzed departure from equilibrium (or from a spatially uniform steady state) was introduced. If the displacement of X

223

from equilibrium, $U = X - X_0$, is positive, then an increase of U is catalyzed at a rate proportional to U; if U is negative, then a decrease of U is catalyzed. In both directions, the change promoted is movement away from the equilibrium value X_0. But in terms of the total concentration X, the enhancement of its increase is catalysis, but the acceleration of its decrease is inhibition. Thus X switches over from being self-catalyzing to being self-inhibiting as it drops below the value X_0. (Note once again that the rate constant for this self-enhancement of the variable U must involve destruction processes for X in the reaction mechanism, because X cannot fall below X_0 as a result of its own *positive* self-interaction.) The Turing model is, for the displacements U and V of two substances X and Y from equilibrium,

U catalyzes itself; U diffuses slowly;
U catalyzes V; V diffuses quickly;
V inhibits U;
(V may catalyze or inhibit itself).

Maynard Smith's (1968) illustration starts with the system at equilibrium; U and V are both zero all along an elongated (one-dimensional) system and are represented by the same horizontal straight line. Now suppose that a small local increase of X occurs somewhere, represented as a positive U peak (Figure 7.1a). Because U catalyzes itself, the peak grows; because U diffuses, the peak spreads sideways. But also, because U catalyzes V, a positive V peak appears; and because V diffuses quickly, this peak becomes broader than the U peak (Figure 7.1b). Now V inhibits U; that is, where V is positive and U is zero, U starts to move to negative values (Figure 7.1c). As explained earlier, when a catalytic variable goes negative, its catalysis switches over into an inhibitory effect. In simpler words, where U is negative, it pulls V down after it (Figure 7.1d). The implication of this diagram is that U and V are beginning to assume wavelike patterns, with the waves for the two substances in phase with each other. It is quite straightforward to continue the argument in the same way for the further sideways spread and see that more half-cycles of the same wave pattern will be generated. As one thinks about this, however, some problems may become evident. First, what on earth is determining the wavelength of these successive half-cycles? Second, if we grant that it looks as if U and V are going to form patterns of indefinite length in the form of waves in phase with each other, what is going to happen to the amplitudes of these waves as time goes on? Will they reach constant values, will they grow, or will the system gradually relax back to equilibrium until another disturbance comes along, like the vibrations of a plucked string? Clearly, Maynard Smith's illustration is a good first step, but far from the whole path we must tread to understand this model. In fact, for different quantitative ranges of values of the kinetic constants, there are several ways the system can go on to behave. Only one of them is the one we are interested in for the explanation of

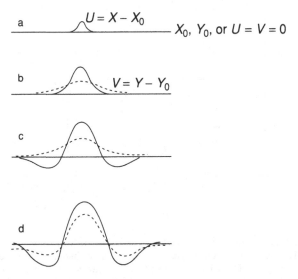

Figure 7.1. Illustration of the formation of standing waves from an initial local disturbance, by Turing dynamics. Redrawn from Maynard Smith (1968).

morphogenesis. Thus far, the illustration has not pointed to any particular advantage or disadvantage in V interacting with itself, as self-catalyst or self-inhibitor. On more detailed analysis, as we shall see later, it turns out to be very advantageous (though not absolutely essential for all examples of morphogenesis) for V to be a self-inhibitor.

7.1.2 Waves in phase as the starting point

Here I have (and take) yet another opportunity to stress that it is of the essence of kinetic theory to view a pattern as a single entity occupying the whole system and behaving dynamically. In Chapter 6, I twice discussed the sine-wave pattern, arbitrarily imposed upon a system as an initial disturbance. For diffusion only, this pattern underwent the Cheshire Cat fade-out, a loss of amplitude without change of form. With self-enhancement added, the fade-out switched over to augmentation above a threshold wavelength.

Similarly, let us take as the initial disturbance in the system of U and V a pair of sine waves of the same wavelength and with shared nodes, that is, either exactly in phase with each other or exactly out of phase with each other. One might object that it is expecting a lot of a system to produce such an initial disturbance. But we have just seen how a localized initial disturbance might change into the pattern of two sine waves. In fact, neither the local disturbance nor the sine waves are very likely to be what nature provides at the start; a jumble of fluctuations of assorted sizes is most likely. It is legitimate to take such a highly idealized initial state because we are heading toward a

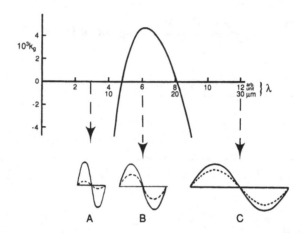

Figure 7.2. Interaction of U and V waveforms to determine growth or decay of pattern amplitude. The larger V is, the larger its negative contribution to the growth of U (inhibition of U by V). Diffusion makes a negative contribution to the growth of both U and V, and this is worst at short wavelength (steep concentration gradients, rapid diffusion). Because V diffuses faster, this effect is more for V than for U, meaning that for the same U amplitude, V will, at long times, settle down to amplitudes in the order $A < B < C$, as drawn. But then the two negative effects on the growth of U are in reverse order. For diffusion, $A > B > C$; for inhibition by V, $A < B < C$. The balance of these effects gives the advantage to B; waves of this length grow, while both A and C decay. (If U only were present without V, only diffusion would limit the growth of U. In that case, A would still decay, B would grow, and C would grow faster, as discussed in Chapters 5 and 6.)

mathematical analysis of a system of *linear* equations. For such, the sum of two solutions is itself a solution. Thus, if we can pick out of the starting jumble a component which is a sine wave of some particular wavelength, we can determine what the mechanism will do to it, with the confidence that when we likewise treat all the other components of the initial disturbance, all we shall finally have to do is to add all the results together.

For our pair of U and V waves (Figure 7.2), two quantities remain unspecified: the wavelength, and the ratio of U and V amplitudes. The latter also represents the ratio U/V at every point along the pattern, because that is obviously constant. (U/V should be distinguished from the activator/inhibitor ratio X/Y, or A/H in the Gierer-Meinhardt model, as discussed in Section 3.3. The latter is a ratio of absolute concentrations. It is high at the peaks and low at the troughs of the waveform. But for the undistorted sine wave, the ratio of displacements from equilibrium U/V is the same at peak or trough or anywhere else.)

Let us take an initial pattern, with wavelength and U/V specified at the outset. How can it change in development? If V were absent, the system would be that of Section 6.2, subject only to Cheshire Cat fade-out or intensification. Only the amplitude can change, not the sinusoidal form or the wavelength. Now if V is added in the same sinusoidal form, its effect on U likewise can change only the U amplitude, but not the shape or wavelength. The same applies to the V waveform, when one considers its self-interaction, the effect of U on it, and the effect of diffusion. All bring about changes strictly in proportion to the current value of V at any point, and therefore change only the amplitude of the wave.

Nothing in all of this, however, suggests that the amplitudes of U and V change in the same proportion. As time goes on, the ratio U/V may adjust to some new value; it is under the control of the mechanism. Everything would become rather straightforward for the ultimate course of development if U/V eventually settled down to a constant value and the two waves then grew or decayed together. Intuitively, one would probably expect this to happen, and the mathematical treatment confirms it (Lacalli and Harrison, 1978) (see Section 7.2). That long-time ratio is, however, wavelength-dependent, and therein lies the essence of how the Turing model works. In place of the simple wavelength dependence of growth rate discussed for one morphogen in Chapter 6, in which longer waves always grow faster than shorter ones [Figure 6.5 and equation (6.45)], there are two *opposing* wavelength dependences in the Turing model. Because of this, both long and short waves may grow more slowly than those of some intermediate length (Figures 3.2 and 7.2).

To see the opposing effects, consider the growth of the U amplitude. It is as described in Chapter 6, except that the presence of V makes an additional negative contribution. All that is needed for the model to work is that, as wavelength becomes longer, that contribution from V becomes larger, so that at some point the overall growth rate constant k_g of U starts to decrease. Why should V increase (i.e., the ratio U/V fall) as wavelength stretches? The reason for this is in the diffusion terms. As wavelength increases, the loss of amplitude by the spreading effect of diffusion becomes less, because the concentration gradients become less. This effect is more for V than for U, because V diffuses faster. Hence, if a wave pattern which has reached a steady-state U/V ratio at some particular wavelength is stretched to a greater length without change in amplitude, V will at once start to grow proportionately faster than U, the U/V ratio will decline, and U will have an additional negative term in its growth rate (Figure 7.2).

This system is akin to an engine (the U system, driving the amplitude increase) with a governor attached to it (the V system). The load against which the engine is working is the spreading effect of diffusion. It is greatest at short wavelength. The engine may then be unable to work even if there is no governor on it (wavelength less than threshold λ_0). If the load is decreased, the engine can work, for instance, to raise the load (Figure 7.3). Suppose now there is a governor not in the modern form of a silicon-chip microprocessor (which gives us no visual analogy because we can't see what it is doing) but in the form of two balls on hinged rods attached to a shaft driven by the engine.

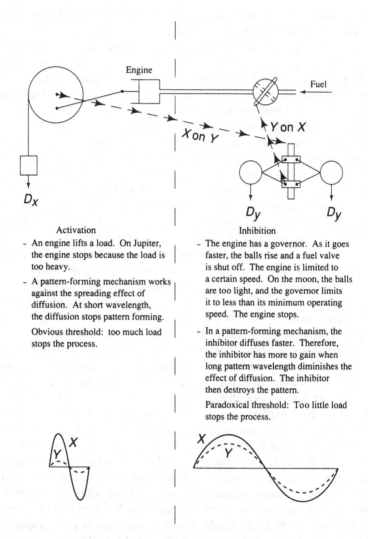

Activation

- An engine lifts a load. On Jupiter, the engine stops because the load is too heavy.
- A pattern-forming mechanism works against the spreading effect of diffusion. At short wavelength, the diffusion stops pattern forming.

Obvious threshold: too much load stops the process.

Inhibition

- The engine has a governor. As it goes faster, the balls rise and a fuel valve is shut off. The engine is limited to a certain speed. On the moon, the balls are too light, and the governor limits it to less than its minimum operating speed. The engine stops.
- In a pattern-forming mechanism, the inhibitor diffuses faster. Therefore, the inhibitor has more to gain when long pattern wavelength diminishes the effect of diffusion. The inhibitor then destroys the pattern.

Paradoxical threshold: Too little load stops the process.

Figure 7.3. Engine-and-governor analogy for U and V (or X and Y) in Turing's model, to show lower and upper thresholds. Most particularly, gravity is the analogue of the "diffusive load" on a pattern, so going to the moon corresponds to increasing the wavelength, which diminishes the effects of diffusion. Upper threshold: In lower gravity, the engine cannot raise the load because the governor shuts the valve.

As the shaft spins faster, the balls rise until a balance is achieved between the centrifugal force driving them out, and therefore up, and gravity pulling them down. The governor is the V system, and the analogue of the diffusive load is the weight of the balls. As the balls rise, they progressively shut down a valve supplying fuel to the engine, or otherwise work something to impede its operation. This is the inhibitory effect of V on U. The governor normally works to limit the speed of the engine, but if the balls rose too far, the valve could be shut completely and the engine would stop.

Let us now transport the engine to the moon, where the gravitational pull is less. The engine has to exert less force to lift the load, but there is also less downward pull on the governor, which may therefore rise to shut the valve and stop the engine. The decreased pull of gravity is the analogue of the decreased diffusive load on a pattern-driving engine when the wavelength is increased. The decrease may help the governor more than it helps the engine, and the engine will stop, though it seems to have an easier job to do (wavelength above the second threshold λ_f). If we were not thinking rather carefully about the governor, we would hardly have expected that an engine which could easily lift a particular load on earth would fail to lift the same load on the moon. This is an example of what Meinhardt (1984) called "counterintuitive properties"; see also the quotation from Meinhardt in Section 10.1.2.

My analogy works well for illustrating upper and lower bounds for operation of a dynamical system, but the reader may find some of its features rather odd. The analogue of a pattern to be formed is a weight to be raised, and the analogue of diffusion is the pull of gravity, both longish stretches for the imagination. These features are, however, specifically intended to encourage a certain attitude: to think of a pattern as a *thing*, and to think of diffusion as something inextricably intertwined in all the dynamics of a system, just as no part of a moving system is without the pull of gravity. Too often, diffusion is thought of in developmental biology only when it is necessary to consider how some material gets from one specific point A to another point B, or how a simple gradient is set up. How diffusion may actually be involved is so much subtler as to seem qualitatively different, though in the end it is still transport down a concentration gradient, and nothing more.

7.2 Turing's equations and the growth or decay of a sine-wave pattern

Turing's kinetic equations are

$$\partial U/\partial t = k_1 U + k_2 V + \mathcal{D}_x \partial^2 U/\partial s^2, \tag{7.1a}$$

$$\partial V/\partial t = k_3 U + k_4 V + \mathcal{D}_y \partial^2 V/\partial s^2. \tag{7.1b}$$

As described in Section 7.1.2, my strategy to show the properties of this model as simply as possible is to take the initial disturbance as a pair of sinusoidal waves in U and V, sharing the same ω or λ, and in phase with each other. Then we may write

$$V = \theta U, \tag{7.2}$$

where the ratio θ is the same at all positions s but may vary with time t. To be sure that θ never becomes dependent on s, we need to see that we can extract from the Turing equations and the initial condition an ordinary differential equation in the time variable for the variation of θ, with no s dependence in it. We may hope to find that as time goes on, θ will settle down to a time-independent value and that the U and V waves can thus be envisaged, at long times, as growing or decaying together with a constant ratio of amplitudes.

As usual, we can dispose of the second derivatives, for sinusoidal disturbance, by writing

$$\partial^2 U/\partial s^2 = -\omega^2 U, \tag{7.3}$$

and likewise for V. The Turing equations now become

$$\partial U/\partial t = k_1 U + k_2 \theta U - U\omega^2 \mathscr{D}_x, \tag{7.4a}$$

$$\partial(\theta U)/\partial t = k_3 U + k_4 \theta U - U\omega^2 \theta \mathscr{D}_y. \tag{7.4b}$$

In the derivative with respect to s, θ can be treated as a constant; but because θ and U are both functions of time, we must use the formula for differentiation of a product of two functions and write

$$\partial(\theta U)/\partial t = \theta(\partial U/\partial t) + U(\partial \theta/\partial t). \tag{7.5}$$

Evidently we can get rid of $\partial U/\partial t$ and set up a differential equation in which the only derivative is the rate of change of θ with time by multiplying (7.4a) by θ and subtracting it from (7.4b), which, on insertion of (7.5), is

$$\theta(\partial U/\partial t) + U(\partial \theta/\partial t) = k_3 U + k_4 \theta U - \omega^2 \theta \mathscr{D}_y U. \tag{7.6}$$

The expression to be subtracted from it is

$$\theta(\partial U/\partial t) = \theta k_1 U + \theta^2 k_2 U - \theta \omega^2 \mathscr{D}_x U. \tag{7.7}$$

The result of the subtraction is

$$U(\partial \theta/\partial t) = (k_3 - \theta k_1)U + (k_4 \theta - \theta^2 k_2)U$$
$$- \omega^2(\mathscr{D}_y - \mathscr{D}_x)U. \tag{7.8}$$

Every term in equation (7.8) is proportional to U, and upon dividing by this quantity, we obtain an equation which has nothing to say about the evolution of the U and V disturbances except what happens to the ratio θ between them. Because we now have an equation for time dependence only, we may as well write it as an ordinary equation. Upon collecting terms in θ and θ^2, this is

$$d\theta/dt = k_3 + \theta[k_4 - k_1 - \omega^2(\mathscr{D}_y - \mathscr{D}_x)] - \theta^2 k_2. \tag{7.9}$$

This may be abbreviated in the form

$$d\theta/dt = c + b\theta + a\theta^2. \tag{7.10}$$

The right-hand side of this expression is the standard form of a quadratic equation. Indeed, that is just what we have if we suppose that θ does approach a long-time limit and seek that limiting value by setting $d\theta/dt = 0$. From this point onward, we have very little calculus to do, but a lot of algebraic juggling with solutions of quadratic equations. Unfortunately, these are expressions in which it is often difficult to see the wood for the trees, especially in terms of the relationship of the behaviour of the model to the numerical values of the six parameters in it. In the present section, the algebraic expressions are set out. In Section 7.3, such simplifications of their significance as seem to be possible are described.

If the two roots of the quadratic equation obtained by setting the rate of change of θ equal to zero in equation (7.10) are called r_1 and r_2, then the solution of that equation for the evolution of θ in time is

$$\theta = (r_1 - r_2)/[1 - e^{-k_2(t - t_0)(r_1 - r_2)}] + r_2, \qquad (7.11)$$

where

$$t_0 = [1/k_2(r_1 - r_2)]\ln[(\theta_0 - r_1)/(\theta_0 - r_2)], \qquad (7.12)$$

θ_0 being the value of θ for the disturbance applied to the system at $t = 0$. According to the values of the k's and $\mathcal{D}$'s, the roots r_1 and r_2 may be real or complex, and hence so may θ. In this respect, the character of θ is the same as that of the fractional growth rate of pattern amplitude [see equation (7.17)]. Real θ describes stationary U and V waveforms (in phase for positive θ, 180° out of phase for negative θ) growing or decaying monotonically. Complex θ describes waveforms with oscillating amplitude. For our purpose of explaining morphogenesis, the ranges of parameters which make θ and k_g real are of the most interest, although it is shown in Section 7.3 that part of the oscillatory region could be connected with morphogenetic capability.

When the roots are real, equation (7.11) shows that at long times, θ tends to either r_1 or r_2, according as the exponent $-k_2(t - t_0)(r_1 - r_2)$ is positive or negative. Thus if k_2 is positive, θ tends to the algebraically larger of the two roots, and if k_2 is negative, θ tends to the smaller root. Both cases correspond to a choice of negative sign in the usual expression for solutions of a quadratic; that is, the long-time limit θ_∞ is

$$\theta_\infty = [-b - (b^2 - 4ac)^{1/2}]/2a, \qquad (7.13a)$$

or

$$\theta_\infty = [b + (b^2 + 4k_2k_3)^{1/2}]/2k_2, \qquad (7.13b)$$

where

$$b = k_4 - k_1 - \omega^2(\mathcal{D}_y - \mathcal{D}_x). \qquad (7.14)$$

Now the fractional growth rate constant k_g of the U pattern is

$$k_g = (1/U)(dU/dt). \qquad (7.15)$$

Thus equation (7.4a), divided through by U, yields an expression for k_g:

$$k_g = k_1 + k_2\theta - \omega^2\mathcal{D}_x. \qquad (7.16)$$

At long times, when θ has become effectively constant, k_g is therefore also constant, and the growth of pattern amplitude is exponential. This limiting value of k_g is found by substituting (7.13) into (7.16):

$$2k_g = k_1 + k_4 - \omega^2(\mathcal{D}_y + \mathcal{D}_x) + (b^2 + 4k_2k_3)^{1/2}. \qquad (7.17)$$

From this equation one of the most important conditions for the Turing model to work emerges at once: the need for two morphogens with different diffusivities. Suppose that $\mathcal{D}_y = \mathcal{D}_x$. Then equation (7.14) shows that b is independent of the diffusivities, because it contains them only as their difference. In that case, the only diffusivity dependence in equation (7.17) is of just the same form as that in equations (6.44) and (6.45), for the optical-resolution model which is incapable of stabilizing pattern: k_g grows monotonically with increasing wavelength. For k_g to pass through a maximum at some finite wavelength, this diffusivity dependence must be offset by a conflicting term, which can come from b only when $\mathcal{D}_y$ and $\mathcal{D}_x$ are unequal. We may examine the passage of k_g through a maximum by differentiating the expression for k_g with respect to wavelength, or, for simpler algebraic expressions, with respect to ω^2:

$$d(2k_g)/d\omega^2 = \alpha(\mathcal{D}_y - \mathcal{D}_x) - (\mathcal{D}_y + \mathcal{D}_x), \qquad (7.18)$$

where α is an abbreviation for

$$\alpha = b/(b^2 + 4k_2k_3)^{1/2}. \qquad (7.19)$$

At a maximum or minimum of k_g, this derivative is zero. Hence,

$$\mathcal{D}_y/\mathcal{D}_x = (\alpha - 1)/(\alpha + 1). \qquad (7.20)$$

Now if α is numerically less than unity (i.e., if α lies between 1 and -1), the ratio of diffusivities is given by equation (7.20) as negative, which is physically impossible. Hence, for k_g to have a maximum at finite wavelength, this range of α is excluded. Equation (7.19) shows that if α is to keep outside this range, $(b^2 + 4k_2k_3)^{1/2} < b$. Clearly, this means that

$$k_2k_3 < 0. \qquad (7.21)$$

Because k_2 is the catalytic constant for the effect of V on U, and k_3 is that for the effect of U on V, and because equation (7.21) means that if one of these constants is positive, the other must be negative, the condition is that if U catalyzes growth of V, then V must inhibit growth of U, and vice versa. This condition is sufficient to ensure that the turning point in k_g versus λ or ω^2 is always a maximum, not a minimum. One may see this in the usual way by differentiating k_g a second time with respect to ω^2 and requiring the second derivative to be negative (Lacalli and Harrison, 1978).

One of the most significant constants of the Turing model is the wavelength of maximum k_g, for that is the pattern spacing between repeated parts. It is

$$\lambda_m = 2\pi(\mathcal{D}_y - \mathcal{D}_x)^{1/2}/[(\mathcal{D}_x + \mathcal{D}_y)(-k_2k_3/\mathcal{D}_x\mathcal{D}_y)^{1/2}$$
$$+ k_4 - k_1]^{1/2}. \qquad (7.22)$$

It can be shown (Lacalli and Harrison, 1978) that for this value of the wavelength, k_g is always real, provided only that condition (7.21) is satisfied. Thus, although oscillatory behaviour can occur in the Turing model, it never happens at the wavelength of maximum growth rate.

The foregoing equations are the basic requirements for using the Turing model. In particular, equation (7.17), with the expression for *b* from equation (7.14), enables one to calculate, for given numerical values of the *k*'s and the $\mathcal{D}$'s, how k_g varies with wavelength (e.g., the graphs shown in Figures 3.2, 7.2, and 7.7). I have not yet discussed, however, how one goes about selecting the numerical values of the parameters. If reaction-diffusion theory had a trade secret, this would be it. For instance, when Lacalli and I did the theoretical work on the regulatory capacity of the model (with an application to slime moulds in mind) for which most of the foregoing account was first devised, our first computations gave oscillations instead of stable pattern. Of the six kinetic parameters, only one had to be changed from our first choice of values, and by only 10%, to stabilize the pattern. But the mathematical analysis was necessary to see what was wrong and what value should be changed. The next section discusses the types of behaviour which a Turing system can exhibit and the ranges of values of parameters for each.

7.3 Turing's conditions

In his long and comprehensive paper, Turing rather casually and briefly set down, with no discussion, a set of relationships between k_1-k_4 and $\mathcal{D}_x$ and $\mathcal{D}_y$ which must be satisfied for the model to exhibit morphogenetic behaviour [Turing 1952, equations (9.4)]. Lacalli and I (1978, 1979) attempted, while giving the otherwise briefer account of Turing's model repeated here, to explain Turing's conditions in some detail. Whatever merit our account may have had as exposition of *how* Turing's model works, it unfortunately conveyed an erroneously pessimistic impression of *whether* it works. Thus Green (1980) wrote that "Lacalli and Harrison have performed the great service of showing how restricted the values of the various parameters . . . must be, if stable waveforms are to be generated. . . ." The problem here is that experimentalists will usually want to think of parameters for dynamics as being rate constants for particular specified chemical reactions. That is not what the Turing parameters k_1-k_4 represent. Their relationship to actual reaction rate constants is different for every specific reaction-diffusion model, and often quite complicated. I believe that if one is going to use two-morphogen reaction-diffusion concepts in discussing morphogenetic mechanisms, one must go as far in the mathematics as to understand and habitually use the Turing parameter space shown in Figure 7.5. One must also be able to relate the Turing constants k_1–k_4 to the actual rate constants of individual steps in whatever particular reaction-diffusion mechanism one is contemplating. The simplest example of this is in Section 6.3.1. In the transformation from equation (6.24) to equation (6.30), it arises that the Turing constant k_1 is not the rate constant of the autocatalytic step, but that of the decay step in a simple two-step mechanism.

In this section I undertake two tasks. The first, in Section 7.3.1, is to show how Turing's conditions may be derived and what they mean in terms of the various kinds of behaviour in a system obeying Turing's equations (e.g., going to spatial uniformity, pattern-forming, or oscillatory). This is done in terms of the values of the Turing rate constants k_1, k_2, k_3, and k_4 and leads to relationships such as

$$k_1 k_4 \geq k_2 k_3 \qquad (7.23)$$

for pattern-forming behaviour to be possible. The need for this and several other relationships between the constants to be satisfied simultaneously can easily give the impression that most arbitrary choices of values for these constants will not give the desired kind of behaviour. Indeed, that is exactly what computer programmers will find if they do not study the conditions, but simple make unguided choices of values.

In Section 7.3.2, I discuss the fact that Turing's model is not a chemical mechanism and that the rate constants in it do not correspond to those of unit steps in a chemical mechanism. As an example of what happens if one tries to work back to the chemical level, I take the Brusselator, which happens to have four steps and therefore another set of four constants, a, b, c, and d. The object of this discussion is to show that the Turing conditions look much less restrictive in terms of a, b, c, and d than in terms of k_1–k_4. For instance, equation (7.23) translates into

$$d/bB \geq 0, \qquad (7.24)$$

where B is a reactant concentration. This is always satisfied if B, b, and d are positive, which they always are, by definition. Numerical ranges at the level of chemical reality are not the same as those at the level of Turing's equations, in which some of the quantities which look like chemical rate constants can be negative.

The problem for the research worker trying to investigate the applicability of Turing's model and all other versions of reaction-diffusion is a very subtle kind of anthropomorphism. One sits down with a model expressed at the level of rate equations and containing several parameters, and it appears that one's choice, *and therefore nature's choice*, of numerical values for these is from the whole range of numbers. Therefore, if nature has managed to select values from a tiny part of that range, *nature has done something arithmetically rather clever*. The words in italics make up the anthropomorphism. Nature did not sit down with a set of rate equations and parameters. These arise from our description of a mechanism that is going on naturally, and that mechanism often does not entail the possibility of the selection of parameters being from unrestricted ranges. Rather, the qualitative features of the mechanism place the parameters, almost automatically, in the appropriate quantitative ranges. The conditions discussed in the present section do not cast any kind of doubt on the model. They provide valuable clues as to what kind of mechanism may be responsible for a system obeying Turing-type rate equations.

The essential distinction here is between two levels of models: rate equations versus chemical reaction equations. The former are the stock-in-trade of the macroscopist seeking to explain morphogenetic process at the scale of the whole organism or that part of it which constitutes a morphogenetic field. The latter are the preserve of the biochemist, enzyme kineticist, and so forth. The macroscopist's parameters are not the enzymologist's rate constants, turnover rates, and so forth. If the former appear very restricted, the proper next step is not to say "therefore nature is not using reaction-diffusion, because it would be so awkward to keep the parameters always inside these narrow ranges." Rather, one should ask: "What kind of chemical mechanism has qualitative features which automatically relieve most of the restrictions?"

7.3.1 The conditions from a computer programmer's viewpoint: Lacalli's (k_1', k_4') space

The Turing conditions can be written in terms of the six parameters k_1–k_4, $\mathscr{D}_x$, and $\mathscr{D}_y$. But upon examination of them it turns out that if one divides the expressions by $(-k_2 k_3)^{1/2}$ and also by $\mathscr{D}_x$, they can be rewritten in terms of only three parameters. These are

$$k_1' = k_1/(-k_2 k_3)^{1/2}, \tag{7.25a}$$

$$k_4' = k_4/(-k_2 k_3)^{1/2}, \tag{7.25b}$$

$$n = \mathscr{D}_y/\mathscr{D}_x. \tag{7.25c}$$

The precise form of these parameters has to be arrived at by doing the algebra. But their nature is easy to rationalize in retrospect in terms of the strategy of operation of the morphogenetic control system. What matters for this is not how fast U self-enhances (k_1). It is how this rate compares with the more indirect self-inhibition of U via the catalysis of V by U (k_3) and the inhibition of U by V (k_2). Clearly, the parameter k_1' gives a measure of this comparison; k_4' similarly compares the direct self-interaction of V with its indirect self-interaction via U. The importance of the diffusivity ratio (n) has already been explained.

For representing as much as possible on one graph, or for making computations with one set of numbers which actually cover wider ranges, it has long been recognized as good practice to use dimensionless parameters wherever possible, and the three parameters here introduced are examples of these. All of the Turing rate constants k_1–k_4 have the dimensions of a first-order rate constant, (time)$^{-1}$, and each diffusivity has the dimensions (length)^2(time)$^{-1}$. But k_1', k_4', and n are pure numbers without dimensions. The divisor which turns the two self-enhancement constants into dimensionless parameters is

$$k_{cr} = (-k_2 k_3)^{1/2}. \tag{7.26}$$

This is the geometric mean rate constant for the two cross-effects, of V on U and U on V. Thus k_1' measures how much faster the self-interaction of U is than the mean cross-effect, and similarly for k_4'; and n measures how much faster Y diffuses than X. Clearly, such relative measures are pure numbers, but they also contain the essence of the interplay which is the pattern-generating ability of the model.

If we want to know how quickly a pattern should grow (i.e., the value of k_g), or what its "chemical wavelength" is (λ_m), then we are seeking quantities with dimensions. It is useful to recast the expressions for these, equations (7.17) and (7.22), partly in terms of the dimensionless parameters, but there is going to be something left over for which the dimensions do not cancel out in these two expressions.

There are six Turing conditions, given later as expressions (7.27a–f), each of which is an inequality. For graphic representation, as devised by Lacalli (Lacalli and Harrison, 1979), each condition can be written as an equation for a boundary in a space in which k_4' is plotted against k_1'. On one side of each boundary, the inequality is satisfied; on the other, it is not. To show on a plane something involving three variable parameters, one of them has to be held constant. Hence we usually set up the diagram for some constant value of the diffusivity ratio n. Provided that $n \gg 1$, the diagram does not vary greatly with n. It is shown in Figures 7.4 and 7.5. The conditions are as follows:

(a) The fact that k_g passes through a maximum at some finite wavelength does not of itself establish that the pattern of that wavelength will grow. It is additionally required that the maximum value of k_g be positive (i.e., at λ_m, $k_g > 0$). Otherwise, the maximum would simply represent the pattern which would decay most slowly, which is not of much morphogenetic interest; a Cheshire Cat endowed with some longevity is still a Cheshire Cat. The required condition is

$$k_4' < n k_1' - 2n^{1/2}. \tag{7.27a}$$

This, as an equation, is the steep line, with slope equal to the diffusivity ratio. It is the most important boundary. To the left of it there is no interesting behaviour for our purpose. To the right of it, an assortment of interesting things may happen, including morphogenesis (Figure 7.7b versus 7.7a).

(b) For selection of a particular pattern spacing, there must be a maximum in k_g at some wavelength less than infinity. Mathematically, this means that

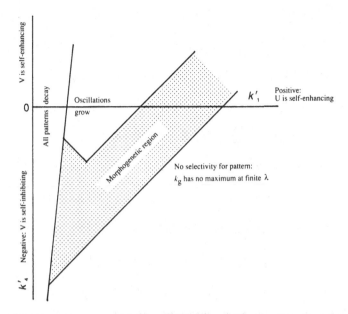

Figure 7.4. Morphogenetic capability of the Turing model as a function of the feed-back constants for U and V. These, k_1' and k_4', are effectively k_1 and k_4 reduced to unit values for the cross-effects of U on V and V on U. See equations (7.25a,b) for precise definitions. The figure is drawn for a diffusivity ratio $n = 10$. This number is the slope of the steep line separating the left-hand region, in which all patterns decay, from the right-hand region, in which oscillatory or pattern-forming phenomena occur. The most important parts of the morphogenetic region are in the quadrant where $k_1' > 0$ (U or X self-enhances) and $k_4' < 0$ (V or Y self-inhibits). See Figure 7.5 for numbering on scales and subdivisions of morphogenetic region.

λ_m must be real. A complex λ_m, with an imaginary component, cannot have real physical significance. The required condition is

$$k_4' > k_1' - (n + 1)/n^{1/2}. \qquad (7.27b)$$

On the diagrams, this is the lowest of three parallel lines of unit slope. All the regions of interesting behaviour lie between lines a and b in Figure 7.5. If the two diffusivities were the same, $n = 1$, line a would also be of unit slope and would in fact coincide with line b. The morphogenetic regions would then have collapsed into zero area on the diagrams. In this way, the need for two different diffusivities can be seen pictorially.

(c) The term "stationary wave," or "standing wave," is often used to describe morphogenetic Turing patterns. Its older and commoner usage is for such things as the vibrations of a violin string. These two classes of phenomena have aspects in common, which makes it possible to use this term to cover both of them. These are (1) that the displacement from equilibrium at

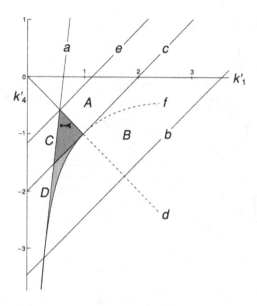

Figure 7.5. Subdivision of the morphogenetic ranges of the two self-interaction parameters. The six boundaries lettered *a–f* correspond to expressions (7.27*a–f*) written as equations, and they separate the regions where these expressions are satisfied or not satisfied when written as inequalities. Shaded regions: *A*: Oscillations grow, but superimposed on nonoscillating pattern which grows faster; likely to be of minor morphogenetic significance, but could be relevant to such cases as the cellular slime moulds (e.g., *Dictyostelium*), in which time-periodic and morphogenetic behaviour occur successively. *B*: No oscillations, but the maximum in k_g versus λ is shallow (Figure 7.7a), and patterns of different complexity might not be well discriminated. *C*: System can oscillate initially if oscillations are somehow initiated, but they decay away, while pattern grows; good morphogenetic region. *D*: No oscillations, and k_g versus λ has a sharp maximum; k_g goes negative at some finite λ above λ_m (Figure 7.7e); excellent morphogenetic region. The point indicated by an arrow in region *C* shows the parameters used in discussing possible application to *Drosophila* imaginal discs (Lacalli and Harrison, 1979; referring to Kauffman, Shymko, and Trabert, 1978) and to *Acetabularia* whorls (Harrison et al., 1984). When the values of k_1' and k_4' for a system are changed so that the steep line *a* is crossed from left to right, the behaviour of the system changes from stable uniform distribution of material to more interesting behaviour. If the region entered is *C* or *D*, this behaviour is morphogenetic pattern formation. The crossing of line *a* is then known as "the onset of the Turing instability."

any fixed time is sinusoidal in distance along the system and (2) the "stationary" aspect: that the nodes of the waveform, where there is no displacement, do not move. The distinction between the two classes of phenomena is that for the violin-string vibration, the displacement at all points other than nodes is periodic in time. A positive wavecrest subsides to zero, becomes negative, thus turning into a trough, and then returns back to a crest again (Figure 7.6a). In the morphogenetic situation, the displacement changes monotonically with

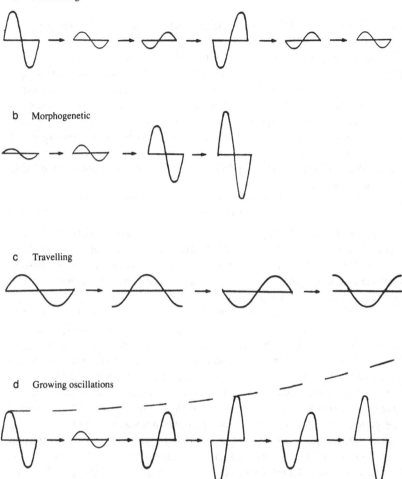

a Violin string

b Morphogenetic

c Travelling

d Growing oscillations

Figure 7.6. Time sequences for the development of various kinds of waves, to clarify terminology: (a), (b), and (d) are all stationary waves, in the sense that the displacement at any point can change with time, but the nodes (points where the displacement is always zero) do not move; (a) is the violin-string behaviour; (b) is morphogenetic behaviour, in which the displacement at any point increases (in the positive or negative direction) monotonically in time, instead of oscillating as in (a); (d) is like (a) except that the amplitude of the oscillations is growing as time goes on; (c) is a travelling wave, moving to the right. Turing waves can show all these kinds of behaviour. Waves (a), (c), and (d) would all correspond to parameter values above line c in Figure 7.5. Wave (a), showing oscillations neither growing nor decaying, would correspond to parameters lying exactly on line d in Figure 7.5. The travelling wave shown in (c) is sinusoidal, and Turing mechanisms can generate such phenomena. These are not the same as the travelling waves produced in the Belousov-Zhabotinski reaction, which have more the character of relaxation oscillations; see Section 10.3.4. For a sinusoidal oscillator mechanism closely resembling the Brusselator reaction-diffusion mechanisms (but without diffusion), see Sel'kov (1968).

time: A crest grows higher, and a trough deeper, continuously, without periodic aspect (Figure 7.6b).

Turing's model can produce both kinds of behaviour in different ranges of the parameters. Mathematically, the time oscillations will occur if k_g is a complex number. Problem solution 6.5.10 is a reminder that complex exponentials are periodic functions; mathematicians usually prefer to write general solutions of differential equations in this form, which is often less cumbersome than using sine and cosine. The boundary between the two types of behaviour is thus simply the boundary between real and complex solutions of a quadratic equation. The condition specifying the "morphogenetic" side of the boundary is

$$k_4' < k_1' - 2. \tag{7.27c}$$

This appears to limit the regions of interest to us to those marked B and D in Figure 7.5; but in Figure 7.4 I have also included region C in the total morphogenetic region, and in Figure 7.5 region A is shaded as of possible significance for morphogenesis. A and C are subdivisions of the time-periodic region, and need further discussion.

(d) The mathematical analysis of the Turing model given in this chapter, however extensive it may appear to the beginner in differential equations, is far from complete. Input has been specified as U and V standing waves in phase with each other. This leaves us discussing only which of the two kinds of standing waves occurs for various values of the parameters, and ignoring all behaviour which does not fall in either category. Computations using parameter values which are "wrong," in the sense of giving the unwanted time-periodic behaviour, easily show that travelling waves as well as standing waves arise. Lacalli and I (1978) fell into this trap in our first computations on regulation of a simple two-part pattern after chopping the system into pieces; this was intended to be relevant to differentiation in the slug stage of the slime mould *Dictyostelium discoideum*. For our first choice of parameter values, two separate travelling U and V waves arose, with the node of each moving across the system and appearing to "bounce" off the boundaries.

From the viewpoint of required conditions to prevent oscillation, however, it does not seem to matter very much which kind of oscillation we are discussing. Choice of parameters to avoid stationary oscillations will also avoid travelling ones, and I continue to classify oscillations only in terms of the former.

The categories of damped and growing oscillations are well known (e.g., respectively, the dying away of the vibrations of a string or a pendulum which has been set in motion but is not being continuously supplied with energy, and the growing positive-feedback howl in a public-address system in which the loudspeakers have been placed behind the microphone). Both can arise in a Turing model. Where oscillation occurs, its amplitude may grow or decay as time goes on. The indicator of which occurs, in terms of k_g as a complex number, is whether its real part is positive or negative. This gives the condition

$$k_4' < -k_1' \qquad (7.27\text{d})$$

to avoid growing oscillations. The boundary is the only line of negative slope in Figure 7.5. It separates region C (damped oscillations) from region A (growing oscillations).

Clearly, the damped oscillations are not a big problem for interference with morphogenetic behaviour. They can never become larger than the initial disturbance, which is not likely to give very much to start them going. Thermodynamic fluctuations do not provide the equivalent of someone starting the pendulum of a grandfather clock, or pushing someone else on a garden swing. Only growing oscillations, which can amplify small rudiments, are likely ever to be seen in biological development. Hence region C is quite legitimately within the morphogenetic region and is a possible region of parameter values for instances in which, experimentally, one sees no hint of oscillatory behaviour at any stage.

All instances of time-periodicity in the Turing model have the characteristic of appearing only above some threshold wavelength. The reason for this can be traced very easily through equations (7.13) to (7.17). Trouble (if one regards complex numbers and vibrating development as trouble) arises if the quantity $b^2 + 4k_2k_3$ is negative, because one has to take its square root. Now k_2k_3 must be negative, so what matters is the relative size of b^2; if it is big enough, it avoids the trouble. Because b contains $\omega^2 = 4\pi^2/\lambda^2$, it goes infinitely positive as λ goes to zero. Therefore, as wavelength decreases, one must reach a point at which $b^2 + 4k_2k_3$ becomes positive. There is no oscillatory behaviour below that wavelength. Hence plots of k_g versus wavelength in the oscillatory regions are like Figure 7.7c,d. Up to the threshold λ, real k_g is plotted as a solid line. It has a maximum, and that is the pattern which grows to dominate the system. Above the threshold, the real component of k_g is shown as a broken line. If it is negative, oscillations die away; if it is positive, they grow. These are, then, disturbances at longer wavelength than the pattern which is becoming established.

(e) The essence of the Turing model is competitive exponential growth of patterns, in which the fastest swamps everything else and appears to be present alone at long enough times. Thus, even a growing oscillation may not be seen if it does not grow as fast as stable pattern. This leads to a further subdivision of the oscillatory region, according to whether its growth rate is greater or less than that of the pattern at λ_m. The condition for oscillation to grow slower than pattern is

$$k_4' < k_1' - 4n^{1/2}/(n + 1). \qquad (7.27\text{e})$$

It is satisfied in the region A, which is therefore of *possible* morphogenetic significance. I have indicated doubt by equivocating about whether the overall

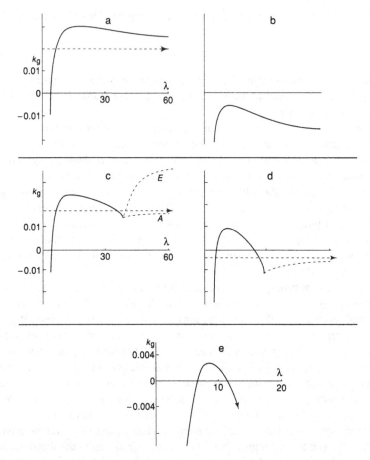

Figure 7.7. Various ways in which the exponential growth rate constant k_g can vary with wavelength, for sinusoidal U and V waves in phase with each other. (a) k_g is real at all wavelengths and passes through a maximum, but rather a shallow one. All patterns of wavelength above the threshold will grow (parameters in region B, Figure 7.5). (b) k_g is never positive; patterns of all wavelengths decay (parameters in region to the left of the steep line in Figures 7.4 and 7.5). (c) k_g has a maximum, but at some longer wavelength becomes complex; the broken lines represent the real component of complex k_g: Line A is for parameters in region A, Figure 7.5; the oscillations do not grow as fast as the pattern at λ_m. Line E is for parameters above line e in Figure 7.5. Oscillatory behaviour builds up faster than pattern. (d) Here again k_g goes complex above some wavelength, but the real component is negative. This means that oscillatory behaviour will die out and is unlikely ever to make itself noticed (parameters in region C, Figure 7.5). (e) k_g has a sharp maximum and goes negative at long wavelengths. This is the band-pass region, D in Figure 7.5, which is strongly discriminating between patterns, and is a likely region of parameters for much morphogenetic behaviour.

morphogenetic region does or does not contain region A, between Figure 7.4 and Figure 7.5. Because oscillations can grow above their initiating disturbance, one might expect to see them transiently, before the organism settles down to proper developmental business, if the parameters are in region A. This could be related to the pulsed behaviour of cAMP in the cellular slime moulds.

(f) Doubters may find their doubts growing monotonically with time about the last-mentioned aspect of the Turing model: that it proposes selection of one pattern from all other possibilities by competition of rates which does nothing to eliminate the slower patterns. They are, in the outcome, still there, though at smaller amplitude than the winner. This statement is, however, strictly correct for all wavelengths of pattern only in a certain part of the morphogenetic region. In another part, pattern decays above a wavelength λ_f. Lacalli and I have stressed the importance of this part of the morphogenetic region by repeated use of parameter values in that region, which includes region C (Lacalli and Harrison, 1979; Harrison et al., 1984) (see Figures 3.2 and 7.2). The curve of k_g versus λ must have the form of Figure 7.7d or 7.7e, as contrasted to Figure 7.7a or 7.7c, for the system to have this "band-pass" discrimination for pattern. The condition to be satisfied is that k_g is negative at infinite wavelength, and the algebra gives

$$k_4' > -(1/k_1').$$
(7.27f)

The boundary is the only one which is not a straight line in Figures 7.4 and 7.5; it is a hyperbola. It appears to squash the most precise pattern-forming behaviour into a tiny region of parameter space, $C + D$, forcing the computer programmer into a very restricted choice of numbers.

But it does not, I stress again as at the beginning of Section 7.3, force nature into a restricted choice, because nature is not a computer programmer and is not directly playing a numbers game. This last condition is the one previously mentioned as expression (7.23), which, for the example of the Brusselator mechanism, means no more than that two of its steps exist.

7.3.2 The conditions from a chemical kineticist's viewpoint: the Brusselator as an example

I raised a question in the preceding section as to whether the Turing conditions are to be regarded as narrowly restrictive or rather easy to satisfy. There is another way to put this question: In Figure 7.5, should we, to cover all likely ranges, be plotting in an area in which the scales of k_1' and k_4' both run up into the hundreds, in which case the area $C + D$ would look tiny indeed? Or, on the other hand, does the area actually plotted cover unnecessary tracts? If, for instance, there are good reasons for k_1' not to exceed unity, about two-thirds of Figure 7.5 could have been omitted, and the remainder drawn on an increased horizontal scale, which would make $C + D$ look much bigger.

The Turing constants k_1-k_4 can have positive and negative values; indeed, one of them (either k_2 or k_3) is required to be negative for the model to work in a morphogenetic manner. This indicates that these constants are not the rate parameters for individual steps in a reaction mechanism; the latter must always be positive. Thus the numerical ranges spanned by their possible values are not the same as those for k_1-k_4. If we have a model specified as a chemical mechanism, the restrictions on its rate parameters must be different from those on k_1-k_4.

By no means has every reaction-diffusion model been written down in chemical-mechanistic terms. The Brusselator has, however, and will serve as an example. The mechanism is

$$A \rightarrow X \qquad \text{(rate constant} = a), \qquad (7.28a)$$

$$B + \ X \rightarrow Y + D \qquad \text{(rate constant} = b), \qquad (7.28b)$$

$$Y + 2X \rightarrow 3X \qquad \text{(rate constant} = c), \qquad (7.28c)$$

$$X \rightarrow E \qquad \text{(rate constant} = d). \qquad (7.28d)$$

The chemical nature and mathematical analysis of this scheme are discussed in detail in Chapter 9. There it is shown that if reactants A and B are supplied at constant concentrations, the intermediates X and Y can settle down to spatially uniform steady-state concentrations X_0 and Y_0, as one would commonly expect for transient intermediates in a reaction. But if X and Y are diffusible, departure from this steady state in the Turing manner is possible, to give spatial patterns of X and Y, which may be expressed in terms of the displacements U and V from X_0 and Y_0. Provided that they are fairly small, U and V develop in time and space according to Turing's equations. The algebra relating the Turing constants to a, b, c, and d is given in Chapter 9, and the results are tabulated in Table 9.1. For the present illustration, we need the last item in that table,

$$k_1'k_4' = -(1 - d/bB). \qquad (7.29)$$

Inequality (7.27f), corresponding to the hyperbolic boundary line f in Figure 7.5, requires that the right-hand side of equation (7.29) be greater than -1. This is clearly satisfied if d, b, and B are all positive. Two of these quantities are actual rate constants for steps in the mechanism; the third is a complete concentration (not a displacement from steady state). All are necessarily positive, and the representative point for the system in (k_1', k_4') space is automatically on the right side of the hyperbola if reactions (7.28b) and (7.28d) exist at all. These are the two reactions in which X is destroyed. If, for the same X concentration, (7.28b) is the faster of these two, then k_1' is positive. Again from Table 9.1, k_4' is automatically negative if steps (7.28a), (7.28c), and (7.28d) all exist.

Consider the following statement: In a particular reacting system with intermediates X and Y between reactants A and B and the final products, all

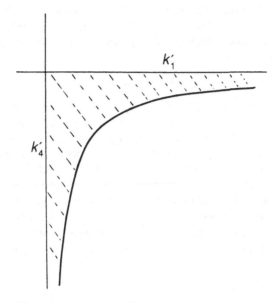

Figure 7.8. An illustration that some of Turing's conditions are not very restrictive when discussed at the chemical-mechanistic level. For the Brusselator mechanism, the representative point in (k_1', k_4') space lies in the shaded region provided that reaction of X with B is faster than its unaided decay.

four steps of the Brusselator mechanism occur; and of the two reactions destroying X, its reaction with B is faster than its unimolecular decay, for the same X concentration. This statement places the system in the region of (k_1', k_4') space shaded in Figure 7.8.

Further carving up of that region, as for instance by the important lines a and d in Figure 7.5, is somewhat more restrictive, but not nearly so much as first appeared in the account in Section 7.3.1. The inequality for which line d is the boundary is, for instance, automatically satisfied if the rate constants are such that the uniform starting concentration X_0 is greater than Y_0. The conditions will not be discussed in further detail here. The above examples are sufficient to show that what may seem abstract, mysterious, and highly restrictive at the algebraic and computer-programming level may turn out to be not much more than common sense whenever one can transfer the analysis to the chemical level.

Finally, the question arises of what quantitative values of spacings between repeated parts in a pattern (wavelengths λ) may reasonably be regarded as capable of establishment by reaction-diffusion. For order-of-magnitude estimates, the message of parameter space is that the Turing rate constants must all have rather similar values. This leads to the possibility of replacing them by the doubling time $\tau_{1/2}$ of pattern amplitude in the regime of exponential growth. From initiation to observable expression of pattern probably needs an increase of pattern amplitude of the order of a factor of 1,000 (Harrison, Snell, and Verdi, 1984), which would take $10\tau_{1/2}$. Lacalli and Harrison (1987) derived the approximation

$$\tau_{1/2} = \lambda^2 \ln 2/2\pi^2\mathcal{D}, \tag{7.30}$$

where $\mathcal{D}$ is the geometric mean diffusivity, $(\mathcal{D}_x\mathcal{D}_y)^{1/2}$. For a very small spacing such as $\lambda = 1$ μm (e.g., xylem rings), $\mathcal{D}$ would have to be 10^{-10} cm^2 s^{-1} to make the pattern development time $(10\tau_{1/2})$ about half a minute. Smaller diffusivity would give proportionately longer development time. For common morphogenetic spacings of order 10 μm, a diffusivity of 10^{-10} cm^2 s^{-1} would allow about 1 hour development time. This quantitative linkage of time and spacing shows that most morphogenetic events are compatible with reaction-diffusion mechanisms, especially if the morphogens are membrane-bound. Turing patterns for non-living chemical systems in aqueous solution (Section 10.3.4) have much larger spacings, because of the high diffusivities.

PART III

Bringing experiment and theory together

> In the actual history of science many of the most fruitful theories have been developed from preconceived ideas of the kinds of laws or theoretical entities that will be discovered to explain the phenomena. The history of the inquiry has to a large extent consisted of using the sharp tools of mathematics and experiment to carve out of these preconceptions a theory exactly fitting the data.
>
> —A. C. Crombie (1959, vol. II, p. 288)

As a schoolboy in England, I was taught the scientific method almost as a kind of credo. Two words were conspicuously (from my present perspective) absent from its description: machinery and models. Science was defined primarily as a function of the human mind, seeking to find order in apparent chaos by a particular kind of formal procedure. In this, the first step is gathering data; the second is making generalizations from data, which are often called laws; the third is explaining those laws on levels different from the observational level of the original data. This is theoretical explanation, and it leads into a sequence of steps in which theory leads to prediction, thence to new experiments, and thence to confirmation, rejection, or detailed refinement of the theory. In this sequence it is sometimes indicated that a theoretical explanation is categorized successively as hypothesis, theory, and law as it becomes more securely established. The progression certainly exists as the principal method of science. But the semantics of the use of the words "hypothesis," "theory," and "law" seems to lack historical validity, and unfortunately gives two different meanings to the word "law," because the established theory is quite a different thing from the generalization from data.

Any semantic confusion in this usage is, however, insignificant compared with the currently popular use of the word "model" indiscriminately for either kind of law and for a theory at any stage in its progress toward establishment or rejection. This usage tends to cast attempts to follow the scientific pathway into a deep fog. The word has also a somewhat pejorative nuance, in that many people think of a model as something distinct from reality. The aim of science is, of course, to find reality in the form of theoretical explanations.

In approaching specific examples, therefore, one should start with the recognition that the scientific method is in full and normal operation in the field of living pattern and form, and that various things now grouped together under the term "model" actually represent different stages along the scientific progression. One should frequently remind oneself, when thinking about a model, that one must be clear where it lies in this progression.

There is very little in this book about the first kind of model, or law, or whatever, arising from the data gathering, that is, the generalization from data. To be sure, such laws exist in the field of developmental biology. An excellent example is the set of rules formulated by French et al. (1976) for limb regeneration in insects and amphibia. These rules, expressed in terms of a polar-coordinate system around the limb and a Cartesian coordinate along it, are quite powerfully predictive for the results of grafting experiments. As a generalization, the scope of the polar-coordinate concept has been remarkably extended by Frankel's (1989, 1990) evidence that it is applicable to a single-celled group, the ciliates. But we still await a definitive explanation of how a living organism goes about setting up a polar coordinate. Generalizations from data have predictive power, but they are all on one level. They do not relate the variables to anything except each other.

This is because generalizations from data are essentially independent of preconceptions. The epigraph from Crombie adds something to the conventional elementary description of the scientific method. In embarking upon their journeys of discovery, scientists commonly have a general preconception of the kind of country to be covered. Up to this point, this book has been largely devoted to expounding, with biologists in mind as the readers, a preconception which many of them do not have but which is widespread among physical scientists thinking about the same experimental phenomena. Turing's equations cannot necessarily be adequately tested against experiment. They compose a roughed-out preconception, which may never exactly fit any precise data. The carving of a theory exactly fitting the data may always involve the use of nonlinear equations. These are introduced in Chapter 9, with an extensive description of the Brusselator model.

Several other nonlinear models are also discussed in that chapter. The general aim is to show that each of them, because of its particular dynamic structure, can serve a particular range of pattern-forming purposes while it altogether fails to serve others. This is the beginning of recognizing reaction-diffusion theory as a class of dynamics containing many species, as do classes of living organisms or classes of molecules such as proteins. Therefore, Part III starts with a brief chapter on classifications.

Many biologists may still, however, be inclined to enquire whether the models discussed here are a class within developmental biology or within a branch of pure mathematics irrelevant to their experimental discipline. Are my titles for Part III ("Bringing experiment and theory together") and for the final chapter ("Approaching agreement?") at best optimistic and at worst quite misleading?

I am encouraged to believe that this is not so, and that my titles are correct, by my current writing of this introduction. I started work on this book in 1984.

Then, and even somewhat later, I had quite a different plan for the contents of Part III. Because I have now written it differently, I have had to rewrite this introduction, and this paragraph is my last task in writing the book. This has led me to recognize that almost none of the material that I have presented in Chapter 10 as evidence for kinetic theory in general and reaction-diffusion in particular could have been written in 1984. Progress is being made toward union of this kind of theory and biological experiment. I believe that the material I have been able to put into Chapter 10 makes the book a more powerful argument for kinetic theory than it could have been if I had finished it sooner.

8

Classifications

The urge to classify is the primary driving force of science. It is the deepest form of the common human desire to make things look tidy. But unlike the domestic analogue which that suggests, the urge to classify operates in realms where one cannot impose order, but can seek to find it if it is already really there (in the deepest sense of the word real, or really, or reality).

Classification can be used in both the analytical and the synthetic parts of the scientific process, the lumping and the splitting. It assists both in the recognition of fundamental common features (e.g., by the setting up of such categories as kingdoms and phyla) and in the drawing of careful detailed distinctions between species and varieties. In the scientific interplay of experiment and theory, classification can appear in different places. It may be the first step beyond data gathering, the organization of data into generalized statements. The larger divisions in the biological classification are of that kind. On the other hand, theories themselves may become complex and diverse enough to require classification. Classification can assist in giving an overview to remove either factual or conceptual confusion. It is most needed where either of these exists in large measure.

Most particularly, classifications of both kinds are needed where experiment and theory are both extensively elaborated, but not well joined together. Both must be properly organized if there is to be a complete plug-in between them. To continue the metaphor of Section 1.2.2, the optic nerve can do its job if the retina and the tectum are each separately already well organized.

Therefore, to assess the current state of kinetic theory of living pattern, and to try to make some contribution to its advancement, I start the last part of this book by attempting some classifications. This is the last chapter in which I try to be in some sense comprehensive. Indeed, earlier in this book I have tried this only in Chapters 1 and 4. Those two and the present chapter can be taken together as my view of the relationship of paradigms and classifications in this subject. Chapters 9 and 10 are quite selective and open-ended, which corresponds to the state of the topic.

8.1   Beginnings of a classification of developmental theories

I am not seeking to propose that kinetic theory should become the be-all and end-all of the explanation of pattern formation, only that it is a *necessary part* of the complete assembly of morphogenetic theories, and should be a very large part. The proper balance will not have been achieved, to my mind, until, in the experimental biology journals and alongside many papers of the kinds which are being written nowadays, there are usually quite large numbers of papers in which the discussion sections use mathematical language and the explanations have been tested by computation.

Table 8.1 is an attempt to map out the main parts of the entire kingdom of developmental theory. The map has not been drawn in equal detail within each of the three "phyla": kinetics, equilibrium, and structure. My purpose is to discuss the first of these. Most biologists will be able to draw detailed maps of the other two (and especially the third) for themselves. Both the "phyla" and the "classes" have been discussed in Chapter 4. But the subdivision of the class "reaction-diffusion" cannot be understood on the basis only of the mathematical introduction in Chapters 6 and 7. The Turing equations express the gist of what a mechanism must have to be able to generate pattern at all. But they are, mathematically, linear equations, which have the unrealistic aspect that their solutions show concentrations of substances growing exponentially without any limitations whatever on the maximum values they may attain. Realistic models, even from very simple putative chemical mechanisms such as the Brusselator, contain nonlinearities. These, for any model worth discussing at all, remove the feature of unrestricted growth and give accounts of patterns which settle down into steady states of concentration distribution. But every specific model has different nonlinearities which give it some quite idiosyncratic properties not shared by other models. These idiosyncrasies hold the promise that reaction-diffusion theory can be fine-tuned to fit many different experimental phenomena, in conformity with Crombie's view of the scientific method quoted as epigraph to Part III.

8.2   The idiosyncrasies of some reaction-diffusion models

8.2.1   *Nonlinearity and the history of reaction-diffusion models*

A mathematician or physicist describing kinetics as being nonlinear means simply that the rate of some process depends on a power different from the first power of the concentration of some substance. For pattern-forming abili-ty, the power should be greater than the first. In Section 5.3.1 I credited Mills (1932) with the original idea of an autocatalysis of kinetic order higher than the first as leading to a two-way departure from equilibrium. This reference seemed to be well known among practitioners of the theory of spontaneous optical resolution, who gave it to me when I first entered that field, and it has been elaborated or independently rediscovered by several of them (Frank, 1953; Seelig, 1971a,b, 1972; Harrison, 1973, 1974; Decker, 1974). Although this has been a continuous line of advance in its own right, it is not clear to what extent it has contributed to the introduction of similar ideas in morphogenetic theory (except for my own rather late entry into the field).

The fact that a combination of positive feedback and diffusion could give an equation in the form of a wave equation in the distance variable was first recognized by Rashevsky (1940). His priority is not acknowledged as often as it should be. Even earlier, Fisher (1937) published an equation for a single diffusible autocatalyst. This has travelling-wave solutions. But the paper of Turing (1952) is far better known, for good reason. [See Murray (1977) for a discussion of Fisher's paper, as well as for one of the numerous opinions in the literature that the concept of reaction and diffusion giving rise to spatial structure starts from Turing.]

Turing's combination of the concepts of autocatalysis, inhibition, and diffusion with a specific interrelationship of cross-interactions between two substances, is a theory having enormously greater power to explain generation of form than any predecessor. It has not yet led to a fusion of chemical kinetics with developmental biology, but it has been the basis for the most extensive theoretical work yet done relevant to that important objective.

Whenever Turing's equations are derived from the rate equations for a putative reaction mechanism, it is easily seen that nonlinearity has two different aspects and functions. The rate equations are written in terms of true concentrations of chemical substances (e.g., X and Y). The corresponding Turing equations are written in terms of departures of X and Y from the spatially uniform steady state (i.e., in terms of the variables U and V). The nonlinearities appearing in the U, V equations are not the same as those in the X, Y equations. They serve different purposes in the possible biological applicability. This is perhaps the primary reason why an experimental biologist interested in using this kind of interpretation of data must understand the mathematical procedure of linearization about the spatially uniform steady state (see Section 6.3 and the full account for the Brusselator in Section 9.1). Without this background, the qualitative features of the theory cannot be properly appreciated.

The first aspect is the general one: Nonlinearity in the chemical kinetics, involving terms in X^2 and so forth in the X, Y equations, is essential to pattern-forming ability. But we saw in Section 2.2 that Mills-type nonlinearity ($D^2 - L^2$) leads to linear growth of the optical asymmetry ($D - L$). And in Section 6.3.1 we saw that a nonlinear term in X^2 gives rise to the Turing-like linear term in U. Thus the *linear* Turing equations in U and V express the essence of the nonlinear chemical kinetics from which they arise.

The second aspect is concerned with the idiosyncratically different behaviours of different models which must be taken into account in all attempts to fine-tune the fit between data and theory. When chemical models with

Table 8.1. *Beginnings of a classification of developmental theories*

Phylum	Class	Order	Family	Genus	Species
	Reaction-diffusion	"One-morphogen"	Pattern transient	Mills (1932) theory of optical resolution	Frank (1953), Seelig (1971a,b), Decker (1974, 1979), Harrison (1973, 1974, 1979)
			Pattern stabilized by fixed sources	Wolpert positional signaller	Zone of polarizing activity
		Two-morphogen (Turing)	Adaptable	Brusselator (Prigogine)	Measuring, Counting
			Headstrong	Gierer-Meinhardt	Hydra grafting, Insect embryogenesis
		Three-morphogen	Travelling waves	Oregonator (Field and Noyes, 1974)	Oscillating reactions
	Mechanochemical	Forces deform cells	Epithelial shape changes	Odell et al. (1981)	Invagination
			Subcellular, cytoskeletal	Ca-regulated	Goodwin and Trainor (1985)
		Forces move cells	Mesenchymal morphogenesis	Oster et al. (1983)	Chemotaxis, Haptotaxis, Random walk
Kinetic	Self-electrophoretic	·············	·············	e.g., Jaffe (1982) ·············	·············
	Complex intercellular signals	Assembly of central nervous system, e.g., Willshaw and von der Malsburg (1976) ·············			

Equilibrium	Cell-as-molecule phase separation	.... Differential adhesion, Steinberg (1970) ....
	Subcellular free-energy minimization	.... Cell division and surface free energy, Thompson (1917), Green and Poethig (1982) ....
Structural		.... "Self-assembly," Inoué (1982) ....

Source: From Harrison (1987).

pattern-forming ability are transformed from the X, Y form to the U, V form, there will always be nonlinear terms in U and V. One ignores these when one uses the linear terms only (i.e., the terms which together make up the Turing equations). Then, one is looking at equations which have the important features of pattern growth, wavelength determination, response to boundary conditions, and so forth, but which also have the unrealistic feature that exponential growth of pattern amplitude goes on forever. This misfit with nature is removed as soon as one adds the nonlinear terms. Pattern amplitudes do not then go on increasing forever. Growth eventually slows down, and patterns reach steady states.

In the development of a pattern, one may then distinguish between an initial linear regime, of which the linear Turing equations give a good account, and a nonlinear regime in which pattern growth slows and finally ceases. For the latter, a host of questions arise. Nonlinear patterns will not be of sine-wave form. How much will they differ from sinusoidal shape? Will the peaks and troughs have the same shape and the same length, or will they differ? For instance, will there be short activated regions and long inhibited ones, or not? Will the number of repeated parts in the steady-state pattern be the same as in the initial pattern of the linear regime, or will there be a drastic change? To put the same question differently, will the system continue to "know" its linear wavelength? By the same token, if a system has reached a patterned steady state, and then some change is made, such as an increase in length of the system or a change in supply of a chemical reactant which affects the linear wavelength, will the steady-state pattern respond as one expects the linear pattern to do? Answers to questions like these are far from obvious, because it has, in general, not yet been found possible to study nonlinear partial differential equations by analytical methods. The method which helps most is the use of the digital computer. A model can be put into it, and one may then regard the model as an experimental system in its own right. One can carry out experiments on the model, just as one can carry out experiments on a developing biological system. This permits some attempt to classify models by their characteristics of pattern development.

8.2.2   Beginnings of a classification
of reaction-diffusion models

In Table 8.1 I have suggested that the reaction-diffusion class might be subdivided into orders on the basis of the number of mutually interacting morphogens in the model. In the end, this may or may not turn out to be the best way of going about the classification. It is not yet clear whether or not there is sufficiently close correlation between numbers of morphogens and differences in dynamics of development. (But one feature is fundamental: Turing's big, original contribution was the addition of the second morphogen, the inhibitor.) This section will be concerned mainly with two-morphogen models, and to some extent with four-morphogen models. The distinctions to be described here go beyond the indications in Table 8.1.

Indeed, two different ways of looking for the characteristics of reaction-diffusion models are suggested here, and they might lead to two different classifications: First, what kind of terms does one see in the equations? Second, what does one learn from computer experiments on the model? Of these, the latter is likely to be more attractive to experimental biologists, because the procedure resembles experimental rather than theoretical work. But biologists are often worried, in relation to reaction-diffusion, whether or not a realistic chemistry is being postulated. For example: "These equations always contain autocatalysis, but we don't know many such processes." Such matters need a hard look at the equations themselves. But in the absence of complete analysis of nonlinear equations, computations can help. Here I discuss the foregoing questions together.

Ideally, a reaction-diffusion model should be postulated first as a set of chemical equations, putative steps in a reaction mechanism. Then the rate equations should be written for this mechanism. This is not always done. Often the model is written as rate equations, and one is left to infer the chemistry. This can lead to misconceptions. Among these, perhaps the most important is that a Turing reaction-diffusion model always requires an explicitly autocatalytic step in the reaction mechanism. Recently, acting as a referee, I recommended acceptance of a paper claiming to show evidence for reaction-diffusion. The other anonymous referee wrote: "A key aspect of all reaction-diffusion mechanisms is the presence of some form of catalysis, typically autocatalysis (hence the 'reaction' in the name)." The editor rejected the paper.

In fact, the "reaction" in the name is not synonymous with autocatalysis. What it signifies is what I have called the kinetic preconception or paradigm: that pattern arises *somehow or other* out of rates of reaction, not out of structural fitting or approach to equilibrium. For the kinetic equivalent of the Turing equations to be obtainable by linearizing the rate equations about the spatially uniform steady state, it is not necessary that the mechanism include an explicit autocatalytic step. There are at least two examples of this in the literature: Murray's (1981a,b, 1989) well-known modelling of mammalian coat patterns, and a three-morphogen model of Meinhardt (1989). In the latter, it is quite obvious in the equations that positive feedback is achieved by an inhibition of an inhibition. This is mathematically equivalent to auto-catalysis, in terms of the kind of dynamics occurring. It is not at all equivalent to autocatalysis in terms of what any one molecular species is doing.

For the scientist whose thinking is molecularly oriented, there are two messages here. First, in conformity with the major theme of this book, it can help in understanding dynamics to try to think about dynamics only and quite avoid the question of which molecules are doing what. Then one can appreciate positive feedback as a "thing," an important "thing" which has to be present in a model to give it pattern-forming ability. When positive feedback is thus separated from the molecular questions, it is quite irrelevant to introduce the term "autocatalysis." The more general term is far superior.

But my second theme is the building of bridges between the dynamic level and the molecular level. It is here that conceptual confusion may easily arise, and classification is often an aid in avoiding confusion. Hence, it may help to distinguish two categories: (1) that in which the positive feedback corresponds to an explicit autocatalysis of one molecular species; (2) that in which there is no explicit autocatalysis, and the positive feedback loop usually has to be traced through the inhibitions.

It is also quite straightforward to devise reaction-diffusion models which do not contain an explicit kinetic inhibitor. This, in fact, is the nature of the mechanism described in most detail in this book, the Brusselator. There is no explicit inhibition in the mechanism shown as equations (6.31a–d) and repeated as equations (9.14a–d). What happens instead is, in step c, the autocatalytic formation of X uses up Y. In the dynamics, this leads to Y playing the role of an inhibitor, because where X is large, the depletion of Y puts the brakes on the rate of formation of X.

Here, then, is another source of quite dangerous confusion. Earlier in this book (Section 2.6 and most of Chapter 7) I discussed Turing dynamics in terms of a true kinetic inhibitor. To do its work of slowing down X production, the inhibitor Y must be present in highest concentration where there is also the most X. For such a substance, the biologist is wrong who says that "I've found an activator at one end of the system; now I'm going to look for an inhibitor at the other end." But if the dynamics involve depletion rather than inhibition, the X and Y patterns will be out of phase with each other, and the right place to look for Y is indeed where there is the least X.

Here, the classification, which Meinhardt (1982) has already pointed out, is into (1) inhibition models and (2) depletion models. The distinction is readily seen if one has either the chemical equations for the putative reaction mechanism or the rate equations for true X and Y concentrations. If, however, one has only the Turing equations for U and V (departures from uniform steady state), it can still easily be seen whether the model is an inhibition or a depletion model. Always, either k_2 or k_3 must be negative, and the other positive [equations (7.1)]. For an inhibition model, k_2 is negative, and computations on these equations will give U and V in phase with each other. For a depletion model, k_3 is negative, and computations will give U and V out of phase.

Thus far, this account has been concerned with properties which affect the basic Turing-like behaviour of a model, that is, in algebraic terms, what is necessary and sufficient for the model to yield the linear Turing equations upon linearization about the spatially uniform steady state. What follows is concerned with nonlinearities which give the nonlinear terms in U and V. These are much more difficult to cope with mathematically. What is known about them is mostly from computer experiments on the behaviour of the model. But these lead to some things which can be seen just by looking at the equations.

First, there is the distinction which I have represented by the words "adaptable" and "headstrong" (Harrison, 1982); see the classification scheme in Harrison (1987) and in Table 8.1. An adaptable model (e.g., the Brusselator) is one which behaves much like the linear model, continuing to "know" its linear wavelength when it is operating in its nonlinear regime, and therefore regulating well to put the right number of evenly spaced pattern repeats within the boundaries of the morphogenetic region. A headstrong model (e.g., the Gierer–Meinhardt model) tends to stabilize peaks once they have been formed and to allow their persistence in the face of production of new peaks in the "wrong" places. Such a model is good for instances in which it is known that interference with the system can lead to teratogenesis. The contrast between adaptable and headstrong models has already been elaborated in Sections 3.2 and 3.3.

Can one look at the equations for a model and tell by inspection which kind it is? There is as yet no definitive answer to this question, but I have a tentative one. The Brusselator equations are given in Section 9.1, both as equations in X and Y [equations (9.15)] and as equations in U and V [equations (9.23) and (9.24)]. In either format, one can see that the nonlinearities in Y or V are exactly equal and opposite to those in X or U. This corresponds to the chemistry of the model. As discussed in detail in Section 9.1.1, the Brusselator is derived from a type of mechanism taught in elementary chemical kinetics courses in relation to balanced radical-chain reactions. These often involve an alternation of steps in which there are two types of radical intermediates. Whenever one is formed, the other is destroyed, and vice versa. Therefore, these alternating steps cannot alter the total content of both radicals together in the system. Likewise, in the Brusselator, the pattern-forming steps exchange X for Y or Y for X but do not change the total in the system. This means that a region of positive U (i.e., excess of X over nonpatterned level) somewhere in the system must be balanced by a region of negative U somewhere else. The total sizes of activated and inhibited regions are tied to each other; they must be equal. (This is for no-flux boundaries.) It seems to me that the high regulatory capacity of the Brusselator is related to this nature of its chemistry, which is seen in the rate equations by the equal and opposite terms in U and V.

By contrast, the rate equations for Gierer and Meinhardt's A and H [Section 9.3.1, equations (9.39)] are quite different. The essence of the pattern-forming ability lies in the fact that the rate of growth of activator H includes a term in A^2/H. The A^2 represents the bimolecular autocatalysis; the $1/H$ is its inhibition. There is no corresponding term in the rate of growth of inhibitor H. This implies a totally different chemistry from that of the Brusselator, related to the fact that one is an inhibition model and the other a depletion model. The nonequivalent dynamics of activator and inhibitor in the Gierer–Meinhardt model give it the power to represent "short-range activation, long-range inhibition" and teratogenesis.

Computer experiments on reaction-diffusion models have often been done for both one-dimensional and two-dimensional morphogenetic regions. Some examples are given in Sections 9.2, 9.3, and 10.1. Two-dimensional work often tests the models in ways which are both more severe and more informative. Especially, the question of whether a model generates a striped or a spotted pattern in two dimensions is of great interest. [A referee for a grant application of mine once scornfully indicated that this problem is so passé that it is a topic for a problem in an undergraduate course in one institution. In the world of reality, I have had a graduate student writing a Ph.D. thesis in physics on one aspect of this problem (Lyons, 1991), and I would confidently expect the question to be a prolific Ph.D. generator through the foreseeable future.]

A general difference in properties between the solutions of linear equations and those of nonlinear equations must here be appreciated. For linear equations, the sum of any two solutions is also a solution. This means that the two physical entities represented by two solutions are noninteracting. They ignore each other (just as molecules in a mixture of two perfect gases behave as if each gas had the container to itself).

If the two solutions are spatial patterns, each with its own development in time, then each will grow as if the other were not there. Suppose that the morphogenetic region is square, with X and Y spatial coordinates, and the patterns are of the form $\exp(k_g t)\sin(2\pi x/\lambda)$ and $\exp(k_g t)\sin(2\pi y/\lambda)$. These are two patterns, crossed at right angles, each like the Cheshire Cat pattern of Figure 6.1. If one of them is a solution of the kinetic equations, with particular values of k_g and λ, then the other will also be a solution, with the same k_g and λ. They will therefore grow at the same rate and together make a spotted pattern.

On the other hand, for nonlinear equations it is not true that the sum of two solutions is also a solution. This is equivalent to saying that the physical entities represented by the solutions do not behave independently of each other, but interact. It is no longer to be taken for granted that two similar patterns, crossed at right angles, will grow at similar rates. Something altogether different may happen, including the effective suppression of one of the patterns, leaving the system striped instead of spotted.

What must be the characteristics of the nonlinearity in a model for it to be good at making stripes without having to rely on strongly asymmetrizing unidirectional gradients or asymmetric boundary conditions? Quite recently, one property has become clear, both from computer experiments (Nagorcka, 1988; Lacalli, 1990; Lyons et al., 1990; Harrison and Lyons, in press) and, with some degree of approximation, from two nonlinear analyses (Ermentrout, 1991; Lyons and Harrison, 1991). The models with the strongest striping tendency are those in which the growth rates of U and V are antisymmetric in those variables (i.e., the growth rates change sign, but not magnitude, when the signs of U and V are simultaneously changed). Some years ago I devised a model of this kind which was an extension of the optical-resolution model (which has that symmetry) and which I described as hyperchirality (Harrison and Lacalli, 1978). It turns out to have a strong ability to form stripes. But in fact, some of the ways in which biologists discuss pattern formation without mathematical analysis contain a concealed assumption which also has the appropriate dynamic symmetry.

8.2.3   When can dynamics be classified as chiral?

A plot of a square-wave pattern of patches of opposite optical resolutions looks the same if it is turned upside down (Figure 2.4c). To the structurally minded scientist, this may seem to have no relevance to biology, except to the speculative question of how spontaneous optical resolution occurred in the remote past. It is well known that it did happen, so that for any chiral biochemical, living organisms nowadays contain only one of the enantiomers. The other is absent, and with it, surely, one half of Figure 2.4c must be absent.

Once again, the essence of my discussion is to put molecular structure completely out of mind and enquire only, What kind of dynamics can give rise to a pattern such as Figure 2.4c? The answer can be put simply in very general terms. The rate of growth of U (upward) at any point depends on the values of U and V and on that property of the shape of the pattern which governs the diffusion rate of U (second derivative in one dimension, Laplacian when generalized to two or three dimensions). Suppose that any local piece of the pattern is inverted through the point $U = V = 0$ (i.e., the signs of U and V are changed without change in numerical values, and the local shape is inverted so that the second derivatives likewise change sign). If the rate of formation of U is thereby changed in sign but not magnitude, and if the same applies to the rate of formation of V, then the positive and negative parts of the pattern will have exactly the same dynamics when they are the same shape. This means that the upward- and downward-facing parts of the pattern will indeed be the same size and shape and will develop, quantitatively, in exactly the same way.

Mathematically, in regard to looking at the equations in U and V to see if this condition exists, one must take note that quadratic terms such as U^2, V^2, and UV do not change sign with U and V, but cubic terms such as U^3, V^3, U^2V, and UV^2 do change sign. If the equations contain only linear terms and cubic nonlinear terms, the pattern will develop the same upward and downward. The peaks and troughs of the pattern are symmetrically related as if increase and decrease in U or V represented resolution to two optical enantiomers, *even if there is no such structural feature to be found in the molecules concerned.* This means that an effective chirality in the dynamics does not require, in molecular structure, the presence of both enantiomers of the same substance.

Consider for instance, the model for optical resolution described in Section 5.3.2 in terms of an array of A's and B's. The letters A and B do not have shapes which are related symmetrically as enantiomers, but they act in the model as a pair of enantiomers would. When the system is analyzed mathematically in Section 6.4.1, we see that the symmetrical relationship between A and B lies in the dynamics of their rates of formation. Equations (6.56a,b) contain the same rate constants, and each equation can be changed into the other merely by interchanging A and B. This example is intended for the final entrapment of the reader who is still trying to think structurally. Look at the equations. See their symmetrical appearance. Recognize that it does not depend on any relationship between the shapes of the symbols A and B. What the equations signify is that a starting material S can be converted into either a final state A or a different final state B with exactly the same dynamics. A and B could be two different molecules, or the symbols could represent cell states, with S as undifferentiated and A and B as two states of differentiation.

In the latter case, the model is for homogenetic induction of cell states A and B. Biologists often postulate induction, but do not often write down the dynamic equations corresponding to the postulate. The absence of the equations effectively hides an important dynamic assumption: Induction to two different states usually is assumed to take place over the same time interval and therefore with essentially the same dynamics, quantitatively. This corresponds to what one sees experimentally, such as (1) in the differentiation of pre-spore and pre-stalk cells at the slug stage of *Dictyostelium discoideum* (which does not, however, give equal space to the two parts of the pattern) and (2) in the formation of pair-rule gene-expression stripes, such as the alternating stripes of expression of *hairy* and *runt* in *Drosophila melanogaster*. This may be a case of chiral dynamics involving quite different proteins in the alternating stripes.

In Section 9.2 I present my hyperchirality model (Harrison and Lacalli, 1978; Harrison, 1979). This uses a structural concept that a protein, itself always of one chirality, might yet exhibit effective chiral asymmetry in two modes of attachment to a membrane. But the proposed dynamics are far more general than the structural suggestion. Dynamics such as those discussed could arise from altogether different structures in which one cannot see any geometrical chiral relationship; see Section 9.2.3.

The essence of the mathematical contrast is that the Brusselator model contains quadratic nonlinearities and in two dimensions tends to give spotted patterns. Force is needed, in the form of unidirectional gradients along the system, to get the Brusselator to make stripes (Lacalli, Wilkinson, and Harrison, 1988). The hyperchirality model has only cubic nonlinearities and makes stripes without the aid of gradients. In living organisms, gradients will tell the stripes which way to be oriented, but they do not have to force the existence of the stripes.

Spotting and striping are general properties of dynamics respectively with and without quadratic terms (Ermentrout, 1991; Lyons and Harrison, 1991). What is more important – the absence of quadratic terms, or the chiral asymmetry which gives rise to that? To my mind, it is the latter, because that can lead us from the mathematics toward cellular and chemical mechanisms.

9

Nonlinear reaction-diffusion models

The term "nonlinear" is common parlance among physical scientists who habitually discuss dynamics mathematically; it is not familiar language to most biologists. Therefore, at the risk of repetition, I call attention again to the fact that, in the most general sense of the term, everything in kinetic theory of pattern formation (including reaction-diffusion, mechanochemistry, and whatever other forms may be devised) fundamentally involves nonlinearity. The reason for this was given as succinctly as is possible by Mills, as quoted in Section 5.3.1. This kind of nonlinearity, however, gives rise to the linear terms in the Turing equations. In this chapter I am considering the roles of the additional nonlinear terms which will always be present in the U, V rate equations for the full representation of the dynamics of any realistic chemistry.

This account is highly selective. Only the two models with which I am most familiar are discussed in detail, in Sections 9.1 and 9.2. One of the models best known to experimentalists, that of Gierer and Meinhardt (1972), is given much less space, for two reasons: first, I have written something about its character in Section 3.3; second, it is not necessary to repeat the extensive account given by Meinhardt (1982). There is no intention here to rank models in order of importance by the amount of space given to them. My purpose in this book is to elucidate general principles which I believe developmental biologists should be using and, for the most part, are not. It will be time for a comprehensive, balanced survey of numerous models when more people have been using them for some time and the biologist habitually thinks about what kind of dynamic mechanisms are present just as much as what proteins are present. The terms in equations have functions in the organism, as groups of amino acids do. Here is an α-helix; it helps to make the protein span a membrane. Here is a cubic term; it helps the organism to make stripes. [Lewis Held (1992) has published a very extensive literature survey, from a viewpoint somewhat similar to mine except that his philosophical orientation is more toward the computability of developmental models than to their expressibility in mathematical form.]

9.1 The Brusselator

9.1.1 Chemical nature of the model (1): elementary two-intermediate schemes

The Brusselator mechanism arises from one of the best-known types of reaction schemes in elementary chemical kinetics. An appreciation of this may help the biologist to understand why physical chemists may persist in believing in type II morphogens although they have proved notoriously difficult to isolate. The hypothetical morphogens X and Y in the Brusselator are in fact kinetic analogues, not of stable reactants and products, but of the unstable intermediates, or free radicals, in elementary schemes for simple organic decompositions. These are not substances that one expects to be able to keep in a bottle, or even to keep for half an hour without a continuously operating source.

Many decompositions of small organic molecules (as well as some simple inorganic reactions) share the common feature of an alternation of steps in which one radical R is destroyed and there is simultaneous production of a different one, R', and vice versa in the alternate step. Standard textbook examples are the gas-phase decompositions

$$C_2H_6 \rightarrow C_2H_4 + H_2 \qquad (9.1)$$

and

$$(C_2H_5)_2O \rightarrow CH_3CHO + C_2H_6. \qquad (9.2)$$

In the first of these, the alternating radicals are $H^\cdot$ and $^\cdot C_2H_5$ in the steps

$$H^\cdot + C_2H_6 \rightarrow {}^\cdot C_2H_5 + H_2, \qquad (9.3a)$$

$$C_2H_5 \rightarrow H^\cdot + C_2H_4. \qquad (9.3b)$$

In the other, they are again $^\cdot C_2H_5$ (written $Et^\cdot$ below) and $CH_3\dot{C}HOEt$:

$$Et_2O + Et^\cdot \rightarrow CH_3\dot{C}HOEt + C_2H_6, \qquad (9.4a)$$

$$CH_3\dot{C}HOEt \rightarrow Et^\cdot + CH_3CHO. \qquad (9.4b)$$

In each of these examples, one of the radicals, with no other reactant, decomposes to generate the other. This corresponds to the reaction $Y \rightarrow X$ which is a step in various of the Brusselator-type schemes.

Pairs of reaction steps such as equations (9.3) or (9.4) have the property that as they occur repeatedly, converting large numbers of molecules of reactants into products, the number of radicals present is conserved. How fast the overall reaction proceeds depends on the steady-state concentrations of radicals, which must evidently be controlled by a balance of production and destruction processes, neither of which has yet been specified in the examples. These intermediates may be produced in a variety of ways, from the principal reactants, or from some other substance present in much smaller quantity and serving only as a radical initiator. Complete reaction schemes therefore vary quite a lot in detail. The closest correspondence to the Brusselator is in the scheme

Initiation: $A \rightarrow R$ + by-product,   (9.5a)

Propagation: $B + R \rightarrow R' + D$,   (9.5b)

                         $R' \rightarrow R + E$,   (9.5c)

Termination: $R \rightarrow$ by-product.   (9.5d)

What is the overall reaction to which this scheme refers? The indication that the first and fourth steps give only by-products suggests that the other two steps occur much more often, and hence that the overall reaction is

$$B \rightarrow D + E.$$ (9.6)

This corresponds to the examples of simple decompositions given earlier. Such processes are called unbranched chain reactions. If, however, one deletes the two indications of by-product formation, it is possible to envisage all four steps going at similar rates, so that A and B are consumed in comparable quantities, and the overall reaction is

$$A + B \rightarrow D + E.$$ (9.7)

The Brusselator is often thought of in these terms, and the product E may be indicated as appearing in the fourth step rather than the third, with no important feature of the mechanism being changed thereby.

Kinetic analysis of such a mechanism is usually carried out by the steady-state method. In this, one writes rate equations for the rates of change of concentrations of the *intermediates R* and *R'*, and one sets each of these rates equal to zero. The chemist commonly calls this procedure an approximation, which it is for the common laboratory situation of a reaction going on in a closed reaction vessel. In that case, A and B are being used up, and because the "steady-state" concentrations R and R' depend on A and B, they cannot truly be steady.

In a flow system, however, it is possible to envisage such a reaction scheme as a small part of a much larger complex of processes in which homeostatic control operates to supply reactants and remove products at such rates that their concentrations stay constant. The steady-state treatment is then exact. But one may expect it still to be only an approximation in the nonhomeostatic conditions of a developing system.

For reactions (9.5),

$$dR/dt = aA - bBR + cR' - Rd = 0 \text{ in steady state,}$$ (9.8a)

$$dR'/dt = bBR - cR' = 0 \text{ in steady state.}$$ (9.8b)

If we add these two equations, the two propagating steps cancel out, and we are left with the rather obvious result that the steady-state concentration of R is determined simply by a balance between its production in the initiating step and its destruction in the termination step:

$$aA - Rd = 0, \qquad (9.9)$$

$$R = aA/d. \qquad (9.10)$$

By substitution in equation (9.8b) we can find the steady-state value:

$$R' = (ab/cd)(AB). \qquad (9.11)$$

But without finding R' we could determine from R the overall rate of the reaction, which is the same thing as the rate of reaction (9.5b), the only source of product D:

$$\text{reaction rate} = -dB/dt = dD/dt = bBR$$
$$= (ab/d)(AB). \qquad (9.12)$$

The role of the third step, reaction (9.5c), is particularly interesting, because this is the one which is altered to an autocatalytic step to turn the whole scheme into the Brusselator. Equation (9.12) shows that the rate constant c of this step does not appear in the expression for the overall reaction rate. We saw earlier [Section 6.3.1, derivation of equation (6.30)] that the rate constant for an autocatalytic step, which is all-important in giving a system the ability to move away from equilibrium, may not actually appear in the rate of departure from equilibrium. We should now be prepared to find the same feature in this more complicated example.

One important feature of such mechanistic schemes is not well shown by the foregoing example. It is quite unsurprising that the overall reaction rate turned out to be first-order in both A and B, but equally simple-looking schemes can give much less obvious dependences. Suppose, for instance, that the overall reaction is

$$B + C \rightarrow D + E, \qquad (9.13)$$

reactant B being consumed in one of the propagating steps, and reactant C in the other. A possible mechanism can be written merely by adding the reactant C to reaction (9.5c) and leaving everything else unchanged. It is then assumed that A is an initiator present in very small quantity compared with B and C.

The steady-state analysis need not be set out in detail. The rate of reaction (9.5c) is amended to cCR', and the treatment is as before, except that c is multiplied by C wherever it occurs. Because c disappeared in the derivation of the overall rate expression, so does C, and the reaction turns out to be first-order in reactant B, but zero-order in reactant C.

9.1.2   Chemical nature of the model (2): the Brusselator itself

The Brusselator mechanism is usually written with six symbols for three pairs of chemical substances: A and B, the initial reactants or morphogen precursors; X and Y, the intermediates or morphogens; and D and E, the final products. Commonly, the entire discussion concerns A, B, X, and Y, and the two products are not mentioned at all beyond their preliminary inclusion in the set of chemical equations. Thus X and Y often may seem to be regarded as both intermediates and products, which amounts to equivocation about their long-term stability.

An experimentalist who discovers some phenomenon which might loosely be called activation or inhibition, and who has any inclination at all to try to relate this to reaction-diffusion, is likely to seek a correlation of the activating and inhibiting substances with X and Y. A major aim of the present account of the Brusselator is to establish, for this example and by implication for other reaction-diffusion mechanisms, that such experimental properties may belong as much to the precursors A and B as to the morphogens X and Y. When interpreting data, one should be very careful as to which of these pairs one thinks one has found. The present section analyzes algebraically how A and B may appear to be activator and inhibitor. The matter is taken up again in the discussion of *Acetabularia* morphogenesis in Section 10.1.

The preceding discussion relevant to equation (9.13) shows us that a relation between two reactants B and C which looks quite symmetric in the overall stoichiometric equation may altogether lack that symmetry in the detailed mechanism. The significance of this for our purpose is that two morphogen precursors may have contrasting control effects, which are relatively easy to predict or explain mathematically, but which are not at all obvious without the algebra. Such effects often involve threshold concentrations for pattern formation.

In this connection, two kinds of thresholds should be distinguished. In Chapter 6 I discussed thresholds which must be passed to ensure that pattern formation does not violate the second law of thermodynamics. To conduct that discussion, one must write every step in the mechanism as a reversible reaction. Such an analysis is complicated enough for the simple optical-resolution mechanism. Models such as the Brusselator are often extensively discussed with the steps indicated as going forward only. It is then tacitly assumed that the equilibrium constant for each step is so large that the thermodynamic criterion is easily satisfied by quite low reactant concentrations, and need not be discussed further.

The other type of threshold is concerned with the satisfaction of Turing's conditions (Section 7.3), that is, the crossing of any of the boundary lines in Figure 7.5. Of these, the two most important are those marked a and b, which enclose all the areas in which anything interesting occurs (i.e., any behaviour of either time-periodic or space-periodic kind). The parameter space of that diagram consists of the parameters k_1' and k_4' derived from the four Turing constants k_1-k_4. In the following account of the Brusselator, I start with four rate constants a, b, c, and d and two reactant concentrations A and B, as in the preceding elementary example of equations (9.5). I go on to show that the Turing constants can be expressed in terms of these six parameters. One of the outcomes of this analysis is that reactant concentrations A and B can easily be chosen to put the system entirely outside the interesting region of Figure 7.5, but that it may then be moved into the region of morphogenetic behaviour by *increasing B* or *decreasing A*. Thus we see that B may appear to be an activator, and A an inhibitor, for pattern formation, although neither is a morphogen in the sense that X and Y are morphogens.

The earliest description of the Brusselator of which I am aware was by Prigogine (1967). More often, the reference given for this model is Prigogine and Lefever (1968), which was the first extensive account of the kinetic properties of the model. Nicolis and Prigogine (1977) gave a very extensive account.

The Brusselator mechanism is

Initiation: $\quad\quad A \rightarrow X \quad\quad$ (rate constant $= a$), $\quad\quad\quad$ (9.14a)

Propagation: $\quad B + X \rightarrow Y + D \quad$ (rate constant $= b$), $\quad\quad\quad$ (9.14b)

$\quad\quad\quad\quad\quad Y + 2X \rightarrow 3X \quad$ (rate constant $= c$), $\quad\quad\quad$ (9.14c)

Termination: $\quad\quad X \rightarrow E \quad\quad$ (rate constant $= d$). $\quad\quad\quad$ (9.14d)

The autocatalytic step (9.14c) is written as being bimolecular in X to imply second-order dependence on the concentration of X as a catalyst. The two intermediates X and Y are considered to be diffusible, with, as usual, Y diffusing faster than X. The rate equations for X and Y, corresponding to those given for R and R' in equations (9.8), but with diffusion added, are

$$\partial X/\partial t = aA - bBX + cX^2Y - Xd \;\; + \mathscr{D}_x \partial^2 X/\partial s^2, \quad\quad (9.15a)$$

$$\partial Y/\partial t = \quad\quad bBX - cX^2Y \quad\quad + \mathscr{D}_y \partial^2 Y/\partial s^2. \quad\quad (9.15b)$$

To find the spatially uniform steady-state concentrations X_0 and Y_0, omit the diffusion terms from equations (9.15) and go through the same procedure used in deriving the concentrations R and R', equations (9.10) and (9.11), from equations (9.8). On setting each of equations (9.15a) and (9.15b) equal to zero and adding them, one sees very easily that

$$X_0 = aA/d. \quad\quad (9.16)$$

Then, upon substitution in equation (9.15b),

$$Y_0 = bB/cX_0 = (bd/ac)(B/A). \quad\quad (9.17)$$

The inverse dependence of Y_0 upon A is a conspicuous difference from the corresponding expression for R' in equation (9.11). This change is obviously caused by the introduction of the X^2 autocatalytic term. A similar effect of this term carries through into the derivation of the Turing constants for this mechanism. In the event, this explains the "upside-down" nature of the threshold condition for the localized pattern in Figure 9.1. There, pattern is produced only where A is *below* some threshold value, which is quite the opposite to the kind of thermodynamic threshold effect discussed in Chapter 6.

Continuing to omit the diffusion terms just for convenience, because we wish to do algebra only on the chemical kinetic terms, we may write the displacements from equilibrium, as usual,

$$U = X - X_0 \quad \text{and} \quad V = Y - Y_0. \tag{9.18}$$

Because X_0 and Y_0 are constants, any time or space derivative of U is equal to the same derivative of X, and likewise for V and Y. Equations (9.15) become

$$\partial U/\partial t = aA - bB(U + X_0) + c(U + X_0)^2(V + Y_0) - (U + X_0)d, \tag{9.19}$$

where

$$0 = aA - bBX_0 + cX_0^2Y_0 - X_0d, \tag{9.20}$$

$$\partial V/\partial t = bB(U + X_0) - c(U + X_0)^2(V + Y_0), \tag{9.21}$$

where

$$0 = bBX_0 - cX_0^2Y_0. \tag{9.22}$$

These equations contain terms in U and V and also some more complicated (nonlinear) functions. The procedure of "linearization about the spatially uniform steady state" consists of retaining only the terms in U and V and ignoring all the rest (i.e., the terms in UV, U^2, and U^2V). If the system is initially in the spatially uniform steady state, $U = V = 0$ everywhere, and starts to depart from it toward small positive and negative values of U and V, the linear terms will at first be much larger than the nonlinear ones. It is then legitimate to ignore the latter. The result is that the equations for rates of change of U and V become Turing's equations. This implies that the departures from spatial uniformity will grow exponentially in time if we have chosen parameter values which let them grow at all. But as U and V depart increasingly from zero, the terms in UV, U^2, and U^2V begin to catch up in magnitude to the linear terms, and the approximation breaks down. This passage into the nonlinear regime has among its properties a slowdown and ultimately complete cessation of the growth of pattern amplitude, so that a stable nonuniform spatial distribution of material is established.

The equations, including the nonlinear terms, but with the diffusion terms omitted, are

$$\partial U/\partial t = (bB - d)U + (a^2A^2c/d^2)V + (2aAc/d)UV + (bBd/aA)U^2 + cU^2V, \tag{9.23}$$

$$\partial V/\partial t = -(bB)U - (a^2A^2c/d^2)V - (2aAc/d)UV - (bBd/aA)U^2 - cU^2V, \tag{9.24}$$

$$\underbrace{}_{\text{linear Turing equations}} \qquad \underbrace{}_{\text{nonlinear terms}}$$

$$\overbrace{\partial U/\partial t = \qquad k_1U + \qquad k_2V,}^{} \tag{9.25}$$

$$\partial V/\partial t = \qquad k_3U + \qquad k_4V. \tag{9.26}$$

The rate constants a, b, c, and d for individual steps of the reaction mechanism are necessarily positive; so are the reactant concentrations A and B. It is evident from the preceding equations that the four Turing constants k_1, k_2, k_3, and k_4 are not rate constants as that term is normally understood in elementary chemical kinetics. First, they are pseudo-constants containing the reactant concentrations A and B, and therefore are constant only in the conditions of homeostatic control of A and B. Second, the Turing constants are not necessarily all positive. In fact, k_3 and k_4 are always negative in this mechanism. Thus, for all possible values of a, b, c, d, A, and B the mechanism automatically assures the satisfaction of two important conditions for morphogenetic capability: k_2 and k_3 are of opposite signs, and k_4 (and hence k_4') is negative. It is not automatic that k_1 is positive. This depends on the comparative values of bB and d and requires a sufficiently high concentration of reactant B.

The reader, having perceived from the foregoing that the Turing equations, with parameters k_1–k_4, are at a long abstraction from the simple-looking chemical reaction scheme, may wonder if the mathematical analysis is unnecessarily cumbersome. It is not, if one happens to possess a computer. Without the translation of terminology represented by equations (9.23) to (9.26), and the consequent possibility of working back from the Turing conditions in Section 7.3 to the values of a, b, c, d, A, and B, one is working almost blind in selecting such values for computations.

This is largely because the computer and its program for solving the equations, plus the input (such as values of parameters) supplied to the program, together are not isomorphous with the living developmental system and are operating quite differently. Solving differential equations with a computer is a different matter altogether from direct simulation. In the latter, the computer does operate in a manner isomorphous with the living process. The model is not converted into equations, but is used directly. For instance, the computations on optical resolution in Chapter 5 (with paper and pencil, leading to Figure 5.6, and with a microcomputer, leading to Figure 5.7) are of that kind. I did, in fact, essentially work blind in setting up the scheme of relative rates used.

Even the conversion of a model directly into a simulation is not always enough to ensure that all necessary conditions are satisfied for the kind of behaviour desired. Only some of the conditions convert, as discussed in Section 7.3.2, into a simple assertion that certain features of the model exist. Others remain a matter of choosing appropriate numerical values for the parameters. For a two-morphogen model, I could not do this without the mathematical analysis.

Prigogine and his collaborators have presented extensive illustrations of the behaviour of the Brusselator, setting $a = b = c = d = 1$. This rather sweeping-looking assumption is in fact not restrictive at all. In effect, the values of the rate constants have been rolled into the parameters A and B. Variation of these two can be used to explore the whole of parameter space, and hence the full range of behaviour of the model. Table 9.1 shows the general expressions for the Turing parameters of the Brusselator, and the

Table 9.1. *Turing parameters for the Brusselator in terms*
of the rate constants and reactant concentrations in the mechanism

Turing parameter	General expression for the Brusselator model	Expressions for unit values of a, b, c, and d
k_1	$bB - d$	$B - 1$
k_2	$a^2 A^2 c/d^2$	A^2
k_3	$-bB$	$-B$
k_4	$-a^2 A^2 c/d^2$	$-A^2$
$k_{cr} = (-k_2 k_3)^{1/2}$	$(bBc)^{1/2}(aA/d)$	$B^{1/2}A$
$k_1' = k_1/k_{cr}$	$(bB/c)^{1/2}(d/aA)(1 - d/bB)$	$(B^{1/2}/A)(1 - 1/B)$
$k_4' = k_4/k_{cr}$	$-(c/bB)^{1/2}(aA/d)$	$-A/B^{1/2}$
$k_1' k_4'$	$-(1 - d/bB)$	$-(1 - 1/B)$

values with this simplification. The latter enable one to see rather clearly the roles of the reactant concentrations in determining whether or not the various Turing conditions discussed in Section 7.3 are fulfilled. For instance, the lower boundary to all regions with morphogenetic capacity in Figure 7.5 (line *b*) has a position which depends somewhat on the diffusivity ratio. But for reasonable values of that quantity, it indicates that values of k_4' below about -2.5 (i.e., numerically larger negative values) are quite unfavourable for morphogenesis. Nicolis and Prigogine (1977, sect. 7.11), following Hersch-kowitz-Kaufman (1975), have discussed how A and B supplies may work to keep the pattern-forming region confined to a part of the whole area of the organism potentially available. This illustration starts from the values $A = 14$, $B = 24$, for which $k_4' = -A/B^{1/2} = -2.857$, a very unpromising value for morphogenesis.

Let us now look at what happens when either A or B is held constant while the other of them is changed. Movement of k_4' upward into the morphogenetic region needs either a decrease of A or an increase of B. If one discovers a pair of substances which seem to be promising candidates for the style and title of morphogen, and one finds that morphogenesis is switched on by decreasing the concentration of one of them, it seems reasonable to label that substance an inhibitor; conversely, a substance which switches morphogenesis on when its concentration is raised may equally reasonably be termed an activator. But these could be the *AB* pair, not the *XY* pair.

9.1.3　Pattern localization: its control
by reactant concentrations

To illustrate pattern localization, Nicolis and Prigogine assume that B is constant throughout the system, but that A is distributed in a steady-state

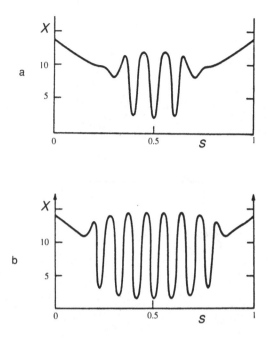

Figure 9.1. Localization of the pattern-forming region; results of computations from Herschkowitz-Kaufman 1975). The boundary conditions are specified at the ends ($s = 0$ and $s = 1$), but pattern-forming is confined to a central part of that region: in (a) about 50% of the overall length, and in (b) about 70%. The model for formation of complex short-wavelength pattern is the Brusselator, with $a = b = c = d = 1$, $\mathcal{D}_x = 1.052 \times 10^{-3}$, and $\mathcal{D}_y = 5\mathcal{D}_x$. The reactant A is also diffusible, with $\mathcal{D}_A = 0.1972$. It is maintained at $A = 14$ at both ends, but falls to a minimum at the centre of the system because of first-order decay (Figure 9.2; compare Section 6.3.1). The reactant B is uniform across the system: (a) $B = 24$; (b) $B = 30$. From Herschkowitz-Kaufman (1975), with permission.

gradient. This has a maximum ($A = 14$) at both ends and a minimum in the middle. Somewhere between the ends and the middle, A is small enough in relation to the constant value of B to move the system into the morphogenetic region. The results of their computations, for $B = 24$ and $B = 30$, are shown in Figure 9.1.

The account given by those authors of the threshold condition which explains this localization is complex and difficult to follow, and it appears to suggest that the condition needed to limit pattern in Figure 9.1b to 70% of the length of the system is different from that which limits pattern in Figure 9.1a to 50% of the length of the system. Both can be understood as manifestations of the same threshold condition in a much simpler way in relation to the Lacalli-Harrison k_1', k_4' diagram (Figure 7.5, redrawn for diffusivity ratio $n = 5$) of Turing's conditions (Figure 9.2).

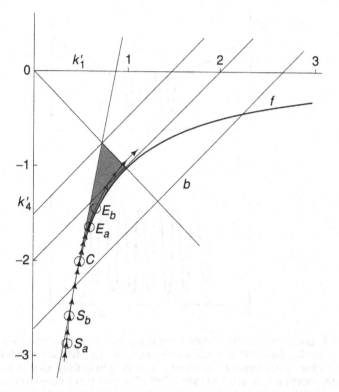

Figure 9.2. The Turing conditions in (k_1', k_4') space, as in Figure 7.5, but redrawn for diffusivity ratio $n = 5$. The important morphogenetic regions C and D are shaded. The line of arrows is the trajectory of the representative point for constant B and decreasing A. For $B = 24$, A falling from 14 to 8.2, the trajectory is from S_a to E_a; for $B = 30$ and the same A limits, S_b to E_b. The curves $B = 24$ and $B = 30$ are so close together that they can be drawn here as one curve.

From the expressions for k_1' and k_4' in the Brusselator model with unit values of the rate constants (Table 9.1), these constants are related by

$$k_1' = -(1 - 1/B)/k_4'. \qquad (9.27)$$

The values of B in the calculations illustrating localized pattern are much greater than unity (24 and 30), so that the constant $1 - 1/B$ is very close to unity (0.958 and 0.967). The curve (9.27) is a hyperbola which lies only a short distance above the hyperbola

$$k_1' = -1/k_4'. \qquad (9.28)$$

The latter is, of course, Turing's condition, equation (7.27f) and line f in Figure 7.5, which gives a two-morphogen model its band-pass character. A hyperbola just above this one passes right through the most important morphogenetic regions (C and D, Figure 7.5 and 9.2, shaded in the latter).

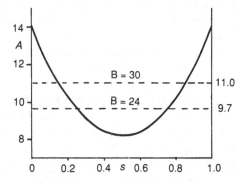

Figure 9.3. Steady-state distribution of A for a source maintaining $A = 14$ at both ends of the system, unit value for the first-order decay constant of A everywhere, and diffusivity of A as mentioned for Figure 9.1. The broken lines indicate that $A < 9.7$ for 50% of system length, and $A < 11.0$ for 70% of system length. These are the extents of pattern in Figure 9.1a,b and were found in Nicolis and Prigogine's computation for $B = 24$ and $B = 30$, respectively. Mathematically, this A distribution makes the Turing parameters k_2 and k_4 position-dependent (see Table 9.1). Analysis of the Turing equations is then a different matter from the analysis of the usual Turing equation with space-independent parameters. To be rigorous about this, Hunding (1989) refers to "Turing patterns of the second kind" when a gradient input is present. In computer experiments, linear analysis based on local values of the graded variable (A, in this figure) seems usually to have given the right indications of what behaviour will be found.

The model distribution of A is shown in Figure 9.3. From one end of the system to the middle, A drops from 14 to 8.2. This corresponds to the combined effects of diffusion and first-order decay:

$$\partial A/\partial t = -kA + \mathcal{D}_A\, \partial^2 A/\partial s^2 = 0 \text{ in steady state,} \qquad (9.29)$$

with $k = 1$.

If $B = 24$ and A falls from its value at $s = 0$ to its minimum at $s = 0.5$, the representative point in Figure 9.2 moves up the hyperbola from S_a to E_a and enters the downward-pointing "horn" of the morphogenetic region close to point C. If $B = 30$, the trajectory is from S_b to E_b. The point C was marked as follows: In Figure 9.3, if pattern is to occupy 50% of system length, the threshold of A must be 9.7 for $B = 24$. Likewise, at $B = 30$, if pattern is to occupy 70% of system length, $A = 11.0$. These two conditions turn out to be almost the same, that is, point C, in Figure 9.2. It is a good approximation to the point of entry to the morphogenetic region.

9.1.4 The joining of models in sequence

Control of a morphogenetic event by a precursor, or input, which has itself acquired nonuniform distribution in a previous morphogenetic event may have very broad strategic significance in biological development. My own research on tip growth in plants (see Section 10.1) provides experimental phenomena which make clear demands of this kind upon theoretical explanation. While a rounded growing tip works to elongate a cylindrical cell, the boundary of this morphogenetic region, where dome joins cylinder, moves in pace with the growth. This movement needs a model, linked to the one for growth. When, at such a tip, a more complex event occurs, such as the formation of a whorl, it is extremely difficult (however diligently one wields Ockham's razor) to model the event with anything so primitive as a single two-morphogen reaction-diffusion model. At least two of these in sequence seem to be needed.

Such joining together of processes in sequence is merely, on a grander scale, what the kineticist always does in writing a multistep mechanism such as the Brusselator. The concept of a chemical reaction being established, with reactants and products and rates of change of the one into the other, one next perceives that the product of one reaction can be reactant for another, and that this can have profound consequences in regard to overall rates of change.

My use of the word "sequence" in this section somewhat confuses the concept of a time sequence and that of a logical sequence of processes, such as steps in a reaction mechanism which are joined in a certain order but which are envisaged as all continuously and simultaneously in operation. Such confusion is usually legitimate and is not troublesome in discussing multistep processes. An effective time sequence is often implicit in a logical sequence. For instance, often the concentration of a reaction intermediate rises, passes through a maximum, and then falls. Thus, at different stages the formation or destruction processes for the intermediate appear to dominate, although all of them actually operate continuously.

The localization of pattern-forming action discussed in the preceding section is our first example of joining entire morphogenetic models in sequence, and it deserves strategic analysis. No less than four complex events are involved:

1. At the two ends of the system, $s = 0$ and 1, localized sources of reactant A have somehow become established. To postulate the presence of such sources with no explanation whatever of why they are where they are jumps right over an important morphogenetic event preceding the one of immediate interest.
2. The sources of A have achieved homeostasis in their operation, so that they deliver A always at the constant concentration $A = 14$ used in the computations. This illustrates that models for development and models for homeostasis must sometimes be envisaged as linked.
3. Reactant A undergoes a morphogenetic process which distributes it in a simple spatial pattern, as shown in Figure 9.3. The symmetry of this reflects the symmetry of placement of the two sources. But the quantitative aspect, the depth of the central minimum, is very important in defining

quantitatively the extent of the pattern-forming region for the next stage in the sequence. The power to do this is conferred by the first-order decay of A, and as stressed in Section 2.1, this is the first step from passive involvement of A as a messenger toward an active role as a morphogen. The reaction-diffusion rate equation (9.29) is linear, but predicts a steady-state distribution of A. This may appear at first glance to contradict my assertion (Section 9.1) that differential rate equations arising from realistic reaction mechanisms will always be nonlinear. This takes us back to stages 1 and 2. No reaction scheme has been written down for the operation of sources of A and for its homeostasis, or steady-state behaviour. If this were done in any realistic manner, the equations would be nonlinear.

4. Reactant A is the start of the Brusselator mechanism, which is the only part of this sequence specified in full mechanistic detail and shown to be capable of generating complex pattern.

In such sequences it is not at all necessary that the earlier stages be mechanistically simpler than the later, as might seem to be implied by stages 3 and 4 of the present example. In fact, stages 1 and 2 may readily need something at least as complicated as the Brusselator for anything like a complete explanation. In Chapter 10 I present my own work on whorl formation, in which a pair of two-morphogen reaction-diffusion models are used in places in the sequence corresponding roughly to stages 3 and 4 of the preceding strategy. The one occupying the position of stage 3 is the Brusselator. In the following stage, I have been able to get a long way by using only a linear Turing model. Thus the more complex model, which is also the one more completely specified chemically, appears earlier than the last stage. This will often be necessary, because one needs for the early stages some representation of outputs reaching a steady state. Otherwise, one has something hopelessly unstable to feed into the last stage as input.

The sequential or hierarchical ordering of the events to be explained is, of course, a consistent theme of embryology; see Figure 5.1 and, for much more information, Slack (1983). But models such as a monotonic gradient and a two-morphogen pattern-former can be linked not only in series but also in parallel; see Nagorcka and Mooney (1982, 1985).

Lacalli (1981) presented the same concept in a somewhat veiled way, because he did not explicitly discuss how the nonuniform distribution of precursor might arise. In a number of calculations on the increase of pattern complexity in a growing one-dimensional system, he used a modified Brusselator model (Tyson and Light, 1973) and, instead of using an ordinary boundary condition, specified a gradient in the parameter b across the system, with either a maximum or a minimum at the centre. He wrote that "in living organisms, boundaries do not always clearly separate the parts of a tissue or cell involved in pattern formation from those that are not. The use of parameter gradients may be the most appropriate means of dealing with such situa-

278   *Bringing experiment and theory together*

tions. Parameter values capable of supporting a pattern in one part of the system would gradually reduce to values incapable of generating pattern in surrounding parts of the system." He gave no model for this spatial variation of parameters except to remark, earlier in the same paper, that "rate constants may contain concealed concentration dependence . . . and could therefore vary across the system."

Lacalli in fact used unit reactant concentrations in his computations, and his *b* parameter is really *bB* in the present terminology. Spatial variation of this across the system could therefore be achieved by a mechanism for nonuniform distribution of the reactant *B*, parallel to the case of nonuniform distribution of *A* presented earlier from Nicolis and Prigogine, and to my whorl-formation model (Harrison et al., 1981; Harrison and Hillier, 1985). A feature of this not present in the examples described earlier is reduction in dimensionality of the morphogenetic field (Harrison 1982). Patterns on a two-dimensional surface, which a growing tip is, effectively, tend to be rather poorly controlled unless they are rather simple. Production of a simple pattern of precursor confines the following complex pattern formation (the whorl) to a ring, effectively a one-dimensional region, in which control of complicated pattern is much easier.

9.1.5   The "adaptable" character of the Brusselator

When a reaction-diffusion model has been put into the computer and values of *X* and *Y* have been set at their spatially uniform steady-state values X_0 and Y_0 (or *U* and *V* have everywhere been set to zero), and some initial random fluctuations have been put on top of this uniformity, then one knows what is going to happen as the dynamics are allowed to run for a short time. The model will initially behave as a linear Turing model, and the separation of maxima in the pattern can be calculated from the linear Turing wavelength. At longer times, when *U* and *V* have grown so that nonlinear terms become significant, there is no a priori reason why the model should still "know" its linear wavelength. But if it has started out with the "right" number of parts in the pattern (i.e., that given by linear analysis), one expects that the model will not easily change the number of parts later. Each part may change somewhat in shape, so that the pattern is no longer a simple sine wave, but the length of a pattern repeat will not change.

Suppose now that a computation has been run long enough to reach a steady state and that some change is then made which, according to linear analysis, should give a different pattern. Will the established pattern start to change, leading eventually to that different pattern, or will it not? No general analytical solution is known to this problem. One has to rely on the computer experiments for answers. In our experience in this laboratory, and also that of Lacalli (1981), the Brusselator has a quite astonishing capacity to find its way from one pattern to another, as linear analysis would predict, when the starting point is in the nonlinear regime. Figure 9.4 shows some plots of *X* morphogen patterns along a one-dimensional system. When the length is one linear wavelength, pattern develops from (a) random input to (b) a steady-state pattern of one central peak. If when that steady state has been reached the parameter values are changed to simulate a sudden stretch of the system to two wavelengths long, development of two peaks occurs by a dichotomous branching of the single peak. Part (c) shows an early stage in the change. But what happens if one similarly stretches the system from one wavelength to three wavelengths? At first, (d), the peak again branches dichotomously, as if to make two peaks. But this is followed, (e) and (f), by change at the boundaries. In this particular computation, a pattern of one central maximum has eventually changed to a pattern with three minima. The pattern unit has been made into three units by an inversion of it. (One could also say that it is still the same way up, but one peak has been split so that it is seen at both ends.) Lacalli (1981, Figure 5), using one of Tyson's variants of the Brusselator model, obtained in the same kind of computation a split of one peak into three without inversion, by repeated branching at the centre. One may look at the stages of either of these computations for a long time and wonder how the mechanism is managing to count to three. It is not at all obvious. The development looks uncannily as if the model is capable of planning ahead by abstract reasoning, as a human designer would.

In fact, what is going on illustrates an important duality in the character of reaction-diffusion mechanisms. In one sense, the whole pattern is an interacting entity. But here, parts of it are changing semi-independently to make a pattern repeat wherever there is a long enough space not occupied by one. In Figure 9.4d there is too long a flat region at each end, so something must happen. In Lacalli's computation, too long a flat region developed in the middle, which therefore had to produce a new pattern unit. That is how the mechanism counts parts. But they all continue to interact with each other, so that they place themselves correctly along the length of the system. This dual character of semiautonomous parts which yet belong to an interacting whole is very relevant to *Drosophila* segmentation (Section 10.2).

The kind of test shown in Figure 9.4 is quite severe, because it requires a readjustment of pattern to start from the extreme limit of nonlinear behaviour, the steady state. But it is by no means the most severe test which can be devised. Lacalli (1981) compared the abilities of the Brusselator (in one of Tyson's modifications) and the Gierer–Meinhardt model to produce a regular hexagonal array of spots in two dimensions. It is general experience from computer experiments (e.g., Murray, 1981a,b) that two dimensions provide much more scope than does one for a pattern to "go wrong."

The Brusselator did produce a regular array; the Gierer–Meinhardt model did not (Figure 9.5a,b). The obvious curvature everywhere, both on the peaks and in between them, in part (a), is in striking contrast to the flat, "long-range-inhibition" regions in (b). This correlates with the contrast discussed in

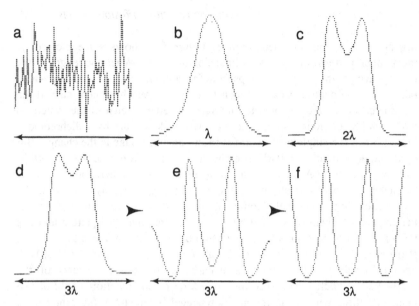

Figure 9.4. The Brusselator for a one-dimensional system. Pattern change from one steady state to another when the length of the system is suddenly changed. (*a*) Random input. (*b*) Steady-state pattern for a system one wavelength long. (*c*) Beginning of the change from (*b*) when the length of the system is doubled. (*d–f*) Stages in the change from (*b*) to a three-repeat pattern when the length is suddenly tripled.

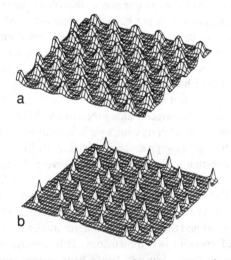

Figure 9.5. Lacalli's (1981) computations of progress toward a regular hexagonal array, from random input, in a two-dimensional system in which the growth mechanism is (*a*) Tyson's modification of the Brusselator or (*b*) the Gierer-Meinhardt (G-M) model. Vertical relief is concentration of the activator morphogen X or U (Brusselator) or A (G-M model). New random input was introduced continuously, and computation was continued until the array of peaks was unchanging on further iteration. From Lacalli (1981), with permission.

Chapter 3 between regulatory behaviour (Section 3.2), for which the Brusselator would be a good mechanism, and teratogenesis, for which the Gierer-Meinhardt model is superior (Section 3.3).

Reaction-diffusion mathematics is usually discussed in the continuum approximation: Chemical substances have concentrations which are continuous functions of position and time, and one may ascribe mechanistically important roles to the slopes and curvatures of these functions, as I did in the preceding paragraph. For computations, the programmer is obliged to approximate these continuous variables by a finite number of discrete points, but is usually careful to use a fairly large number of points for spatial position within each wavelength of the pattern.

Nature is not always so careful. If the points in a computation are supposed to represent cells or nuclei in the real system, there may be very few of them to a wavelength. Continuum diffusion may be replaced, for instance, by cell-to-cell exchange via gap junctions, with each cell being effectively a well-stirred compartment. Wolk et al. (1974) and Wolk and Quine (1975) pointed out, in relation to the spacing of heterocysts in the cyanophyte *Anabaena,* that the appropriate unit for diffusion distance in multicellular organisms is the cell. Even in a syncytium, with a diffusion continuum between nuclei, the localization of crucial steps in developmental chemistry at the nuclei may amount to coarse compartmenting of a pattern unit. The syncytial blastoderm of *Drosophila melanogaster* is a good instance. At the time when the segmentation pattern of the insect is first generated as a set of stripes of expression of some pair-rule genes, each stripe is only three or four rows of nuclei wide; see the cover illustration on the book by Gilbert (1988).

What happens if one does a one-dimensional computation on pattern formation by a reaction-diffusion mechanism, but uses only three points per wavelength? Given a set of 22 points (i.e., a region 7 wavelengths long), will the mechanism produce 7 pattern repeats, or will it, because of the coarse subdivision and lack of continuity of derivatives, totally lose track of the size of pattern unit it is "supposed" to generate? In this laboratory, we tried this computer experiment with the same Tyson modification of the Brusselator which Lacalli used in most of his computations (Harrison and Tan, 1988). The result (Figure 9.6) was that the model sometimes went slightly astray, giving a small defect somewhere in the pattern which led to a total of 6 or 8 units instead of 7 (Figure 9.6a, defect between 4 and 5 units on the distance scale), but in general showed a strong continued "knowledge" of its linear wavelength and the number of pattern units it should therefore produce, despite the coarse subdivision.

A second, and equally important, feature of this result is that the pattern unit, having only three points, could have only two or three concentration levels. This computation produced the pattern unit with two levels: two points at a low concentration of X, and one at a high concentration. This means that the Brusselator is a good mechanism for producing on–off switching be-

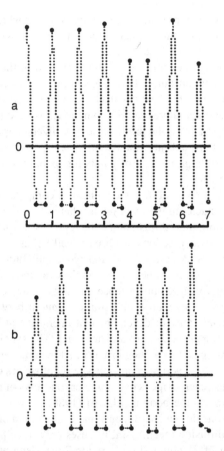

Figure 9.6. Reaction-diffusion pattern in a coarsely compartmented region. Results of a computation with Tyson's modification of the Brusselator, as used extensively in the work of Lacalli (1981), from Harrison and Tan (1988). Our experience in using both the original Brusselator and the Tyson modification is that, for most of the properties of interest for biological pattern formation, they are very similar. Parameter values were chosen to give 7 Turing wavelengths along the system, with 3 points per wavelength. The solid horizontal line marked 0 is the spatially uniform steady state. Plots are of concentrations of the X morphogen. (*a*) Initial input is random noise above and below the line 0. The computation was started with 12 points per wavelength, reduced to 3 in the later stages. (*b*) To remove pattern defects such as that between 4 and 5 wavelengths in (a), the initial input was a sine-wave pattern with the right wavelength. With only 3 points per wavelength, the computer starts with a sawtooth pattern of three concentrations. This turns out to be unstable. The computation changes it to an "on–off switching" pattern of two concentration levels. The pattern produced in these computations suggests the expression pattern of a segment-polarity gene, such as *engrailed*, in *Drosophila* (see Section 10.2.1) and is my only attempt so far to model segment polarity. For different but related ideas on applying reaction-diffusion to this problem, see Russell (1985).

haviour, which is generally much more discussed by biologists than is continuous concentration variation. The two are not necessarily different in chemical mechanism. Reaction-diffusion can do both.

Why is the Brusselator so good at behaving quasi-linearly in its nonlinear
regime? This question calls for algebraic analysis which has not yet been
achieved by mathematicians for this instance or for nonlinear dynamics in
general. Speculatively, I suggest that an important feature may be the close
match between the nonlinearities in rates of growth of U and V. They are in
fact equal and opposite, so that if we enquire about the overall rate of growth
of the sum of the two morphogen concentrations $(U + V)$, by adding equations (9.23) and (9.24), we have, with diffusion terms added,

$$\partial(U + V)/\partial t = -Ud + \mathcal{D}_u \partial^2 U/\partial s^2 + \mathcal{D}_v\, \partial^2 V/\partial s^2. \tag{9.30}$$

Consider a system of length l, from $s = 0$ to $s = l$, with no-flux boundaries at
both ends. The boundary condition means that the slopes of the pattern (first
derivatives of U and V with respect to s) must be zero at both ends. Let us
compare the excess of U above the spatially uniform state $(U = 0)$ in the
positive half-waves with the deficiency of U in the negative half-waves. For a
linear waveform (i.e., a cosine wave), excess and deficiency would cancel
each other out because the positive and negative halves are the same size and
shape. They are not the same shape for nonlinear Brusselator patterns, but
what about size, that is, integral of U along the system? At steady state:

$$\text{Integral } I = \int_0^l U\, ds = (\mathcal{D}_u/d) \int_0^l (\partial^2 U/\partial s^2)\, ds$$
$$+ (\mathcal{D}_v/d) \int_0^l (\partial^2 V/\partial s^2)\, ds = (\mathcal{D}_u/d)[(\partial U/\partial s)_l$$
$$- (\partial U/\partial s)_0] + (\mathcal{D}_v/d)[(\partial V/\partial s)_l - (\partial V/\partial s)_0]. \tag{9.31}$$

Because all the first derivatives are zero at both ends, the integral vanishes.
This means that however a Brusselator pattern develops, it must always keep
the area of the peaks equal to the area of the troughs. There can be no long-
range inhibition unless the inhibited state is very close to $U = V = 0$. This
seems rather unlikely. When the model is operating with pattern-forming
ability, it is not going to leave any region essentially unpatterned. A region of
flat concentration profile would have to be displaced some significant distance
from the unpatterned, spatially uniform steady state. Hence the Brusselator
does not produce long unpatterned regions and always behaves as if every part
of the pattern is looking to right and to left and interacting with what is there.

9.2 The hyperchirality model

9.2.1 *Big hands from little hands*

The word "chirality" means "handedness," and the concept is brought into
our minds by our right and left hands. However much we think of a molecule
as a real object, we remain accustomed to drawing a distinction between the

molecular and macroscopic scales, and we do not readily think of our hands as two enantiomers. But they are, if the shapes are smoothed by a reasonable spline fit (see the discussion of large-scale symmetry in Sections 5.1 and 5.2). Yet at the microscopic extreme, we find that both right and left hands are built of L-amino acids only, assembled into polypeptide chains and protein polymers up to the size of cytoskeletal elements still in one chirality only. Yet at some spatial scale between those polymers and the hand there is a symmetry-breaking which permits those polymers to make up the two enantiomeric hands. What is the spatial scale concerned, and what is the significance of this symmetry-breaking in relation to mechanism of pattern formation?

Whatever that spatial scale may be, its significance is that at that point something other than geometrical self-assembly must take over control of pattern, to override the propagation of a single chirality to larger spatial scales. In the unicellular ciliates, such as *Tetrahymena* and *Paramecium*, highly organized cell surface structures are made up of arrays of microtubules. The latter are all of one chirality. But there are many instances of formation of mirror-image doublet cells which have, effectively, left- and right-handed versions of the organized structures. Here, the spatial scale of the symmetry-breaking event can be rather precisely identified as being between 1 and 10 μm. Frankel (1989) has given a comprehensive account of these organisms from the viewpoint of pattern formation, and he maintains that there is evidence for "hidden global controls of pattern formation." In another review (Frankel 1990) he introduces the major problem with my phrase "big hands from little hands," first used in 1979 (Harrison, 1979). Another interesting aspect of this geometry has been discussed by Brown and Wolpert (1990), and it involves two different meanings of "handedness." We may use the word as a synonym of chirality, to mean a shape which does not superpose on its mirror image, and we may then consider bilateral symmetry as the presence of both handednesses on the two opposite sides (e.g., right and left hands). Otherwise, we may use "handedness" in the sense of saying that someone is left-handed (i.e., that there are limitations on the bilateral symmetry). If we look on a fine enough spatial scale, we may find that right and left halves are not mirror-symmetry equivalent because they are both made of molecules of one chirality. In some sense, the different relationship of these "little hands" to the two "big hands" may confer some kind of advantage on one of the big hands. This is likely to happen only if the chirality of the little hands is somehow translated to a larger scale, statically. Again the question of the spatial scale arises. Brown and Wolpert argue that molecular chirality may be translated into an effective orientation of a whole cell, which they represent by the asymmetric form of the letter F. If these cells are then organized in a bilateral symmetry, on one side of the organism the open side of each F will point medially; on the other side, they will all point laterally.

Similar thinking led me to my "hyperchirality" model for reaction-diffusion (Harrison and Lacalli, 1978). In this, some geometry involving protein tetramers and their orientations on a cell membrane is suggested. Once again, I see a danger that the molecularly minded scientist may give most attention to the details of this speculative chemistry. They are not the point. Similar dynamics might arise from many different models, in which the change from one chirality present to two chiralities present would occur at different spatial scales. What matters is the symmetry of the dynamics, in which four morphogens reduce effectively to two.

9.2.2 Dynamics of the model

A mechanism for spontaneous optical resolution is discussed in Section 6.3.3. The gist of it is the autocatalytic production of two substances A and B out of the same substrate S with exactly the same kinetics. The most obvious structural situation which would lead to such symmetry in the kinetics is that in which A and B are related as two optical enantiomers. But, as discussed in Section 8.2.3, any such structural aspect is a secondary consideration. The important thing is the kinetic symmetry between the two processes. The experimental biologist may readily arrive at either of two contradictory conclusions about the possible value of this kind of model. On the one hand, the biologist will say that on the molecular level, both enantiomers of any given structure are hardly ever present in a living system. On the other hand, the biologist will agree that, especially on the cellular level, establishment of two new states of differentiation often takes place on exactly the same time scale. I conduct the following analysis as if pairs of enantiomers are involved and use subscripts D and L for the two enantiomers of the same substance: X_D, X_L and Y_D, Y_L. These are supposed to be formed out of two precursors: A for the X enantiomers and B for the Y enantiomers.

The spatially uniform steady state corresponds to the racemic mixtures $X_D = X_L$ and $Y_D = Y_L$. The two measures of departure from uniformity, U and V, therefore correspond to optical asymmetries:

$$U = X_D - X_L \quad \text{and} \quad V = Y_D - Y_L. \tag{9.32}$$

If it is recognized that X_D and X_L may be symmetrically related in *kinetic* reality but not in *structural* reality, it will then be appreciated that the morphogen variable U can be a very far cry from a concentration of one chemical substance.

For these four morphogens, X_D, X_L, Y_D, and Y_L, in two symmetry-related pairs, the activator–inhibitor dynamics of a Turing model are achieved by a crossover in stereospecificity of catalysis in which X catalyzes formation of Y the "right way round" (D catalyzes D), but Y catalyzes formation of X the "wrong way round" (D catalyzes L). This makes the asymmetry of Y an effective inhibitor of the asymmetry of X. These two asymmetries are U and V, and the model leads to Turing equations with the signs as for an inhibitor model (k_2 negative, k_3 positive), not as for a depletion model such as the Brusselator.

The catalyses are all represented as being bimolecular and second-order in both the substrate A or B and the catalyst X_D, and so forth, and hence involving squared concentrations of all these. Thus, the autocatalysis of X_D appears in the equations as $k_{XX}A^2X_D^2$. As in the optical-resolution models of Sections 5.3.2 and 6.3.3, the catalytic species are supposed to occupy sets of sites limited in number. If we are thinking of processes at the plasma membrane of a cell (or assembly of cells), we shall call these sites receptors for X and Y. It is assumed that there is one set of sites for X, each site having the same binding characteristics for X_D or X_L, with the total concentration of these sites being P_X. A similar set of sites is postulated for Y, with total concentration P_Y. If each set is always saturated with X or Y, but with variability in the ratio X_D/X_L or Y_D/Y_L, then at all times,

$$X_D + X_L = P_X \quad \text{and} \quad Y_D + Y_L = P_Y. \tag{9.33}$$

For the rate equations to represent this condition, they must contain displacement terms showing the removal of X from the X sites at a total rate equal to the rate of formation of X, but at an X_D/X_L ratio corresponding to that on the sites at any moment. Terms with P_X or P_Y in them in the equations are these displacement terms. An interesting feature is that the model has no explicit effect of Y on itself. But when the algebra has been done to yield the Turing equations, there is a k_4 term, and k_4 is always negative, as required to put the constants in a morphogenetically promising region of parameter space. If one traces it through the algebra, one finds that this effective self-influence of Y arises from the displacement terms.

In summary, the catalytic interactions are

X_D catalyzes its own formation;
X_D catalyzes formation of Y_D, which catalyzes formation of X_L,

and the same two statements with D and L interchanged throughout.

With all the terms for catalysis, displacement, and diffusion together, the rate equations for formation of the X enantiomers in a one-dimensional region (distance coordinate s) are

$$\partial X_D/\partial t = k_{XX}A^2X_D^2 + k_{XY}A^2Y_L^2$$
$$- (X_D A^2/P_X)[k_{XX}(X_D^2 + X_L^2) + k_{XY}(Y_D^2 + Y_L^2)]$$
$$+ \mathcal{D}_X \partial^2 X_D/\partial s^2, \tag{9.34a}$$
$$\partial X_L/\partial t = k_{XX}A^2X_L^2 + k_{XY}A^2Y_D^2$$
$$- (X_L A^2/P_X)[k_{XX}(X_D^2 + X_L^2) + k_{XY}(Y_D^2 + Y_L^2)]$$
$$+ \mathcal{D}_X \partial^2 X_L/\partial s^2. \tag{9.34b}$$

On subtracting (9.34b) from (9.34a), with the substitutions

$$U = X_D - X_L, V = Y_D - Y_L, k_{1a} = k_{XX}A^2P_X/2,$$
$$k_{1b} = -k_{XY}A^2P_Y^2/2P_X, k_2 = -k_{XY}A^2P_Y,$$
$$k_5 = k_{XY}A^2/2P_X,$$
$$k_1 = k_{1a} + k_{1b}, \tag{9.35}$$

we obtain for the rate of growth of the hyperchiral asymmetry of morphogen X

$$\partial U/\partial t = \{k_{1a}[1 - (U^2/P_X^2)] + k_{1b}\} U + k_2V$$
$$- k_5UV^2 + \mathcal{D}_X \, \partial^2U/\partial s^2. \tag{9.36}$$

For U and V small enough to allow neglect of the nonlinear terms, this expression reduces to the Turing equation (7.1a), and k_2 is negative (i.e., V destroys U).

Proceeding in like manner for Y, we amend equations (9.34) by interchanging the letters X and Y throughout, replacing A by B, writing $k_{YY} = 0$, and, in the cross-catalysis term in k_{YX}, interchanging L and D, so that X_D catalyzes production of Y_D. Subtraction of the resulting equations, with substitutions as in equations (9.35) and with the additional substitutions

$$k_3 = k_{YX}B^2P_X, k_4 = -k_{YX}B^2P_X^2/2P_Y,$$
$$k_6 = k_{YX}B^2/2P_Y, \tag{9.37}$$

we obtain for the rate of growth of the hyperchiral asymmetry of morphogen Y

$$\partial V/\partial t = k_3U + k_4V - k_6U^2V + \mathcal{D}_Y \, \partial^2V/\partial s^2. \tag{9.38}$$

For small X and Y, this reduces to the Turing equation (7.1b), and k_4 is negative. If the receptor concentrations P_X and P_Y are similar in magnitude, then k_4 is approximately $-\frac{1}{2}k_3$, and if, additionally, the cross-catalysis constants k_2 and k_3 are similar in absolute magnitude, then the absolute value of either of them is roughly k_{cr} and $k_4' = k_4/k_{cr} \approx -\frac{1}{2}$. This places k_4' within the limits of the most generally useful morphogenetic region of parameter space (see Figure 7.5).

The hyperchirality model arose from my initial interest in both optical resolution and morphogenesis, as represented particularly by semicell development in the desmid *Micrasterias*. It now appears to me less likely that this is the appropriate reaction-diffusion model for the desmids. We now have a much better model for the desmids which uses the Brusselator with an additional feedback loop into its reactant A (Harrison and Kolář, 1988). But the hyperchirality model, in two-dimensional computations (Harrison and Lyons, in press; Lyons and Harrison, 1991), turns out to have great power to produce striped rather than spotted patterns (Figure 9.7). This is the property discussed in Section 8.2.3, and there indicated as belonging to mechanisms which have only cubic nonlinearities (or, more generally, only nonlinearities which are odd functions of the simultaneous sign changes of U and V). Inspection of equation (9.36) shows nonlinearities only in the forms U^3 and UV^2, and only U^2V in equation (9.38).

a

b

c

d

9.2.3 A structural model, and a wider dynamic significance

A well-known feature of proteins is their quaternary structure, which is quite often a dimeric or tetrameric association of identical subunits. When such structure is present, the polymerization is commonly a necessary condition to give the protein its function (e.g., as an enzyme). This makes such proteins very good candidates to be recognized as morphogens, because this structure–function relationship should produce nonlinear dynamics. My attention was first drawn to this kind of geometry by looking at diagrams of the quaternary structure of the soybean lectin concanavalin A (Figure 9.8a). Here, each subunit is to be seen as a left-hand mitten, the four being joined at β pleated sheets which are their palms.

My structural model of hyperchirality used protein tetramers on a cell surface and the concept of "flatland" chirality (Gardner, 1964, 1982). The latter is illustrated by the shape of four joined rectangles in Figure 9.8c. If this shape is an object so strongly attached to a flat surface that it is confined to two spatial dimensions and cannot be taken out into a third dimension and turned over, then it is chiral with respect to reflection in a mirror line in that surface. By a distortion of the concanavalin A quaternary structure, I devised a geometry which could be attached to a surface in two ways to make the two flatland enantiomers (Figure 9.8b). The hyperchiral property belongs to the general outline of the tetramer, a spline-fitting which does not see the geometry of the primary, secondary, and tertiary structure (i.e., a "big hands" view). Figure 9.8d shows such flatland chiral structures acting as stereospecific templates for the autocatalytic assembly of more tetramers. The structural relationships are those which would lead to the kinetics of equations (9.34).

I present this model with some trepidation as to whether it will assist or hinder the advance of kinetic theory. The danger is that the molecularly

Figure 9.7. *(opposite)* A computation, from Lyons and Harrison (1991), of the ability of the hyperchirality model to form stripes. The array is square and has no gradients on it to force the formation of stripes. Periodic boundary conditions were used both between top and bottom and between left and right edges. The final pattern, which looks like four zigzag stripes (looking at light or dark only), is thus actually a single stripe. Light colour represents high values of $U = X_D - X_L$; but this is of no significance. The model is chiral, and the significances of D and L can be arbitrarily changed. (a) Random input, $t = 0$. (b) $t = 10,000$. (c) $t = 40,000$. (d) $t = 100,000$. Time is in arbitrary units, representing number of iterations of the computation. If, instead of the hyperchirality model, a Brusselator had been used, a regular spotted pattern like that of Figure 9.5a would have been obtained. We have recently done a series of computations (Lyons and Harrison, not yet published) with increasing amounts of a quadratic term added to otherwise hyperchiral equations. As the coefficient of the quadratic term is increased, the steady-state pattern after a long computation changes, going from striped, through splotchy patterns of spots partly joined into stripes, and finally to spots when the quadratic term is large. From Lyons and Harrison (1991), with permission of Elsevier Science Publishers.

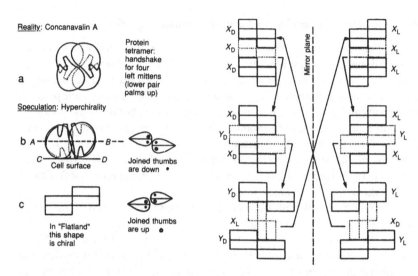

Figure 9.8. Left-hand side: A speculative model of a tetrameric protein structure attachable to one side of a cell membrane in two different ways which are chirally related in two-dimensional flatland: (*a*) the actual arrangement of four subunits into a tetramer in concanavalin A; (*b*) a hypothetical distortion of this geometry; (*c*) an idealization of this structure as seen in plan view, looking down on the cell surface, as an arrangement of four rectangles with the same flatland chirality. Right-hand side: Autocatalytic assembly of subunits into tetramers, for two proteins X and Y in a flatland-stereospecific way. The longest arrows indicate the crossover in which Y_L catalyzes formation of X_D, and vice versa, so that the asymmetry $(Y_D - Y_L)$ inhibits growth of the asymmetry $(X_D - X_L)$. Hyperchiral kinetics are not necessarily related to anything remotely like this structural model; but for recent indications of possible relevance of cell surface structures to chiral symmetries, see Brown and Wolpert (1990) and Frankel (1990). From Harrison (1979), with permission.

oriented biologist will find the specific structural proposal marginally possible, but very unlikely, and will dismiss the dynamic discussion from consideration because it seems to be tied to an improbable structural hypothesis. But the dynamic model is not so tied. The essential symmetry is in values of rate constants, not necessarily in geometrical structure at all. Biologists are fond of postulating that cells might differentiate in two quite different ways *on the same time scale*. That is hyperchiral kinetics.

In the higher vertebrates, each side of the visual cortex of the brain is connected to both eyes. (The optic chiasm is differently arranged from that in lower vertebrates, in which right eye connects to left brain, and vice versa.) In the Primates, the information from both eyes is found to be received in a set of alternating parallel ocular-dominance stripes, and in these instances the mix of neurons from the two eyes connecting to one side of the cortex is close to equality, 50/50 on a percentage basis. By contrast, in the cat the mix is 70/30, and the pattern of ocular-dominance regions is spotted. Also, for the macaque monkey, which is the Primate on which most experiments have been done, monocular deprivation by blindfolding one eye for a few days after birth leads to disruption of the striped pattern.

These observations correlate with the kinetic concept of hyperchirality. Somewhere in the formation of connections of the optic nerves to the cortex via the lateral geniculate nuclei, neurons from the two eyes have acted as distinguishable from each other in the left-versus-right sense, but otherwise in kinetically equivalent ways. This is hyperchirality, and it is good for making stripes when the left/right mix is 50/50. But it involves dynamics on the grand scale of the whole metabolism of cells and the electrochemistry involved in making stable synapses. Structurally, we are a long way from my putative protein tetramers on a cell surface when we think of ocular dominance; kinetically, we may be in the same place (Harrison and Lyons, in press).

This example illustrates particularly well the basic philosophy of the kinetic-theory approach to pattern formation. One may reject totally the structural model of Figure 9.8 as being unrealistic, unknown, and irrelevant to living things; but one still needs to know how dynamics with the equivalent of a left/right asymmetry are expressed in equations lacking quadratic terms and generate striped patterns.

9.3 Brief comments on other models

The brevity of this section is not intended to convey any information about relative importance of models. I have dealt with the Brusselator at length, first, because the originators of the model have neglected to reach out to biologists in their presentations of its properties, second, because it is one of the best. to illustrate some important general principles simply, and, third, because I have used it extensively and regard it as promising for many biological applications. The hyperchirality model is my own, and in the dozen or so years since I devised it I have done less than most modellers in advertising the merits of my own model; and it has a special merit, namely, its strong tendency to form stripes.

The following accounts can be much briefer because all of the originators have done a thorough job of relating their models to biological problems, and their own work can be consulted. The following sections are not much more than a list of models, with a few comments to indicate where these models may fall into place in the classification of kinetic theory. The times are not ready for an exhaustive taxonomic survey of dynamic models.

9.3.1 The Gierer–Meinhardt model

The first paper of Gierer and Meinhardt (1972) discussed a number of models, and the book by Meinhardt (1982) presented even more. I think it legitimate, however, to refer to the model in the singular because these authors are particularly responsible for one pair of dynamic equations, and many of the other models they have discussed are essentially variants of these. The special character of the Gierer–Meinhardt model is given to it by the way in which the inhibitor effect is put into the rate of growth of activator. This has been discussed in Section 3.3, and the question of whether or not anteroposterior gradients in insect embryos are Gierer-Meinhardt patterns is further alluded to in Section 10.2.

The dynamics which may be regarded as *the* Gierer-Meinhardt model are

$$\partial A/\partial t = \rho_0\rho + c\rho(A^2/H) - \mu A + \mathcal{D}_A\, \partial^2 A/\partial s^2, \qquad (9.39a)$$

$$\partial H/\partial t = c'\, \rho'A^2 - \nu H + \mathcal{D}_H\, \partial^2 H/\partial s^2. \qquad (9.39b)$$

Here, A and H are morphogen concentrations corresponding to Turing X and Y. They may be taken as the initials of the German words for activator and inhibitor, but it has also been noted that the first names of the originators of the model are Alfred and Hans. ρ_0, c, μ, c', and ν are constants. ρ and ρ' are functions of position s representing "source gradients" which are taken as preformed and unchanging in time.

The reader of this book may be interested in two aspects of any specific reaction-diffusion model which are contrasted and sometimes even in conflict: First, does the model give a good fit to some particular range of developmental phenomena? Second, from a pedagogic viewpoint, is the model an easy one to approach and to appreciate within the general framework of Turing dynamics, so that the research worker who is trying to use reaction-diffusion can make progress?

As to the first question, I have indicated elsewhere in this book (especially in Section 3.3) that I see the Gierer–Meinhardt model as being exceptionally well suited to the explanation of some phenomena, but inferior to other models such as the Brusselator or hyperchirality model for some others. The things with which it does particularly well are as follows:

1. Amplification of a shallow gradient and representation of nonregulatory events following grafting of pieces which retain their fragments of the gradient, as in *Hydra* (Gierer and Meinhardt, 1972) (see Section 3.3.2).
2. Formation of a small region of morphogenetic activity within a much larger region, the greater part of which does not manifest such activity ("short-range activation, long-range inhibition").
3. Teratogenesis of the kind in which peaks fail to maintain the proper spacing between them which linear Turing analysis would indicate (Section 3.3.1).

The Gierer–Meinhardt model seems to me, however, to be definitely inferior to the Brusselator in giving an account of formation of patterns of repeated parts with quantitative control of either the number of parts or the spacing between parts (i.e., morphogenesis that performs well the job of either measuring or counting). And both of these models are inferior to the hyperchirality model in respect of the tendency to form stripes rather than spots in a two-dimensional region. A comparison which the reader might find instructive is the use of these diverse two-morphogen models to tackle the problem of *Drosophila* segmentation (Section 10.2). Meinhardt, on the one hand, and Lacalli and I, on the other, are enthusiastic about the possibilities of reaction-diffusion within the complex hierarchy of events in *Drosophila*; but the places we identify as most likely to involve two-morphogen activity are different.

As to the pedagogic question, two things are somewhat unfortunate: First, Gierer and Meinhardt themselves, though often briefly acknowledging Turing's priority, do not to my mind stress nearly enough the value of seeking to understand two-morphogen models always within the Turing framework. This involves linearization of the equations about the spatially uniform state and use of parameter spaces. In my laboratory, when people are working on reaction-diffusion computations, one tends to trip over parameter-space diagrams like Figure 9.2 all over the place. Second, the Gierer–Meinhardt model happens to give rather cumbersome expressions when one does the linearization. The formulae for Turing constants k_1–k_4 are not easy to work with, whereas those for the Brusselator are (Table 9.1). Of course, if the Gierer–Meinhardt model is looking like the right explanation for some phenomenon, one must live with that complexity. The point is that the Gierer–Meinhardt model is not a good start for a beginner.

The transformation of Gierer–Meinhardt parameters to Turing ones is

$$k_1 = (\mu/M)(c\nu - \rho_0 c'\rho'), \tag{9.40a}$$

$$k_2 = -c\nu^2\mu^2/M^2\rho, \tag{9.40b}$$

$$k_3 = 2M\rho/\mu, \tag{9.40c}$$

$$k_4 = -\nu, \tag{9.40d}$$

where

$$M = c\nu + \rho_0 c'\rho'. \tag{9.40e}$$

The reader who desires some algebraic exercise might try the following: Derive the expressions for k_1' and k_4'. Consider a system in which ρ and ρ' are gradients monotonic in the spatial variable s. They may both decrease in the same direction, or they may be opposed. For both cases, as one moves along the system, how does the representative point for the dynamics move in (k_1', k_4') parameter space? Could the crossing of boundaries in that space on moving up or down a gradient be related to the "short-range activation, long-range inhibition" property of the model?

The term A^2/H in equation (9.39a) is chemically unusual. Mechanisms containing inhibition more often give expressions such as $A^2/(H + K)$. The distinction is important when a new peak is attributed to inhibitor loss, as in the double-abdomen deformity (Section 3.3). I know of only one mechanistic scheme which, to a certain approximation that is exact for the steady states, reduces to the Gierer–Meinhardt rate equations: that of Babloyantz and Hiernaux (1975). We have started some computations with this model to study its dynamics far from the steady states.

A basic concept of the Gierer–Meinhardt model, as its originators have used it, is that it features changes on two different time scales: Chemical reactions between molecules are envisaged as occurring rapidly, whereas changes in cellular states are seen as occurring so much more slowly that they can be taken as static on the chemical time scale. That is, A and H can change quickly, but the source gradients ρ and ρ' are fixed. This corresponds to the wealth of very convincing experimental data, both for *Hydra* and for insects, that gradients are persistent even in grafted pieces. But the head-activator and head-inhibitor identified in *Hydra*, and therefore looking like promising candidates for the title of Turing morphogen, are activator and inhibitor for cellular differentiation as neurons. These cells are more concentrated in the head region and could therefore constitute the source gradients. It seems to me that any attempt to bring these substances into the Gierer–Meinhardt dynamics must involve a substantial modification of the model involving feedback into the source gradients. The dynamic preconception, and the kind of mathematics needed to follow it through, would remain unchanged; and the starting point could be the Gierer–Meinhardt equations. There is opportunity for much work by many scientists in reaction-diffusion. This paragraph defines a field big enough for quite a few Ph.D. projects.

9.3.2   Murray's model

Murray, like Gierer and Meinhardt, has worked with more than one model. In a recent paper (Shaw and Murray, 1990), his group has become the first, within my knowledge, to combine reaction-diffusion and mechanochemistry in parallel as putative parts of the same patterning process. This is the formation of patterns on vertebrate skin (hair follicles, scales, etc.) in which they have assumed mechanochemistry in the dermis interacting with reaction-diffusion in the epidermis.

Murray's name is, however, probably most specifically associated with a model for striped and spotted mammalian skin markings which has led to the biomathematical theorem that "it is not possible to have a striped animal with a spotted tail; the converse is quite common" (Murray, 1981b). The special feature in this work is a set of reaction-diffusion equations in which the kind of pattern obtained is rather sensitive to the shape of the region in which it forms. I copy these from equations (4) of Murray (1981b), with some changes in notation to simplify the parameters. [Bard (1981) used different reaction-diffusion equations for the same phenomena.]

$$\partial A/\partial t = \alpha(A_0 - A) - \rho SA/(1 + S + KS^2) + \mathcal{D}_A \nabla^2 A, \qquad (9.41a)$$

$$\partial S/\partial t = \gamma(S_0 - S) - \rho SA/(1 + S + KS^2) + \mathcal{D}_S \nabla^2 S. \qquad (9.41b)$$

These equations represent a specific model in which two substrates S and A of a membrane-bound enzyme diffuse to the membrane at rates represented by the $(A_0 - A)$ and $(S_0 - S)$ terms. The denominator $(1 + S + KS^2)$ combines a Michaelis–Menten binding with an inhibition by S when it is large. That latter is the KS^2 term. I write the Laplacian operator in general form, whereas I wrote it in one dimension for the Gierer–Meinhardt equations, simply because all Murray's investigation of these dynamics involves two-dimensional regions, whereas Gierer and Meinhardt have done extensive work in one dimension.

Murray's equations have two features which have not appeared previously in this account: first, the absence of an explicit autocatalytic term and, second, the presence of a saturation term in the concentrations, that is, the $S/(1 + S)$ term, which is the Michaelis–Menten form until the KS^2 term becomes significant. Murray has studied the patterns which these dynamics will produce on surfaces of a variety of shapes approximating the skins of animals on their trunks, legs, and tails. He has found interesting correspondences with some of the things that actually happen in mammalian coat patterns.

As for every nonlinear model, the main burden of discovering its properties is currently carried by the computer. Algebraic analysis lags far behind. Murray has particularly stressed that one-dimensional patterns computed to steady state often are only moderate distortions of those expected from linear analysis and seen in the early stages of a computation, but that this is definitely not so for two-dimensional patterns. These often change quite substantially between linear and nonlinear regimes.

The diversity of patterns which this model can produce and its sensitivity to the shape of the region seem to me to indicate a striping tendency greater than that of the Brusselator but less than that of the hyperchirality model. Meinhardt (personal communication to T. C. Lacalli) has suggested that a saturation term helps in stripe production. Lacalli (personal communication) has suggested that saturation will flatten sharp peaks and hence will make a model which otherwise would behave in the "short-range activation, long-range inhibition" manner instead produce rather comparable flattish peaks and troughs. The patterns thus produced would to some extent mimic those made by kinetically chiral models. Perhaps the behaviour of Murray's model is an illustration of this.

9.3.3   The Sachs–Mitchison model, and an acknowledgment to Rashevsky

Turing was not the first to invent a model for kinetic production of periodic pattern. That honour belongs to Rashevsky (1940), an American pioneer of theoretical biology. Turing is justly celebrated for the elegant simplicity and probable great importance of the concept of activator–inhibitor mutual interaction leading to pattern. Rashevsky's idea was different. He proposed that a membrane permeability might be self-enhancing. A few years earlier, indole-3-yl acetic acid (IAA) became the first growth-promoting substance in plants with a known chemical identity. [For a good, concise review of the first half-century of discovery in the field of substances which control plant growth, see Wain (1977).] This has become the type substance of the class of substances known as auxins, and the singular noun "auxin" is often used for IAA itself.

In the 1940s and 1950s it became clear that auxin was involved in the formation of veins in plants. Sachs (1969) suggested that the efficiency of transport of auxin increases with its flux. Mitchison (1980, 1981) formulated this concept as a mathematical model. His equations show auxin flux J from one cell to the next as the sum of two terms, one representing passive transport of neutral IAA molecules across the plasmalemmas, and the other representing active pumping of IAA anions:

$$J = q(C_b - C_a) + pC_b,   (9.42)$$

where p and q are constants, and the concentrations C_b and C_a are those of auxin at the basal end of one cell and the apical end of the next. This is equation (1) of Mitchison (1981), with notation to the physico-chemical conventions of J for flux and C for concentration. The essence of the pattern-forming model is that the polar (active-transport) part of the flux is envisaged as autocatalytic in J, not C:

$$\partial p / \partial t = \epsilon(kJ^2 - p),   (9.43)$$

where ϵ and k are constants. The theme of nonlinearity, shown in the mathematically simplest way by a squared term, which has been present in most of the models discussed in this book, is here again in the J^2 term. By elaboration of this kind of dynamics, Mitchison has been able to model vein formation, including a strange phenomenon sometimes observed in which a closed loop of vein is formed, with circular transport around it. This peculiarity is somewhat distantly related to the main vein-forming activity, both in nature and in the model, but the ability of the model to produce it is evidence in favour of the model.

A strong distinction should be borne in mind between equation (9.43) and all previous postulations of autocatalysis mentioned in this book. Because the *flux* is autocatalytic, the model can lead to fast-transport paths (veins) in which the *concentration* of auxin is lower than in the surroundings. This leads to two rather obvious questions: First, what is the mechanism whereby either each channel becomes more efficient or the number of channels between two cells increases as a consequence of the rate of pumping? This matter has not been addressed in any account I have read of autocatalytic flux. When I try to think of mechanisms, I find myself nearly always putting in something in which the presence of auxin is producing some kind of activation. This is inappropriate when the active veins are going to be regions of low concentration. Is it possible that the essential step is autocatalytic assembly of the quaternary structure of auxin pumps on the plasmalemmas, so that the flux dependence is a symptom of the pattern formation event, not part of its etiology?

The second question, which seems to arise readily whenever an auto-catalytic flux is mentioned, is the possibility of a geophysical analogue in the formation of the drainage patterns of rivers and their tributaries. Such analogies are of doubtful value. In the geophysical system, neither the magnitude of the total flux nor the local directions of parts of it are under such strong feedback control as that of a set of self-organizing pumps. Perhaps a better analogue might be human settlement, with the associated networks of roads and railways being the veins. By contrast to river systems, in which there are no uphill flows, the movement of people and goods to where they are already more concentrated is perhaps too much in the latter analogy.

10

Approaching agreement?

Là où je cherchais de grandes lois, on m'appelait fouilleur de détails. [Where I was looking for great laws, they called me a digger for details.]

—French-Canadian saying quoted by A. J. Libchaber
at 13th International Liquid Crystal Conference, 1990,
session on Pattern Formation

How precise a fit should one expect between experiment and theory at various stages in the scientific enterprise? Where experiment rests upon clearly established "great laws" (i.e., preconceptions or paradigms), one knows what to measure, and one may seek ever-higher precision, as in modern spectroscopy. When one is dealing with applied science, whether it be the clinical testing of a drug or the construction of an engine, precision is again of the essence. (I have heard it suggested that, conceptually, the Elizabethans could have devised a steam engine, but they had no lathes to give a good enough fit of piston to cylinder to make it work.)

It is quite a different matter when the "great laws" are not firmly established. How shall one improve the precision of measurements when one is not yet sure which properties may be the significant ones to measure precisely, and the worker in the next laboratory clearly has a different opinion on this? Major concepts have arisen from partially correct data. For instance, Dalton's atomic theory arose from the law of constant composition. The controversy between Dalton and Berthollet, with the latter maintaining that the composition of a compound by weight was variable, is very well known. In retrospect, we know that both were correct for different ranges of compounds and that the berthollide nature of many metal oxides can be accounted for by variable mixes of, say, Fe^{2+} and Fe^{3+} in the crystal. But would we ever have got around to using those symbols without Dalton's insistence on constant composition? In the same vein, the best-known instance of wasted effort in science is the extreme concentration, late in the nineteenth century, on determining atomic weights to ever-increasing accuracy, before it was known that the existence of isotopes makes those atomic weights variable well outside the limits of accuracy which were being achieved.

The foregoing is not intended as a polemic against precision and in favour

298

of sloppiness, either in experimental work per se or in the fit which one regards as acceptable between experiment and theory. I am an experimental scientist who worries about details in the laboratory (see Section 10.1). The matter is one of perspective. For instance, the quotation from Crombie which I have used as epigraph to Part III is immediately followed in his book by this sentence: "A good example of this is the atomic theory, first seen as scientific material of this kind in the 17th century and eventually reduced to exact empirical form by John Dalton in 1808." But in what sense was the atomic theory of 1808 a "theory exactly fitting the data," as specified in Crombie's previous sentence? The atomic weight scale was in serious confusion up to the 1860s; indeed, I have even seen a paper from the 1890s in which the formula of water was given as HO. And I have mentioned earlier the dispute between Boltzmann and the positivists in the 1890s, in which many scientists perceived the atomic theory as a "tottering edifice."

In the sections of this chapter, I try to assess the present extent of two kinds of fit between biological experiment and kinetic theory of pattern formation. The first is between the predicted and observed dynamics of pattern formation, sometimes with a little knowledge of what chemical substances are patterned but without knowledge of what substances do the patterning (i.e., are morphogens). The second is the bridge between macroscopic dynamics and biochemistry, which does require morphogen identity. In the following sections I make no attempt at a comprehensive account of all the experimental evidence. The primary objective of this book is to conduct an examination of the differing philosophical attitudes of two groups of scientists who should be getting together and are not. What I present in this last chapter is my own view of what features of the evidence are important in this enterprise, and it is to be read as commentary.

The examples chosen for discussion illustrate the generalization that I am quite happy, in the first instance, to consider kinetic mechanisms as promising possibilities both for unicellular and for multicellular phenomena of pattern formation. Unicellular instances are important to consider as a counterbalance to the common tendency to identify pattern in terms of gene switchings in multicellular assemblies. For other accounts of unicellular pattern formation, see Frankel (1989) on the ciliates and Harold (1990) for a perspective covering diverse microorganisms.

10.1 *Acetabularia* and some desmids

I start with the only organism on which I have been doing extensive experimental work in my own laboratory: the marine siphonous green alga *Acetabularia acetabulum* (or *A. mediterranea*, a name in disfavour among the taxonomists, but to my mind better because it gives significantly more information than the repetitive *A. acetabulum*). Single cells grow into cylindrical structures of enormous size (up to 4 cm long and 0.4 mm in diameter), mainly by action at a dome-shaped growing tip at the opposite end of the cell from its single nucleus. From time to time, complex structures of whorl form are produced at the growing tip. Among these, the best-known are the reproductive structures ("caps"), formed once only in the life cycle after several months of vegetative growth. During those months, however, a whorl of hairs forms at the growing tip every few days.

10.1.1 Choice of organism and of developmental event to study

Acetabularia has been advocated as a very suitable organism for the study of some aspects of cell biology (Puiseux-Dao, 1970; Berger et al., 1987), but has not attracted the experimental attention of a large number of biologists. There are several good reasons for this, in regard to the usual kinds of experimental programs in biology. Because the organism is uninucleate throughout its vegetative growth, pattern formation in the whorl-producing events cannot be linked to patterns of differentiation and gene-switching at the nuclear level. Thus, many may reasonably presume an irrelevance of pattern-forming mechanisms in *Acetabularia* to those in higher plants and animals. There is a dearth of mutants, and the long life cycle hinders the search for them (Green, 1976), so that *Acetabularia* seems the very antithesis of *Drosophila* (or, in the plant kingdom, *Arabidopsis*) for a program of work intended to throw light on the relation of development to genetics. There is no strict control of the number of parts in a whorl pattern. Vegetative whorls can have from 3 to 35 hairs. The extremes are uncommon, but a normal healthy population provides numerous examples having from 7 to 21 hairs (Harrison et al., 1984, Figure 3a).

Why, then, did I choose this organism for developmental studies, when most biologists have failed to respond to admonitions to do so? First, my preconception that the same kind of dynamics may be expressed at a variety of levels of organization and by diverse chemical reactants leads me to believe that study of pattern formation in a unicellular plant may readily throw light on what is going on in multicellular animals. The unicellular plant may indeed be a minimal system, showing how much an organism can manage without (e.g., differentiation, patterned gene-switchings) and yet still have all that is needed for generation of complex pattern.

Second, as a classical physical chemist I like straight lines; that is, I like to change one variable continuously and see the response as a continuous quantitative change in some other variable. Unlike the biologist, I tend to avoid the study of discontinuous variations (i.e., on–off switchings). From this viewpoint, a pattern of a variable number of parts within wide limits is attractive. After a few months of preliminary study of *Acetabularia* in 1978, I decided that the vegetative whorls were definitely more interesting to me than the cap, because of the greater variability in the number of parts in the former.

Third, it is a commonplace that biologists would like to study phenomena mainly in vivo if possible; but to a chemical kineticist, it is an absolute imperative that change must be studied as it occurs. At the growing tip of *Acetabularia,* one can observe quantitative changes in dimensions hour by hour in the formation of one whorl, and one can show directly that the number of hairs in the next whorl formed a day or so later is usually not the same.

The first result that led me to continue studying *Acetabularia* whorls as a long-range project was that under fixed environmental conditions the spacing between adjacent hairs in a whorl is constant; that is, one whorl makes twice as many hairs as another because its diameter and perimeter are twice as large. Diameter is sloppily controlled. Spacing of hairs is precisely controlled.

In discussing these data with biologists who have not heard of them before, I find that a misunderstanding very commonly arises. Biologists often think at first that I am measuring the longitudinal distance from one whorl to the next along the main axis of the cell. That variable, together with longitudinal growth rates, does not interest me. (Being persuaded by biologists to try some longitudinal growth measurements, I did so and found them hopelessly idiosyncratic from one cell to another, in growth conditions in which the hair spacing gave well-behaved results which I find significant.)

The object, then, is not to study anything to do with a set of events in a time sequence, such as formation of successive whorls, but to concentrate on the properties of a pattern which arises as one entity, all its parts appearing simultaneously. Every whorl-forming event in *Acetabularia,* cap or vegetative whorl initiation, is impressively simultaneous. All the initials appear at once.

What should one study as independent variables, what should one control precisely, and what can one afford to control rather more sloppily? My cultures are maintained in growth chambers just as biologists usually do it. Temperature is controlled to ± 0.5K, and light is between fixed limits of intensity, on and off on a 12 : 12 cycle. But my experiments have been done as a physical chemist of the 1930s and 1940s would do experiments on chemical systems: in water baths controlled to ± 0.02K, with lights above to give more than a minimum intensity on a 12 : 12 cycle, but open to the laboratory so that large amounts of stray light on indeterminate schedules are in no way excluded. This very precise control of temperature and sloppy control of lighting are choices opposite to what biologists would be most likely to do. My procedure is not intended to deny that it has been clearly shown (Schmid, Idziak, and Tunnermann, 1987) that whorl formation can be initiated by a flash of blue light in *Acetabularia* otherwise grown in red light. The implication of my procedure is that quantitative control of hair spacing is quite a different matter from whorl initiation. In my work, whorls can be initiated at any hour of the day or night, and the time of initiation does not affect the hair spacing. But temperature does, and so does the calcium concentration in the culture medium. These, again, are the kinds of variables the classical physical chemist would think of first.

10.1.2   Predictions, and assessment of the significance of results

Two philosophical matters relating to the dynamics of scientific progress are mentioned as general problems here and are discussed in relation to data on *Acetabularia* whorls. The first of these is also quite relevant to the present state of work on *Drosophila* segmentation, which is discussed in the next section. The question is, On what scale of time, number of people and projects involved, and generality or detail of experimental data does one expect the progression from experiment to theoretical prediction to further experiment to operate? The second question is, When a theory concerns complex phenomena, so that its predictions are not simple but need an extensive literature for their description, is it straightforward for all scientists to assess properly whether or not some particular data support the theory? Meinhardt (1984) wrote that "without knowledge of the underlying principles on which pattern formation is based, the chance would be high that a successful experiment would be incorrectly interpreted as a failure."

To my mind, the most fruitful predictions at the present stage are quite unlikely to be those which seek to guide an experimental program from step to step in close detail for one particular organism. Rather, the useful predictions will probably be those of a general nature which must eventually be verified in a variety of organisms if the theory is to be recognized as having any general validity. For instance, it may be predicted from reaction-diffusion theory that the chemical wavelength, or distance between adjacent repeated parts of a periodic pattern, should be controlled by the concentrations of immediate precursors to the morphogens in the chemical mechanism (e.g., the A and B of the Brusselator, where X and Y are the morphogens). This control should be of such a nature that spacing becomes smaller as the precursor concentration increases. It may be a simple inverse variation, or an inverse dependence on the square root or one-quarter power of the precursor concentration; for example, for the Brusselator an approximate form of equation (7.22) for large diffusivity ratio n leads to

$$\text{chemical wavelength } \lambda = 2\pi(\mathcal{D}_x\mathcal{D}_y/k_{cr}^2)^{1/4}, \qquad (10.1)$$

and if we use in this the value of k_{cr}^2 for a Brusselator with unit values of rate constants a, b, c, and d, as given in Table 9.1, then

$$\lambda = 2\pi(\mathcal{D}_x\mathcal{D}_y/A^2B)^{1/4}. \qquad (10.2)$$

Thus the spacing between adjacent repeated parts should be proportional inversely to $A^{1/2}$ and $B^{1/2}$. The same approximation applied to the hyperchirality model (Harrison and Lacalli, 1978, equation 47) indicates, for the two precursors A and B of that model, inverse proportionality both to $A^{1/2}$ and $B^{1/2}$.

I am not aware of any theory of pattern formation other than reaction-diffusion which predicts this kind of concentration dependence. Therefore, if such a dependence is found in any pattern-forming process, I regard it as quite strong evidence for a reaction-diffusion mechanism. The substance producing the concentration-dependent effect is not a morphogen, but an immediate precursor of a morphogen in the biochemical mechanism.

The preceding two paragraphs contain, to my mind, a definite and unequivocal prediction which is a proper part of the scientific dynamic at its present stage in developmental biology. But to many experimental biologists it may appear quite otherwise, in fact not a prediction at all. This is because there are three things totally unspecified in the prediction; the organism, the developmental process within the organism, and the specific chemical substances involved in that pattern-forming process. What the prediction says is very general: that among all the processes being studied in developmental biology, some (and probably a large number, eventually) will be found in which pattern spacings decrease as the concentration of some chemical substance is increased; and the variation will have a simple functional form, related to the reciprocal of a power of the concentration (probably the first power or a fractional power). This is a testable prediction within my understanding of how the scientific method is supposed to work. But it doesn't tell anybody precisely what experiment to do at the bench tomorrow, which seems to be the stringent requirement of many experimental biologists for something to be classed as a testable prediction.

For instance, a major project in relation to hair spacing in *Acetabularia* whorls in my laboratory has been to study the quantitative effect on spacing of the calcium concentration in the artificial-seawater culture medium (Harrison and Hillier, 1985; Harrison et al., 1988). Figure 10.1 shows the variation found. It is of the kind predicted earlier. But it was not with this expectation that I started the study. The motivation for looking at calcium concentration was that I believed the morphogenesis to have its primary control in the rate of plasma membrane extension. This involves the rate of fusion of membrane vesicles with the membrane, and calcium is a likely controller of that process. In the event, we found a variation in accordance with my general prediction and, to my mind, pointing toward an extracellular role for calcium, rather than to intracellular control of membrane fusion. Thus the month-to-month dynamic of what we were doing in the laboratory was driven by something quite other than the prediction we ended up verifying, though that prediction was my own.

The reader may, however, object that my prediction is for lines showing not just a linear variation of spacing with some inverse power of a concentration, but direct variation (i.e., lines that go through the origin of coordinates). Those in Figure 10.1 do not. I return to this point in the next section.

Meanwhile, one should always recognize that everything one does within the scientific method, whether experimental or theoretical, is part of a continuity and therefore both answers a question and asks another. Evidence such

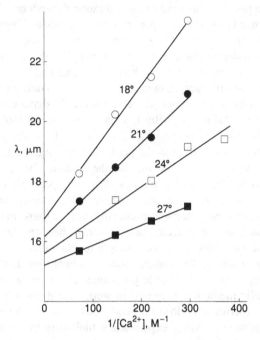

Figure 10.1. Dependence of spacing between hair initials in whorls of *Acetabularia mediterranea* upon $1/[Ca^{2+}]$ in the culture medium (artificial seawater, Shepard's medium), showing linearity of each plot and systematic variation of its slope and intercept with temperature. These latter were interpreted in such a way as to give binding constants and thermodynamic properties for binding of extracellular calcium to a cell surface receptor which then becomes a morphogen precursor. A problem for the reader: The Brusselator expression (10.2) indicates dependence of spacing on $A^{-1/2}$ and $B^{-1/4}$, not a simple dependence on reciprocal of concentration. What needs to be changed in the Brusselator mechanism to make one or other of these a simple inverse dependence? Is the required change reasonable in relation to the possible nature of morphogens? See Section 10.4 for my answer.

as Figure 10.1 is consistent with a prediction of reaction-diffusion theory, but does not definitely prove that the mechanism of pattern formation in *Acetabularia* whorls is of that type. It does, however, present a challenge to all future theories. They should address the question of how this concentration dependence arises.

More directly, the data point to a particular role for calcium in the morphogenetic mechanism, and ask a question: Can one find a spatial distribution of calcium corresponding to that role? And first, what is that spatial distribution? Here, Meinhardt's remark that ". . . a successful experiment would be incorrectly interpreted as a failure" is very relevant. In my laboratory, we looked for calcium distributions during the whorl initiation event by using chlorotetracycline as a fluorescence chelate. We found a sharply defined annulus of membrane-bound calcium at the location of whorl initiation, and it was uniform, not breaking up into the whorl pattern until the initials were already well established morphologically (Harrison et al., 1988, especially Figure 3C–E). This is just as expected on the basis of reaction-diffusion theory. For instance, in the Brusselator mechanism, patterning of a whorl would be envisaged as arising first in the morphogens X and Y. But according to equation (10.2) the spacing between adjacent hairs (λ) should be governed by concentrations of precursors A and B and should be uniform only where those precursor concentrations are high and uniform. If we had found that calcium was never distributed uniformly around the whorl-forming region, but formed a pattern of peaks where the hairs were going to be right from the start, we would have been faced with a real problem. I know of no theory in which that kind of distribution is compatible with a λ-controlling role.

An anonymous referee of my next research-grant application made a complimentary remark about this work, but added in passing that it is a pity that the calcium was not found in the whorl pattern at an earlier stage. This referee had fallen right into the trap of which Meinhardt warned. If one does not look carefully at the mathematics of the theory, it might seem that one is looking for something distributed in the whorl prepattern, and it is then a pity if one does not see it before morphological expression. But the mathematics requires one to distinguish X and Y from A and B and to think of the different things that these two pairs of substances are doing. A whorl pattern at an early stage for something which otherwise has an important attribute of A or B (i.e., spacing control) would be evidence that one has got onto the wrong theory and needs to think again.

10.1.3 Morphogens and mechanisms

Physical chemists who were brought up on gas-phase kinetics expect very simple reactions to have very complex mechanisms, with multiple roles for very few elements. The gas-phase combination of hydrogen and oxygen to form water has more than a dozen steps, among which two steps show why the reaction occurs explosively over certain ranges of concentration. All the steps involve only two elements. Such a chemist, faced with the problem that calcium has something to do with *Acetabularia* morphogenesis, will probably not ask, (a) Is calcium a morphogen? (b) What is the role of calcium in whorl morphogenesis? Rather, the question will be, (c) Which of the multiple roles of calcium is involved in the particular aspect of morphogenesis we are considering at the moment?

For the formation of *Acetabularia* whorls, two groups have been following parallel lines of work, both using kinetic theory but in different forms, and both involving calcium. My approach has already been outlined up to the evidence of Figure 10.1. Those lines, however, do not go through the origin of coordinates. This suggested to me that the morphogen precursor is not free calcium ion but a membrane-bound form of it involving a calcium receptor on the outside of the plasma membrane. If the spacing λ is inversely proportional to the concentration of bound calcium, then the plots of Figure 10.1 become plots of (1/bound) versus (1/free). These are analogous to the familiar Lineweaver–Burk plots of Michaelis–Menten enzyme kinetics. There, the rate of reaction is proportional to the concentration of bound substrate, so that (1/rate) plotted as ordinate is proportional to (1/bound). Of course, the concept which I discuss here of spacing being inversely proportional to a concentration arises from the idea that spacing, however static it may look, is a manifestation of reaction rates.

On this basis, I was able to extract quantitative values of the binding constant from (intercept/slope) of the plots, of order 10^3 M^{-1}. Such values are common for the binding of calcium to a pair of carboxylate ligands and a few neutral donors (Williams, 1976, 1977), including the weak calcium-binding sites of some proteins (e.g., the calcium-binding site of concanavalin A, where the anionic ligands are on the side-chains of Asp 10 and Asp 19). The binding constant increased with temperature, indicating endothermic entropy-driven binding ($\Delta H^0_{298} = 88$ kJ mol^{-1}, $\Delta S^0_{298} = 356$ J mol^{-1} K^{-1}). Endothermic binding is in fact quite common for calcium (e.g., with phthalate, malonate, or lactate as ligand).

This line of physicochemical reasoning points toward an unidentified protein on the plasma membrane as the essential morphogen precursor, with a weak calcium binding needed to activate it. Thus the calcium is a somewhat incidental player in the morphogenesis. The possibility that a protein has the leading role makes it easy to see, in general terms, that the genome can exercise control of species-specific differences in cell surface architecture. This has astonishing diversity in the Dasycladales.

The parallel line of work is that of B. C. Goodwin with his group and collaborators. They have found that whorl morphogenesis can be switched off by decreasing the calcium level in the culture medium below about one-fourth of the normal level, or by using a calcium ionophore (Goodwin and Pateromichelakis, 1979; Goodwin, Skelton, and Kirk-Bell, 1983). These data point clearly to an involvement of calcium in morphogenesis, and the effect of the ionophore suggests that intracellular calcium is involved. These workers have gone on to develop a mechanochemical theory of whorl morphogenesis using the cytoskeleton and intracellular calcium (Goodwin and Trainor, 1985; Brière and Goodwin, 1988). This type of theory is of course within the general category of kinetic theory which I am advocating in this book. Also, it is quite concordant with current biological thinking that extracellular receptors on the cell surface and the intracellular cytoskeleton should be linked together in the performance of a single biological event. My concept of calcium triggering a membrane-bound protein into becoming a precursor of a Turing morphogen is, however, the only theory so far published which accounts for the physicochemical data on control of hair spacing by calcium concentration in the medium.

No other workers seem to have studied calcium distribution at closely spaced intervals during whorl initiation. But at a later stage, when the hairs have grown extensively, Reiss and Herth (1979) showed by chlorotetracycline fluorescence that calcium was concentrated just below the point of emergence of each hair from the main stalk of the cell. Cotton and Vanden Driessche (1987) followed that up by showing that at the same developmental stage, calmodulin was concentrated in a similar region. This points toward involvement of intracellular calcium at some stage in hair development.

Vanden Driessche (1990) has reviewed these various observations on calcium and morphogenesis in *Acetabularia*, particularly from the viewpoint of characterizing intracellular calcium as a second messenger. She mentions at some length my concept of membrane-bound morphogenetic calcium, but from my evidence mentions only two qualitative points which I regard as quite minor. The usual philosophical gap, or paradigm difference, between physical scientists and biologists is once again apparent. I regard the quantitative evidence in Figure 10.1 and the thermodynamic data derived therefrom as being the burning issue. These are what point so strongly to an extracellular role of calcium. Although the reference is given, the significance of these data is not mentioned in Vanden Driessche's account.

What may be the nature of the morphogen precursors which are activated by this calcium binding? It was suggested to me by an anonymous referee that autophosphorylating protein kinases offer a promising possibility. Lisman (1985) devised a putative bistable switching system in which the nonlinear (bimolecular) step was intermolecular phosphorylation of this kind. Goodwin and Pateromichelakis (1979) found a strong gradient in phosphate incorporation along the stalk of *Acetabularia*. They mentioned that protein kinases are known to be present in *Acetabularia* and to increase apically during cap formation (Pai et al., 1975). Vanden Driessche (1990) refers to this work and to known relationships (e.g., in maize) between protein phosphorylation and auxins. Perhaps the Turing morphogen pair in whorl morphogenesis is a protein kinase and a small-molecule plant hormone, whether among the auxins or in some other category. A connection between extracellular calcium binding and intracellular bimolecular auto-phosphorylation, via integral membrane proteins, may eventually serve to put these diverse observations and theories together in a consistent picture in which very little will have been discarded of all the current ideas.

10.1.4 Three feedback loops

As discussed earlier, my scheme for hair-whorl morphogenesis has two stages. In the first, an annular pattern is formed. This defines the region in which whorl pattern is set up essentially one-dimensionally, that dimension being distance around the perimeter of a circle. The concept here is much more general than the idea of using reaction-diffusion in each of these patterning processes. I have pointed out (Harrison, 1982) that wavelike patterns are particularly badly controlled on a hemispherical surface and that reduction of dimensionality in a preliminary stage might be expected to be a common feature of developmental hierarchies. Is there any more direct evidence for the existence of two stages?

First, there is a little phylogenetic evidence. In the Palaeozoic *Rhabdoporella*, believed to be the ancestor of all the modern Dasycladales, the arrangement of hairs was random. Organization into whorls arose in the Mesozoic (Fritsch, 1956; Herak, Kochansky-Devidé, and Gušić, 1977). This suggests that for these two stages, ontogeny reverses phylogeny. The first stage was absent in the earliest dasyclads. (Presumably they did not have the trick of from time to time boosting an input, S in Figure 10.2, so that it changed the A pattern from apical maximum to annular.)

Second, the observation of Goodwin and Pateromichelakis (1979) that ionophore A23187 switches off whorl formation but allows tip growth to continue seems to indicate that these two processes have different mechanisms.

Third, there is a contrast between main whorl formation and subsequent branching of each hair. The former needs a switch-on event (Harrison et al., 1984; Schmid et al., 1987). But once the main whorl has formed, the subsequent branchings always occur, on a fairly regular time schedule. This is compatible with the second stage, once activated, remaining active in the hairs.

Most of the foregoing considerations (except for the parenthetic remark about S and A) can be taken in the spirit of general systems theory, and could be correct if the mechanisms do not involve reaction-diffusion at all. In Harrison et al. (1988), a version of Figure 10.2 was shown with stages I and II thus numbered, but without the equations suggesting two reaction-diffusion processes in series, the X of I being the A of II, and therefore designated as A in Figure 10.2.

Tip growth is a remarkably difficult phenomenon to explain even in the absence of branching at the tip. For it to happen at all as a means of elongating a cylinder, the tip must be self-limiting in extent; that is, the morphogenetic activity which sustains the dome-shaped tip and elongates the cell must be able to pull its own boundaries up after it. If the boundaries should cease to move, cell surface extension would "blow a bubble," with the fixed boundary as the end of the bubble pipe. Such morphogenesis happens in other dasyclads at particular stages (e.g., *Cymopolia,* while producing hair whorls, makes reproductive structures by forming large spherical gametangia on the ends of certain hairs).

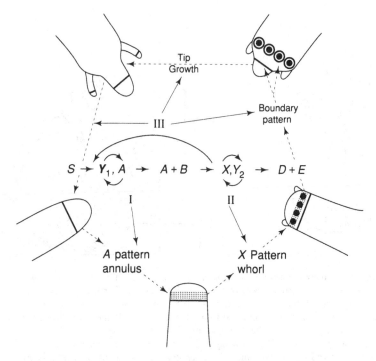

Figure 10.2. Three feedback loops for whorl morphogenesis in *Acetabularia*. Loops I and II are two Brusselators in sequence, the X morphogen of the former being designated A because it becomes the A input of the latter. Loop I has the necessary patterning power to concentrate A into an annular region where loop II can produce the whorl pattern. Both patterns were observed for membrane-bound calcium during whorl initiation (Harrison et al., 1988). Loop III is the means by which the morphogenetic tip draws its own boundary around itself. It is necessary to give tip growth its essential character as a cylinder-elongator. In branching tip growth, it must be active to form a new boundary round each hair, so that each of these may grow as a cylinder. Loop III gives scope for new ideas, both as to its biological mechanism and as to its mathematical formulation. In the modelling of sequential coupling of loops I and II done in my laboratory by G. D. Zeiss [Harrison et al. (1981); computation of whorl morphogen distribution shown again in Harrison and Hillier (1985), Harrison et al. (1988), and Nagorcka (1989)], loop I was modelled with a Brusselator, but loop II was modelled with a linear Turing model with concentration-dependent parameters as one might expect them to be for a Brusselator, hyperchirality model, or others. In modelling a sequence, one must put in some specific nonlinearities in the earlier stages because one needs to take them to the steady state. Only the last stage can be modelled more simply.

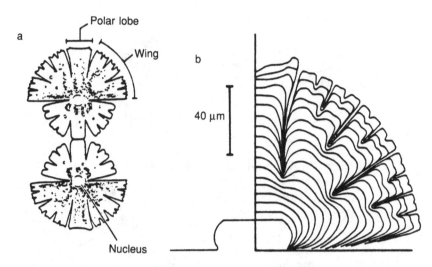

Figure 10.3. Morphogenesis in *Micrasterias rotata*, a desmid alga common in fresh-water ponds in widespread locations. (*a*) About 3 hours after mitotic division, each daughter cell is about 60% of the way through growing a new half-cell. (*b*) Growth profiles for the semicell margin at 10-min intervals, 20°C. In the third dimension, this cell is quite thin (without the indentations, its shape is somewhat like that of an ordinary biconvex lens). The growing region may be approximated as an edge (i.e., the profile drawn in these diagrams). More precisely, it is a set of growing tips. Each of these does not remain circular like *Acetabularia* tips. They become somewhat flattened into the plane of the diagram. From Lacalli and Harrison (1987), with permission.

10.1.5   *Branching tip growth in some desmids*

Another group of algae, the placoderm desmids, display both tip growth, with branching, and "bubble-blowing" at various stages of semicell morphogenesis. Figures 10.3 and 10.4a–e show some typical cell shapes. When mitosis occurs, the daughter cells separate at the isthmus (i.e., the narrow junction between the two halves, horizontal in the diagrams). Each daughter cell is closed at the isthmus by a flat septum of cell membrane and wall. This develops into the shape of a new half-cell by growing out into a "bubble" which in some species stays fairly simple in shape (Figure 10.4a–c) and in others becomes much more complex by repeated dichotomous branching (Figures 10.4d,e and 10.3). Each lobe of these complex shapes advances by tip growth, with progressive isolation of an increasing number of mor-phogenetic regions from each other.

Morphogenesis in *Micrasterias* was studied extensively by Kiermayer (1970) and later by Lacalli (1973, 1975a,b, 1976). The latter work gave me my introduction into the whole field of morphogenesis. Reaction-diffusion modelling has been based, as in my work on *Acetabularia*, on the concept that the prepatterns are formed at the cell surface, and that one of the patterned

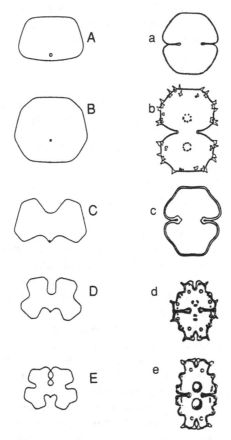

Figure 10.4. Right-hand column, (a)–(e): Line drawings of the shapes of some fully developed desmids. From West and West (1905). (a) *Cosmarium nitidulum;* (b) *Xanthidium armatum* var. *irregularius;* (c) *Cosmarium hammeri* var. *homalodermum;* (d) *Euastrum divaricatum;* (e) *Euastrum rostratum.* Left-hand column, (A)–(E): Some computed morphologies in two dimensions. From Harrison and Kolář (1988). The growth region is a closed loop, initial state circular (not shown). A Brusselator patterns X and Y around the loop. Every small region of the perimeter grows at a rate proportional to X. The algorithm for converting length increase unequivocally to shape change implies cell-wall mechanics which tend to preserve local directions. These calculations illustrate the need to use "feedback loop III" in the model (Figures 10.2 and 10.5) to obtain repeated dichotomous branching. (A) and (B) are from a computation with no such loop. Beyond stage (B), the shape soon again becomes circular, as it was initially, but now much larger. Stages (C)–(E) are from a computation in which persistently low X ultimately switches off the morphogenesis. The Brusselator then produces repeated branching. In (E) the branches are slightly overlapping in the middle. The suggested comparison of the computations with the several desmid shapes is to be understood in terms of the shape of the upper half-cell of the real desmid shape.

morphogens governs the rate of extension of the surface. This assumption is strongly supported in *Micrasterias* by the fact that cytoskeletal inhibitors do not prevent morphogenesis. Kiermayer was the first to suggest that the rate of membrane extension by fusion of vesicles from the Golgi was of primary importance in shape development. Lacalli had this in mind in choosing the organism for extensive study. Without at first being aware of the earlier suggestions, I came around to the same idea upon reading Staehelin and Giddings (1982), who, in this same alga, identified rosette-shaped structures carried by Golgi vesicles to the plasmalemma as the active cellulose polymerases. Each rosette produces a microfibril of fixed length. Therefore, it seemed to me, cell surface extension must go in lock-step with the addition of polymerase rosettes to the plasmalemma. Control of the rate of that process, at the membrane itself, should be the essential shape-forming event.

In the first application of reaction-diffusion theory to this example (Harrison and Lacalli, 1978), we used the hyperchirality model, which I had just spent a lot of time devising. I no longer think that this is the most appropriate form of reaction-diffusion for *Micrasterias*. The strength of hyperchirality lies elsewhere, in the stripe-forming ability associated with its special symmetry properties (Sections 9.2 and 10.2). Lacalli (1981) gave an extensive discussion of reaction-diffusion modelling for desmids and other unicellular algae, mainly in terms of a modification by Tyson of the Brusselator model. Harrison and Kolář (1988) used the standard Brusselator in a model which is the closest so far published to being complete in the sense of containing all components needed for repeatedly branching tip growth.

Complete modelling is very demanding because the cell shape is formed by a sequence of events on a continuously changing shape (Lacalli and Harrison, 1987). In *Acetabularia*, one can demonstrate that a chemical model will give a whorl prepattern on a fixed dome shape. Feedback loops I and II (Figure 10.2) have to be linked together, but one can proceed a long way without having to compute the actual morphological growth of the whorl or to devise a mechanism for feedback loop III, the delineator of new boundaries for morphogenetic regions. In *Micrasterias*, there is no avoiding growth and loop III in the model. For the latter, reactant *A* of the Brusselator turns out to be very useful in the modelling. In other instances, I have used *A* to model developmental hierarchies: *Acetabularia* (Section 10.1.4), hierarchy from loop I to loop II; *Drosophila* (Section 10.2.1), hierarchy from maternal-effect through pair-rule gene activation (Lyons et al., 1990). For the desmids, I have used the same control of morphogen *X* by input *A*, but have closed a loop by feedback from *X* to *A* (Figure 10.5).

This feedback is described in the legend to Figure 10.5. It involves a postulated aging effect in the cell surface, in which old surface fails to supply reactant *A*. A continuous decrease in *A* with increasing age α was put into our

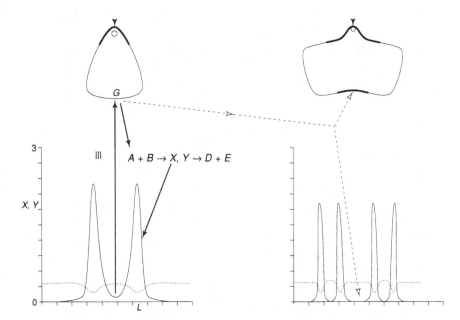

Figure 10.5. My first, and rather primitive, model for feedback loop III. The upper diagrams are loop morphologies, as in Figure 10.4 (*A–E*). The lower diagrams are plots of Brusselator *X* and *Y* against distance *L* round the loop, from the top as $L = 0$ (upper diagrams, black arrowheads). If the loop is not growing, its material (cell wall and membrane) is aging in real time. If it is growing, the added material is taken as zero age at time of addition. An average age α for a small region of the loop can be found by an integration over time of the amounts of material added at various times. This will be less than real-time age. Where *X* is high and growth rapid, the cell surface is kept continually rejuvenated. Where *X* is low and growth is slow, the surface ages more rapidly. Without detailed discussion of mechanism, it is assumed that supply of reactant *A* depends on the age and is completely cut off at a threshold value α_{th}. *X* goes to zero irreversibly in those regions, shown by heavy lines in the upper diagrams. At point *G*, this is about to happen. From then on, the shape development does not take the path back to a circle shown in Figure 10.4 (*A* and *B*). Instead, the morphogen peaks are nicely separated, and each splits into a pair. The right-hand stage in this diagram arises in place of Figure 10.4 (*A*), and further repeats of the same feedback through α take the shape into the sequence of Figure 10.4 (*C–E*).

computations in the simplest possible way, the expression for the rate *aA* of step 1 of the Brusselator mechanism being

$$aA = a_0(\alpha_{th} - \alpha)/\alpha_{th}. \tag{10.3}$$

Here, α_{th} is the threshold age at which supply of *A* totally ceases. Equation (10.3) is used for $\alpha \le \alpha_{th}$. For all greater ages, *aA* is taken as zero.

An age effect is likely to be quite plausible to botanists, but the properties I assume for it are unconventional. In higher plants, especially at the onset of rapid elongation of cells, it is evident that mechanical factors are involved in growth. Botanists are likely to expect an age effect to be formulated in terms of mechanical hardening. This is not what I have done. *Micrasterias* grows by formation of a thin, flexible, temporary wall (primary cell wall) which has randomly oriented microfibrils. The thick, rigid, secondary wall with arrays of parallel microfibrils is laid down only after semicell morphogenesis is complete and the cell has already reached its final shape. This is a very different kind of growth from higher-plant cell elongation. I am inclined to expect aging effects in *Micrasterias* to involve changes in chemical components of the membrane more than mechanical properties of the wall. For instance, primary-wall formation involves the accumulation in the membrane of polymerase rosettes which, for unknown reasons, have become irreversibly inactive after formation of a definite amount of cellulose. Also, when vesicles are conveying this enzyme to the plasmalemma, what else may they be conveying in addition? Whatever the precursor A to morphogen X is, it may readily be conveyed in this way. If X catalyzes cell surface extension by catalyzing vesicle fusion, then there is a feedback from X into A.

This account is intended to attract the attention both of physical scientists and of biologists. For the theoretician, the feedback loops of Figures 10.2 and 10.5 afford much scope for mathematical analysis. My rudimentary model for the aging effect [equation (10.3)] and hence for loop III is just a pointer to what could be a large and almost untouched field of modelling. There is also the matter that loops I, II, and III together form a loop on a grand scale. Another feature of the present model which is quite rudimentary is the kind of mechanical interaction implied by the algorithm which we used to translate surface extension unequivocally into a uniquely determined shape. I believe that this represents reasonably well the mechanical properties of a fairly flexible wall with a membrane inside it, but there is scope for a lot of mathematical analysis here. Finally, simultaneous prepatterning and shape change must be commoner in plants than in animals, because plants continue to develop throughout their lives. Very little has been done either of analysis or of study by computation of what could be a large field here.

10.2 *Drosophila* segmentation

Segmentation of insects in general and *Drosophila melanogaster* in particular has been treated by many of the persistent practitioners of reaction-diffusion theory: Meinhardt (1977, 1986, 1988), Nagorcka (1988), Kauffman (1977, 1981), Goodwin and Kauffman (1990), Harrison and Tan (1988), Lacalli et al. (1988), Lacalli (1990), Lyons et al. (1990), Hunding, Kauffman, and Goodwin (1990). Also, without any reference to *Drosophila*, Herschkowitz-Kaufman (1975) used the Brusselator model to demonstrate that production of pattern could be confined to a central part of a morphogenetic region (Section 9.1.3). The pattern of X and Y generated by the Brusselator mechanism had seven repeats occupying somewhat over half the overall length of the region between the two *A* sources. This is remarkably similar to the initial appearance of the *Drosophila* segmentation pattern.

To my mind, reaction-diffusion is a very promising type of mechanism to account for insect segmentation. It can represent the hierarchical ordering of the several classes of segmentation genes in a manner formally similar to the two-stage model discussed for *Acetabularia* in Section 10.1, albeit in a different geometry. It can address the question of how pattern formation in two dimensions is controlled to form stripes rather than spots (Sections 8.2.3 and 9.2). Its nonlinear dynamics correlate well with increasingly frequent references to bimolecularity in the actions of regulatory proteins. None of this is intended to signify that reaction-diffusion has yet been proved definitively to be the mechanism for any particular stage in the segmentation sequence. But neither has any other mechanism, and reaction-diffusion is one of the most exciting possibilities, which ought to be engaging the attention of substantial numbers of experimentalists and theoreticians interacting with each other.

In the past few years there has been, among drosophilologists, a brief period of incipient interest in kinetic theory (Akam, 1987), followed by renewed doubt (Akam, 1989). From others among them, I have met attitudes ranging from indifference through disappointment to overt hostility. The following two sections explore two aspects of this gulf between attitudes.

10.2.1 Communication versus "no crosstalk"

A living organism is, above all, a display of large-scale organization. At some level of developmental processes, this requires long-range interaction. Physical scientists tend to start from the assumption that these long-range influences are built into most levels of the processes which generate complex pattern. Developmental biologists tend to suppose that in a hierarchy of events leading to complex pattern, only the top level of the hierarchy involves long-range interaction. All the others can be understood on a local basis. Two essentially equivalent terms denote this kind of hierarchical concept: "gradient reading" and "positional information." In my classification, such models use type I morphogens at the top of the hierarchy and do not use type II morphogens anywhere. By contrast, the kind of mechanisms which I regard as most promising require type II morphogens for a pattern of repeating parts to be formed at all.

The general outline of the experimental facts and the hierarchical order which they require of a mechanism of either of the foregoing types is not in dispute. Some 4 hours after fertilization, at the start of interphase 14 in the sequence of divisions of the fertilization nucleus, the *Drosophila* egg has about 8,000 nuclei in a syncytium. They form a single layer just inside the surface of the egg, which is roughly a prolate ellipsoid 0.5 mm long. Adjacent nuclei are a distance of order 10 μm apart. Interphase 14 is a long pause in the division sequence. Within that period, the egg becomes, first, biochemically segmented and, second, multicellular. The latter event occurs by multiple invaginations of the plasma membrane to envelop each nucleus and the surrounding cytoplasm in a separate plasma membrane. But while the egg is still syncytial, some proteins, such as the product of the *fushi tarazu* gene, are found to be produced in a pattern of seven stripes, each about three to four rows of nuclei wide, transverse to the anteroposterior direction (Gilbert, 1988, cover illustration). The genes concerned are called "pair-rule" genes, because two gene products are so related that they form alternating stripes; for example, in the intervals in which *fushi tarazu* protein is not present, *even-skipped* protein is found.

(The Japanese word *fushi* has the primary meaning of the repeated nodes seen very obviously on a bamboo cane. By metaphorical extension, the word is used as an abstract noun for both space-order and time-order. Thus, *fushi tarazu* may be translated "not enough segments" or "not enough rhythm." Many *Drosophila* genes have names suggesting a deficiency. Each was first recognized as existing because of the deformity of a mutant insect lacking the particular gene activity concerned. Physical scientists seeking to contribute to explanations of biological development should recognize that the chemical phenomena which they are considering – gradients and more complex patterns of chemical substances – are still rarely known directly. Much more commonly they are inferred from studies of developing morphology. Instances of direct observation, such as *fushi tarazu,* remain exceptional even in such huge bodies of information as that now existing for *Drosophila.*)

The pair-rule genes are the third level of a hierarchy of five classes of genes: maternal-effect, gap, pair-rule, segment-polarity, homeotic. In brief:

1. Maternal-effect genes. These genes are transcribed in the mother. The mRNA is transferred to localized regions at the anterior or posterior end of the egg and does not move. It is translated into proteins which can move. One of these, *bicoid,* is now known (Driever and Nüsslein-Volhard, 1988) to be distributed along the egg in a gradient which, for a large fraction of the length, has the exponential form expected for a type I morphogen (Figure 2.1b).

2. Gap genes. These genes are active along parts of the egg much longer than one stripe of the 14-stripe pattern of two alternating pair-rule genes, and they appear to control the expression of the pair-rule genes. Thus the absence of gap gene activity leads to defects in about three *fushi tarazu* stripes and hence to major morphological defects, as described by names of mutants and genes such as *hunchback, Krüppel* (cripple), and *knirps* (dwarf). [For a summary, see Carroll and Scott (1986).]

3. Pair-rule genes. These genes are expressed first in quite long domains, but by interphase 14 have achieved the pattern of seven stripes of expression of

one gene alternating with seven of another. There are four such pairs of genes.The most widely studied is *fushi tarazu*, but it appears to be a passive recorder of a pattern first appearing in the pair *hairy* and *runt*. If there is within the pair-rule system a two-morphogen pattern generator, it is most likely to involve the control systems of *hairy* and *runt* with some feedback from *even-skipped*, the partner of *fushi tarazu*. The region occupied by a stripe is referred to as a parasegment, because the pair-rule stripe pattern is staggered with respect to the ultimate morphological segmentation (i.e., a segment corresponds to the posterior part of one stripe and the anterior part of the next).

4. Segment-polarity genes. It has long been known from work on insects other than *Drosophila* (e.g., *Rhodnius*) (Locke, 1967) that each segment of an insect contains a repeat of the same anteroposterior gradient, so that along the whole length of the insect the gradient has sawtooth form. Segment-polarity genes are those which somehow specify this gradient by being expressed only in the same localized region of each segment. In *Drosophila*, the expression of these appears to begin mainly late in interphase 14, after cellularization. Two well-known examples are *engrailed* (a heraldic term for a scalloped border with the spikes pointing outward from the bordered region) and *wingless*. Both are normally expressed only in the posterior region of each segment. Absence of *engrailed* leads to a mutant morphology in which, roughly, the posterior part of each segment develops abnormally as a mirror image of the anterior part. A defect in *wingless* can sometimes affect only the posterior part of the second thoracic segment, from which the wings arise. This gene is homologous to a vertebrate gene *int-1* (55% identity from *Drosophila* to mouse, 99% from mouse to human), which is an oncogene (i.e., a gene for some part of a signal transduction pathway which becomes drastically overexpressed in cancers).

5. Homeotic genes. The term "homeosis," coined by Bateson (1894), describes an abnormality in which one organ is replaced by a likeness of another (e.g., a leg-like structure grows where an antenna ought to be). The modern definition of a homeotic gene in an insect is that it is one which determines the identity of a particular segment.

These five classes of genes appear to be involved in a hierarchical ordering of control processes. [See Akam (1987) for a review of the gene classes.] These processes may be described as follows:

1. establishment of monotonic anteroposterior gradients along the whole embryo,
2. demarcation of fairly long "cardinal regions" [terminology of Meinhardt (1986)] along the embryo,

3. formation of the 14-stripe pattern,
4. formation along each segment of a similar anteroposterior gradient,
5. specification of an individually different identity for each segment.

The preceding generalities are, I believe, not in dispute and are also unlikely to be changed over the next few years by new evidence, even though it
is being obtained very rapidly by many workers. Controversy surrounds both
the nature of the pattern-forming control processes and the manner in which
one should go about trying to establish this. On the latter point, I maintain that
the first approach should be to the generalities. Fine-tuning to the impressive
array of known details is probably going to take decades. Many drosophilologists would, I think, argue that progress beyond the current level can
be made only by increasingly meticulous attention now to all those details.

The question of type I (Wolpert) versus type II (Turing) morphogens arises
even at the top of the hierarchy, in the establishment of the anteroposterior
gradients. The fundamental distinction is that a Wolpert positional gradient
arises from a preexisting localized Saunders source (and is therefore not really
the top of a hierarchy; that honour belongs to the source-positioner, e.g., the
nurse-cell/egg interaction in *Drosophila*). A Turing morphogen pair has the
power to locate its own sources. The evidence that maternal-effect mRNAs
are Saunders sources for Wolpert gradients of their translation products is very
strong in *Drosophila*. But the double-abdomen deformity in *Smittia* correlates
well with a Turing mechanism and even allows one to start the fine-tuning of
theory to experiment by recognizing that the nonlinear structure of the Gierer-
Meinhardt equations is more appropriate to this case than is the structure of
the Brusselator (Meinhardt, 1977, 1982) (see Section 3.3.1, Figure 3.12).
Meinhardt (1986), as copied in amended form by Harrison and Tan (1988),
has also used this two-morphogen mechanism for the *Drosophila* gradients,
despite the evidence for fixed sources. Do insect eggs in fact have multiple
anteroposterior gradients, some of one type and some of the other, and therefore capable of diverse kinetic events (just as the many proteins in a cell have
diverse functions)?

Beyond the gradients, developmentally (or below them, hierarchically), lies
their interpretation in the expression patterns for gap genes and pair-rule
genes. Herein is the greatest divergence between the current views of reaction-diffusion theorists and experimental drosophilologists, as summarized in
the title of this subsection: communication versus "no crosstalk." The approach taken in my laboratory builds on the work of Herschkowitz-Kaufman
(1975). We start the hierarchy by using the reactant A to represent the combined effects of both anterior-high and posterior-high maternal gradients (respectively, *bicoid* and probably *nanos,* though the posterior effects seem more
complex, and *oskar* has also been mentioned in this connection) (Irish,
Lehmann, and Akam, 1989). Is it reasonable to set up a model by rolling the
effects of two or more different substances together into a double-ended

gradient of a single substance *A?* I have been severely criticized by drosophilologists for doing this, and the procedure certainly runs quite counter to
their energetic efforts to disclose the variety of detail in the actions of the
several substances forming the anterior and posterior gradients. The next
section addresses this issue.

My gradient of *A* is simply the sum of two type I morphogen distributions,
of the steady-state exponential form in distance along the egg. The anterior-
high part is modelled on the known *bicoid* distribution. The posterior-high
part can be modelled more freely, because less is known experimentally about
it. Given these gradients, what happens down the hierarchy? At the lower
levels, especially the pair-rule level, we depart from both the experimentalists
and the approach of Meinhardt (1977, 1986) by using two-morphogen reaction-
diffusion as the pattern-former. This involves communication by diffusion of
X and *Y* from nucleus to nucleus through the intervening cytoplasm. With the
Brusselator mechanism and appropriate choices of parameter values, we can
produce one long plateau of morphogen concentration or a pattern of several
equally spaced peaks. The former might represent a gap gene distribution of
activation, the latter a pair-rule pattern. Eventually, we want to tie these two
kinds of computations together so that the whole hierarchy can be put into one
program. To do this will involve using two Brusselators in sequence, with a
morphogen of the first (gap gene pattern) becoming a reactant input to the
second. This is a strategy similar to feedback loops I and II in Figure 10.2 for
the *Acetabularia* whorl pattern.

That will be a major project needing a long-term approach in many steps.
For the nonce, our first step (Lyons et al., 1990) is to assume a gap gene
product distribution. Just as reactant *A* is used for all maternal-effect gene
products, so we use *B* for all gap gene products, and we model mutants by
taking pieces out of the otherwise uniform *B* distribution to correspond to
known regions of activity of particular gap genes. In this way we have been
able to model the defects in pair-rule pattern produced by null mutants for the
gap genes *hunchback* and *Krüppel* (Figure 10.6).

To do this, however, we did not have to drop *B* to zero. A diminution from
30 to 15 arbitrary concentration units was quite enough. For this we have
received the criticism that we are modelling a haplo-insufficiency rather than a
null mutation, and that it takes total absence of a gap gene product to erase
pattern. We are unrepentant. This work is a first attempt. After much more
theoretical work by several groups publishing many papers, the fine-tuning
of theory to experiment may progress, perhaps, into continuing with two-
morphogen reaction-diffusion but using a different model from the Brusselator; and all quantitative aspects could change. Our work seeks to point the
way to a promising-looking path.

The no-crosstalk model which is currently much preferred by many drosophilologists does not involve any communication between nuclei. Each
nucleus is envisaged as responding independently to concentration levels of

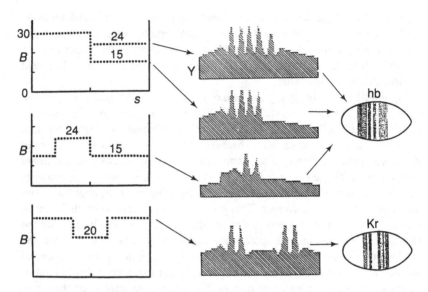

Figure 10.6. One-dimensional Brusselator modelling of the *fushi tarazu* expression patterns at cellular blastoderm stage (interphase 14) in two *Drosophila melanogaster* gap gene mutants, *hunchback* (hb) and *Krüppel* (Kr). Modified from Lyons et al. (1990). A reactant had the double-ended gradient shown in Figure 9.3. With constant $B = 30$, this gives a 7-peak pattern of Y, corresponding to the minima of X in Figure 9.1b. In the present figure, Y concentration patterns are shown (cross-hatched) for distributions of B with distance s along the egg as shown on the left. Idealized experimental data in the egg shapes on the right are adapted from Carroll and Scott (1986).

the primary maternal-effect gradient. This response first produces the "cardinal regions" (Meinhardt's terminology) of the gap gene activities. These regions have some overlap, and the changing ratio of gap gene product concentrations in the overlap regions is seen as a means of turning the primarily monotonic positional information into a set of repeating stripes of pair-rule gene activities (Figure 10.7) (Edgar, Odell, and Schubiger, 1989; Pankratz et al., 1990; P. M. Macdonald, personal communication).

Whatever the details of intermediate stages in translation of a monotonic gradient into a striped pattern, however, the precision with which the gradient must be read remains the same. This may be estimated from the *bicoid* data of Driever and Nüsslein-Volhard (1988), with an uncertainty of about ±30% arising principally from the question of how to correct for the background intensity shown in their graphs. The exponential concentration gradient is of course linear in free energy (chemical potential) (Figure 2.2), and I estimate the change between adjacent rows of nuclei as about $k_B T/25$ per molecule, or 0.1 ± 0.03 kJ mol^{-1}, in other words about 1/25 of thermal noise. This is not nearly enough to switch on a gene in one row of nuclei while switching it off

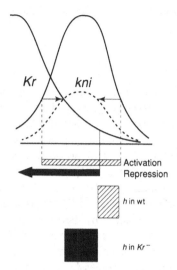

Figure 10.7. A detail of the no-crosstalk gradient-reading model at present more favoured than reaction-diffusion by many experimental drosophilologists. Names of mutants and their related genes: *h, hairy; Kr, Krüppel; kni, knirps;* wt, wild-type. From Pankratz et al. (1990), © Wiley-Liss, with permission from Wiley-Liss, a division of John Wiley and Sons, Inc.

in the next with the precision needed to control the 14-stripe pattern, by an interaction of one positional-information molecule with one binding site on nuclear DNA. The pattern would simply get lost in the noise.

There are various ways out of this difficulty. All of them involve the operation of some kind of amplifier. Therefore, they all require careful *mathematical* consideration. The kinds of models which the experimental drosophilologists are formulating require different mathematics from reaction-diffusion, but they do not permit full expression with *less* mathematics. No model for the pattern-forming phenomenon has been adequately formulated until its mechanism for overcoming thermal noise, to allow such astonishingly precise gradient reading, has been presented. I present this consideration as a challenge to the no-crosstalk theorists. Reaction-diffusion theorists have done their work on this aspect. For instance, the first paper of Gierer and Meinhardt (1972) contained a computation in which a very noisy gradient was amplified into a much steeper and much smoother one (see Figure 5.3). In two-dimensional computations in my laboratory (Figure 9.7) (Lyons et al., 1990) it is routine to put in thermal noise at every iteration.

Here, I am by no means maintaining that it is impossible to find an amplifier which will overcome noise in a no-crosstalk model, only that this essential part of the modelling has not yet been done. Edgar et al. (1989) have modelled a bistable switching system which is an amplifier. But they have not yet shown fully how this could be used to fine-tune the gradient reading to the required extent.

10.2.2   Carving out a theory

Crombie's phrase (from the quotation used as epigraph to Part III) suggests a correlation in procedure between scientific and artistic enterprises, which are indeed the two faces of human creative activity. Modern scientific theory is sufficiently elaborately detailed that one should take as its analogue classical representational art. But what kind of art should be envisaged: painting, sculpture, or whatever? If one thinks of the painting of a vast mural, one may envisage many workers simultaneously painting the details of diverse small sections. The whole surface on which to do this is available from the start. The artists do not have to produce it by their art.

The word "carving," however, suggests an art with different constraints, in which the sculptor must rough out a human shape from a block of stone before the carving of details of a face becomes feasible. Much of the art lies in the roughing out. Contemporary sculpture often says important things about the human form without going into the details. I contend that Crombie has chosen his word with great precision. Scientific theory is an art in which the surface on which to elaborate the details is not available a priori, but is itself a product of the enterprise. The double helix of DNA, which is the broad ground on which so much detail is being elaborated by molecular biologists, is itself a product of similar work. In longer perspective, before the middle of the nineteenth century Liebig had roughed out the ground for protein science: "There is, indeed, no part of an organ possessing a form or structure of its own, the elements of which are not derived from the albumin of the blood. All organized tissues in the body contain a certain amount of nitrogen" (Leicester, 1974). But nitrogen itself is one detail of the elaboration, half a century earlier, of the system of chemical elements.

What has all this to do with the production of 14 stripes by an interphase-14 *Drosophila* egg? In essence, I believe that all theories of this phenomenon should be appreciated today as being at the roughing-out stage. They should be judged on how they are coping with the broad generalities. Thus the procedure which I have used for reaction-diffusion, to lump all the maternal effects into a double-ended gradient of one substance, should be seen as normal procedure. It should be anticipated that the single precursor A will eventually be split down into two or more substances. Lacalli (1990) has pointed the way toward this in his discussion of four-morphogen models, which are basically two models of the two-morphogen type coupled in parallel. My hyperchirality model in Section 9.2 (Harrison and Lacalli, 1978; Lyons et al., 1990) is a four-morphogen model which, because of a particular symmetry relationship between the rate constants, can be viewed alternatively as a two-morphogen model [the four being X_D, X_L, Y_D, Y_L, and the two U and

V, as in equations (9.34) through (9.38)]. This illustrates that the count of morphogen concentration variables in the dynamic equations may readily be different from the number of chemical substances concerned. These differences do not matter in the roughing-out stage of the theory.

The no-crosstalk gradient-reading theories, on the other hand, require also some general mathematical discussion to establish their feasibility as a class of models. The crucial question was raised in the preceding section. Can these models be provided with amplifiers which will provide the very precise readings of concentration levels required and which at the same time will fight against thermal noise in those concentrations, without the models turning into reaction-diffusion models?

10.3 From slime moulds to salamanders

In this section I group together a number of phenomena which are biologically and chemically quite disparate. Their common feature is that each represents, to my mind, a quite significant advance toward the establishment of kinetic theory in general and reaction-diffusion in particular as being relevant to real pattern-forming processes. These advances are as follows: identification of what is very probably a Turing morphogen pair; observation of the expected intermediate stages between randomness and order in Turing dynamics; evidence for an activation-inhibition-communication mechanism in a vertebrate mesoderm; and two contrasted instances of Turing patterns in nonliving chemical systems, one spotted and the other striped.

These examples also illustrate, however, the difficulties in establishing morphogenetic mechanisms with certainty, especially in regard to the roles of the molecular scale and the cellular scale. (I mentioned this earlier in Section 4.4.1.) What is an activator morphogen, Turing and Prigogine *X* or Gierer–Meinhardt *A*, supposed to activate? For the stalk-cell-pathway activator DIF-1 in *Dictyostelium* and the heart-formation activator in *Ambystoma*, as also for the head activator in *Hydra* (Schaller, 1973; Schaller and Gierer, 1973; Bodenmüller and Schaller, 1981), the activator is known to govern a particular kind of cellular differentiation. Whenever the differentiated cells are found to produce more of the activator than do other types of cells, the existence of a positive feedback loop has been established. This is promising as a component of a kinetic pattern-forming model. But it is not the Turing k_1 term, nor the third step of a Brusselator mechanism, on a strictly *molecular* scale. The feedback goes through the cellular scale.

The reader may have noticed an incorrect identification between two things in the preceding paragraph. It was put in as a deliberate mistake, to highlight the difficulty of investigating kinetic self-organization, possibly the principal reason why the field is advancing so slowly. The autocatalytic rate constant k_1 of the Turing equations is *not* the rate constant of the autocatalytic step (*c*) of the Brusselator. In Table 9.1 I have listed Turing k_1 as Brusselator $bB - d$, an

expression not containing the rate constant of the explicitly autocatalytic step
at all. This was mentioned earlier in Section 6.3, following a simpler bit of
mathematics in which I showed that for a simple model of autocatalysis and
decay, the Turing k_1 constant is the decay constant [Figure 6.4, equation
(6.30)]. If this decay is brought about by an enzyme, addition of that substance could appear as an activation.

The foregoing may seem like a counsel of despair. If experimental ap-
pearances can be so deceptive in relation to real mechanism, what hope is
there of putting the two together? Beyond the stages already reached, as
indicated in the following sections, what is the next step? In fact, I remain
altogether excited and optimistic about this. The next step for the biologist
should be the first on a totally new staircase. It starts, in this book, at the
beginning of Chapter 6, and the ascent of it involves gradually introducing
into our discussions of developmental mechanisms all the mathematical language presented from there through Chapter 9. To my mind, without that
language there is great danger of setting up a tottering edifice of false conclusions; with it, there is hope of greatly accelerated progress in the field. The
following rather brief note on a few promising examples should be read in that
light.

10.3.1 A morphogen pair in Dictyostelium discoideum?

The differentiation of two types of cells in the "slug" stage of the cellular
slime mould *Dictyostelium discoideum* was mentioned in Section 4.4.1, especially Figure 4.13a. It is known that both cell-sorting and differentiation of
cells in situ in a manner sensitive to their surroundings occur in the slug stage.
Therefore, both cell-as-molecule and purely molecular versions of kinetic
theory may be considered. The linear Turing model was applied by Lacalli and
Harrison (1978) to the problem of regulation of this two-part pattern in the
face of variation of slug size or interference by cutting a slug into fragments
during this patterning process. This work, showing quite good regulatory
capacity, was presented to argue against the indication of Bard and Lauder
(1974) that the Turing model lacks regulatory capacity. This had tended to
diminish the enthusiasm of biologists for Turing models (e.g., Cooke, 1975),
though Bard (1981) was soon in favour of these models again.

There are three candidates for the title of Turing morphogen in *D. discoideum:* cAMP, DIF-1, and ammonia. It has been known for nearly a
quarter-century that cyclic adenosine monophosphate (cAMP) is the chemotactic signaller to which separately moving amoeboid cells respond in the
aggregation stage, to come together to form the slug (Konijn et al., 1967).
cAMP was identified in 1960 and is best known and described in most
biochemistry texts as a "second messenger," that is, a compound which acts
intracellularly to transmit and amplify the signals delivered to the outside of
the cell surface by "first messengers" [i.e., hormones, especially epinephrine

(adrenaline)]. But in *D. discoideum*, cAMP can be secreted by the cells within which it acts, and it can act on the outside of those cells as a first messenger itself. This is a positive feedback loop, and it makes cAMP a possible X morphogen.

The other two compounds, DIF-1 and NH_3, are, however, the ones currently most strongly indicated as a probable type II morphogen pair. Schindler and Sussman (1977) showed that ammonia plays a role in the patterning. DIF-1 is a lipid-like molecule which appears to be an activator for the differentiation of cells as pre-stalk cells (Town, Gross, and Kay, 1976; Gross et al., 1981). Gross has proposed these two as a Gierer–Meinhardt morphogen pair, the short activated region being that of the pre-stalk cells, and the long inhibited region that of the pre-spore cells. Speculatively, Gross et al. (1988) have linked the entire postulated dynamics to pumping of calcium into calcium-sequestering vesicles (Figure 10.8). To my mind, this diagram should be taken as a systems-theory schematic in the same spirit as my diagram of three feedback loops in *Acetabularia* (Figure 10.2), but with substantially more knowledge, for *Dictyostelium*, of the actual molecules involved. This should be the takeoff point for the mathematical advance.

10.3.2 Stages of Turing kinetics in **Polysphondylium pallidum**

Figure 2.7c shows how a pattern with a definite wavelength may be expected to develop out of random noise which contains a mixture of many wavelengths, when a Turing mechanism is operating. Direct observation of intermediate stages would assist greatly in establishing the reality of such dynamics. In my laboratory, we observed something different from this, but closely related to it, in *Acetabularia* whorls: the shift from one hair spacing to another via a spread of intermediate values when the temperature of a culture is suddenly changed and spacings are measured in whorls formed through a period of several hours after the change (Figure 10.9c).

E. C. Cox and co-workers have reported much more direct observations of the Turing dynamics in the slime mould *Polysphondylium pallidum* (Figure 10.9a,b). Unlike *D. discoideum, P. pallidum* does not culminate its life cycle by forming a single ball of spores atop a single stalk. Instead, a number of whorl masses are segregated at equally spaced intervals along the main stalk. Each of these develops from 1 to 8 secondary tips around its equator. These grow out into a whorl of secondary stalks on which the spore masses (sori) finally form. Byrne and Cox (1986, 1987) made several antibodies, one of which, called anti-Pg101, detected a pattern of antigen expression on the whorl masses which appears to correspond to the prepattern for whorl formation (Figure 10.9a). The distribution of immunofluorescent stain around the equator of the tip was measured quantitatively (Byrne and Cox, 1987) and converted by Fourier transformation into a "power spectrum" (Figure 10.9b). This procedure resolves the pattern into intensities of sine-wave components

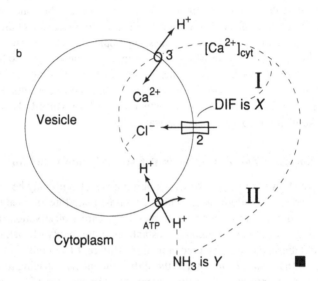

Figure 10.8. (*a*) The structure of DIF-1. (*b*) Schematic of feedback controls in *D. discoideum* patterning, from my notes of a lecture by J. D. Gross (1989), with addition of broken lines representing my understanding of the feedback loops which make DIF an *X* morphogen and ammonia a *Y* morphogen. The large circle is a calcium-sequestering vesicle. 1, 2, and 3 are transmembrane structures, respectively a proton pump, a chloride channel, and a proton-calcium antiporter. To sequester calcium, the vesicle must be acidified so that the antiporter has an adequate supply of protons to function. This acidification is done by the proton pump, but its proper functioning depends on the chloride channel being open to supply counter-ions for the protons. The feedback loops I and II involve unspecified processes in the rest of the cell, in which the cytoplasmic calcium level controls production of both DIF and ammonia. DIF opens the chloride channels, while ammonia combines with protons and hinders the proton pump. Gross et al. (1988) describe this mechanism with a diagram similar to parts of (*b*).

with wavelengths corresponding to various numbers of peaks around the equator. As expected for Turing kinetics, out of many peaks of comparable intensities, one grows to dominate the system. Byrne and Cox (1987) and McNally and Cox (1989) performed reaction-diffusion calculations showing a good match to this observed behaviour.

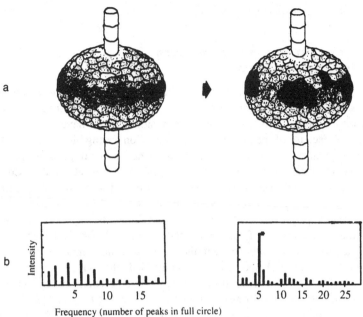

a

b

Intensity

5   10   15

Frequency (number of peaks in full circle)

5   10   15   20   25

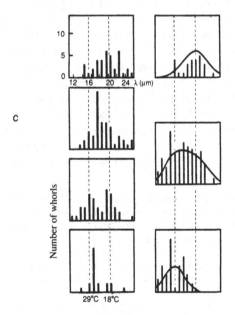

c

10

5

0

12 16   20   24 λ (μm)

Number of whorls

29°C 18°C

Wavelength = circumference/frequency

Figure 10.9. (a) and (b) Whorl formation around the equator of a whorl mass in *Polysphondylium pallidum*, from Byrne and Cox. (a) A whorl mass with dark shading showing the whorl prepattern as revealed by anti-Pg101 immunofluorescent staining, at two different stages in whorl mass development. (b) For samples at approximately

10.3.3   Complex kinetics in the mesoderm
of Ambystoma mexicanum

It has long been recognized that some features of vertebrate development, such as control of the number of somites, occur very early, suggesting the existence of prepatterns. But reaction-diffusion models have been largely discounted because of the rather general impression among biologists that this kind of mechanism cannot cope with adjusting the number of parts in a pattern precisely when the sizes of individual embryos are somewhat variable. Cooke (1975) has succinctly summarized this general line of thinking. The problem here has been the primitive nature of the reaction-diffusion models, which have remained primitive much too long because not enough people have been doing the detailed work of elaborating them. I have shown (Harrison, 1982) that when one takes into account the ways in which inputs such as the Brusselator reactants *A* and *B* may be controlled, the number of parts in a pattern produced by reaction-diffusion can be made constant for changing system size, or variable in any way one might want it to vary (see Section 10.4). This should not have been left for me to do, thirty years after Turing's paper.

In view of the general scepticism, any evidence pointing toward reaction-diffusion in the vertebrates is welcome. It has long been known that formation of some organs from the mesoderm involves an inductive effect from the endoderm. A specific example is the heart in some amphibians. Evidence in some species has shown that the inductive effect is active until very shortly before the heart begins to beat, giving the impression that the mesoderm responds rather passively to endodermal instructions and does not have a big job to do by itself in specifying the heart-forming tissue.

It is therefore very significant that in one species of salamander, the axolotl *Ambystoma mexicanum,* the inductive effect has been found to be complete by the end of the third day of development, whereas the beating heart is not

Figure 10.9 *(cont.)* the same stages, the power spectra of stain intensity. Each bar in the histogram shows the intensity of a sine-wave component of the pattern having, all round the equator of the whorl mass, the number of peaks shown by numbering on the scale of abscissae. This is therefore a scale of spatial frequency, or reciprocal of wavelength. As expected for Turing dynamics, the spectrum first has many lines of comparable intensity; but later, one grows to dominate. In the case shown, that one corresponds to a whorl of five secondary stalks. (*c*) The change in wavelength (hair spacing) in *Acetabularia* whorls when the temperature is suddenly changed from 18°C to 29°C. From Harrison et al. (1984), with permission. Here the abscissae are wavelengths in micrometres, not reciprocals of wavelength. Right-hand column: top, expected statistical distribution from data at a constant temperature of 18°C; bottom, the same at 29°C; middle, half-way stage, sum of top and bottom graphs. Left-hand column: distributions of spacings in whorls formed at various times after the temperature shift, from top to bottom: 0–2 h, 2–4 h, 4–6 h, 6–8 h. (a) and (b) From Byrne and Cox (1986, 1987), with permission of Academic Press.

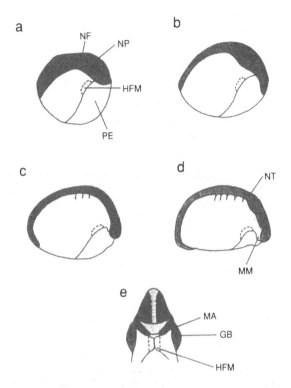

Figure 10.10. (*a*)–(*d*) Diagrams of the right side of *Ambystoma mexicanum* embryos at four stages of neurulation, showing the advancing edge of the flank mesoderm and the position within it of the heart-forming mesoderm (HFM). PE is the pharyngeal endoderm which supplies the inducing signal. NF, NP, NT: neural fold, plate, and tube. MM: mandibular mesoderm. (*e*) A view of the ventral side of the anterior part of the embryo, anterior upward, at a later stage (29) when the two edges of the mesoderm have just met and the two HFM regions are in contact. MA: mandibular arches; GB: gill bulges.

operative until the end of the fifth day. The two-day gap makes it possible in this species to do experiments establishing that in that two-day gap between end of induction and start of heartbeat, something rather complicated is going on in the mesoderm itself. This probably does not mean that heart formation in the axolotl is fundamentally different from that in other amphibians, or for that matter in other vertebrates, from mouse to human. The earlier induction is probably quite a trivial feature, but it opens a window to the experimentalist which is closed in other species (Smith and Armstrong, 1990; Smith, 1990). Figure 10.10 shows the position of the inducing pharyngeal endoderm and several sequential positions of the leading edge of the flank mesoderm through the developmental stages of most interest.

It has been known since 1926 that the heart field (i.e., the region of

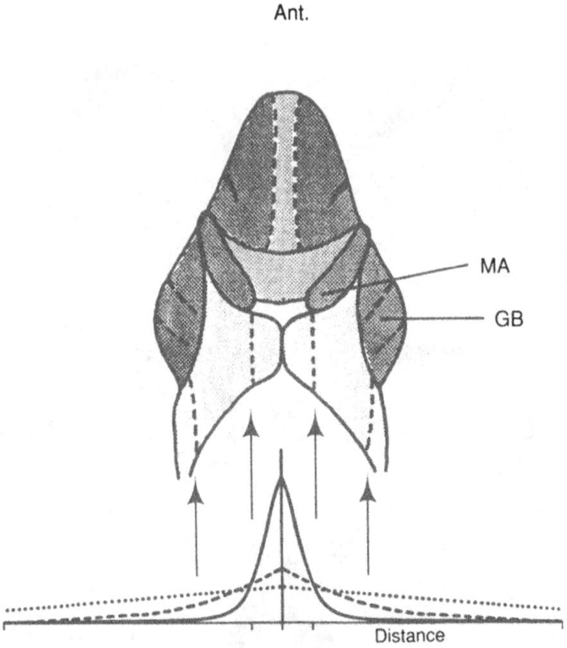

Ant.

Figure 10.11. Top: The same ventral view of a stage-29 embryo shown in Figure 10.10e, with the addition of indications of the relative extents of the heart-forming mesoderm (inner pair of dashed lines and arrows) and the heart-field mesoderm (outer pair of dashed lines and arrows). Bottom: A Gierer-Meinhardt diagram of short-range activation, long-range inhibition from the operation of their model on a shallow source gradient. Solid line, activator concentration. Dashed line, inhibitor concentration. Dotted line, source gradient. In heart formation in *Ambystoma mexicanum* and many other species, the source gradient is to be taken as the inducing effect from the mesoderm. Diagrams combined from two in Smith (1990), with permission.

mesoderm capable of responding to the inducing signal) is larger than the formed heart will be (Figure 10.11). Sater and Jacobson (1990) have discussed the progressive restriction of the heart field in *Xenopus laevis* and come to no definite conclusions on the cause of it. An obvious possibility is the progressive operation of an inhibitor in that part of the heart field which does not become heart-forming. Where does the inhibitor come from? Jacobson (1960) and Jacobson and Duncan (1968) studied heart induction in amphibians, especially *Taricha torosa*, and found in that species an inhibitory effect from the neural tissues (i.e., a source of inhibitor outside the heart field). But Smith and Armstrong found no such effect in *A. mexicanum* and concluded that inhibitor production is intrinsic to the heart-field mesoderm itself.

A strong line of evidence for this arises from the existence of a mutant, *cardiac-lethal* (*c*), in which most of the events preliminary to heart formation occur normally, but the heart primordium fails to beat. All the necessary muscle proteins arise in the heart-forming region, and this can even roll up into a tube, the shape of the normal primordium when it first beats. The specific defect in the *c* mutant is failure of the muscle proteins in each cell to organize into sarcomeres. Earlier work suggested that the defect was absence of inductive signal from the pharyngeal endoderm. But that was based on an assumption that the induction occurred in *A. mexicanum* at the same developmental stage as it does in *T. torosa*. That turned out to be incorrect. Following their precise determination of the much earlier inductive stage in *A. mexicanum*, Smith and Armstrong went on to show conclusively that the inductive signal in *c* is quite normal and that the defect is in the mesoderm. It can be corrected by contact with normal mesoderm.

They discussed their data in terms of the interaction of a diffusible activator and a diffusible inhibitor, both produced within the mesoderm. In this interpretation, the *c* mutant either underproduces activator or overproduces inhibitor. They propose that these two substances take part in Turing kinetics (Armstrong, 1989; Smith and Armstrong, 1991). They regard the restriction of the heart-forming region within the larger heart field as an instance of Gierer–Meinhardt "short-range activation, long-range inhibition" (Figure 10.11). But no detailed attempt has yet been made to match their data by computations with any specific reaction-diffusion model. Armstrong and I are collaborating on this project.

This, of course, implies that I regard the reaction-diffusion hypothesis as highly promising for this developmental event. But it is not yet certainly proved. The problem of certain proof is a common one when reaction-diffusion is postulated, as I mentioned in the prefatory remarks to Section 10.3: What is an activator morphogen supposed to activate, or an inhibitor inhibit? Most commonly, activator and inhibitor are detected by their effects on something else, in this instance the assembly of muscle proteins into functional sarcomeres. But proof of two-morphogen reaction-diffusion needs a knowledge of the effects of activator and inhibitor on themselves and on each other. These are not at all easy to observe directly, and they are not the same in all models which lead to Turing kinetics (e.g., the X morphogen is not necessarily explicitly autocatalytic).

There is also the possibility that mechanochemical effects are involved. Pieces excised during microsurgical experiments will roll up in reproducible directions, indicating that there are patterned mechanical stresses in the embryo.

These uncertainties do not bother me. The clear indication from Smith and Armstrong's data is of kinetically generated pattern, involving complex kinetics in the heart-field mesoderm and requiring investigation by the methods of kinetic theory, that is, discussion in terms of partial differential equations and computer experiments on the model which those equations describe.

10.3.4   *Turing patterns in nonliving chemical systems*

Patterned distributions of chemical substances arising during chemical reactions have long been known, such as the Liesegang rings formed during precipitations occurring when either two gases (NH_3 and HCl, for instance) or two solutes in liquid solution diffuse into each other from opposite ends of a container. Oscillatory behaviour and travelling waves are known in chemical systems, especially in the well-known Belousov–Zhabotinski reaction, which is cerium-catalyzed oxidation of malonate by bromate in aqueous solution; see Tyson (1985). This reaction has attracted substantial attention from chemists, among whom it is one of the best-known instances of a "dissipative structure" in the Prigogine sense; the first textbook mention of this concept in physical chemistry appears to have been by Moore (1972). The Belousov–Zhabotinski reaction has been analyzed kinetically as a three-morphogen model in the "Oregonator" scheme of Field and Noyes (1974), the morphogens X, Y, and Z being bromite ($Br\ O_2^-$), Br^-, and Ce^{4+}.

This instance, however, is rather on the fringes of the conceptual field of this book. As Winfree (1987) has pointed out, travelling-wave phenomena are most often studied in systems having a single diffusivity or, as is common in aqueous solutions such as the Belousov–Zhabotinski (B–Z) mixture, several substances with essentially the same diffusivity. Such systems do not have the large disparity between two diffusivities which is a characteristic feature of the Turing "diffusion-induced instability." If the B–Z reaction has biological analogues, they are most probably excitable systems such as neural systems and the heart muscle. In these, each excitable element has three states: resting, excited, and refractory. The third of these has no analogue in Turing dynamics. By the same token, travelling waves in excitable media have the character of relaxation oscillations rather than sinusoidal oscillations. For the B–Z reaction, Kopell and Howard (1973) performed an experiment which cast doubt on the existence of diffusive coupling in the travelling-wave phenomenon.

It is therefore noteworthy that in 1990 two reports were published concerning patterns in chemical systems which seem much closer to real instances of Turing patterns. Further, one of these is a hexagonal array of spots, such as the Brusselator readily produces (Castets et al., 1990), and the other is an array of parallel stripes of alternating character (Tabony and Job, 1990), as seen experimentally in *Drosophila* pair-rule patterns and by computation in the behaviour of the hyperchirality model. The spotted pattern arises in a solution occupying a narrow rectangular piece of polyacrylamide gel. One face of the gel strip is in contact with a solution containing sodium chlorite ($NaClO_2$), the other with a solution of malonic acid. Both solutions contain potassium iodide. The chemical system is thus related to that of the B–Z reaction, but oxidation of iodide to iodine is involved, and the patterns are observed with the aid of a starch-like stain for elemental iodine. Both stripes and spots have now been seen in this reaction (Ouyang and Swinney, 1991).

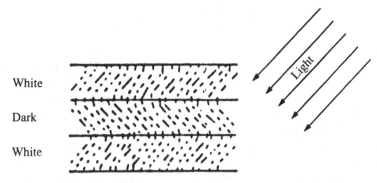

White

Dark

White

Figure 10.12. Diagram of the arrangement of microtubules in two nematic liquid-crystal phases oriented orthogonally to each other and forming a pattern of alternating stripes in arrays of microtubules self-organizing in vitro. Alternate stripes are in a certain sense of macroscopic geometry mirror images of each other, like stripes which are formed by the hyperchirality model. This phenomenon may fall into the "big hands from little hands" category (Harrison, 1979), for which examples of living organization are the mirror-image doublets in ciliates (Frankel, 1989, 1990). From Tabony and Job (1990), © Macmillan Magazines Ltd., with permission from *Nature*.

The pattern first appears as a set of stripes parallel to the faces at which the solutions are supplied and therefore somewhat similar to Liesegang patterns, but the significant observation is that each stripe breaks up into a series of equally spaced spots. This, as the authors stress, is a symmetry-breaking perpendicular to the gradients of the reactants. Further, the spots formed out of successive stripes are staggered in position so that the whole set of spots in two dimensions forms an essentially hexagonal array. The nearest-neighbour distance is about 0.2 mm.

The pattern of alternating stripes is closer to biological realities, being an array of microtubules self-organizing in vitro. The sample cell is $40 \times 10 \times 1$ mm, roughly the size of a 1-mm-thick layer on a microscope slide. The solutions contain tubulin, guanidine triphosphate (GTP), acetyl phosphate, and acetate kinase. Microtubule formation and stability after assembly require the hydrolysis of GTP to GDP and inorganic phosphate as the free-energy-liberating process and hence thermodynamic driving force. Tabony and Job observed that arrays of microtubules were formed as stripes which, in appropriately directional light, appear alternately light and dark. Each stripe is an oriented nematic liquid-crystalline phase with microtubules lined up parallel to each other. But in one stripe they are lined up at 45° to the direction of the stripe, in the next at 135°. Thus alternate stripes have orthogonal orientations of microtubules (Figure 10.12). The pattern was characterized by study with ordinary light, polarized light, and small-angle neutron scattering. ^{31}P-NMR was used to monitor the reactions involving phosphate, to show that pattern formation is indeed being driven by an irreversible chemical reaction.

10.4   Measuring, counting, regulation: *Acetabularia* versus *Drosophila* versus *Dictyostelium*

Reaction-diffusion theory continues to be rejected by many biologists as a promising possibility on the same basis that led to its rejection by Waddington (1956): that it requires a rather precise fit between the Turing wavelength and the size of the system. A much more recent and very precise statement of this objection to the theory is that of Cooke (1981): "A decisive limitation of reaction-diffusion pre-patterns is that numbers of peaks or singularities generated are sharp functions of the spatial extents of tissues in which they operate, unless the reaction and diffusion parameters can have their values adjusted within tissue by appropriate feedback from the overall size." The first part of this sentence specifies the ability of reaction-diffusion mechanisms to measure; the second part specifies in general strategic terms what is necessary in a mechanism if it is, instead, to count.

In the account from which this is quoted, Cooke evidently sees the necessary feedback from overall size as rather unlikely and indicates that reaction-diffusion is not a promising possibility in relation to somite determination. Likewise, when I tried a few years ago to publish a brief note indicating the probable nature of such feedback and the versatility which it gave the Turing theory in stabilizing pattern over a wide range of system size, I was sternly taken to task by an anonymous referee for making "*ad hoc* additions" to the Turing model.

It is important, I believe, to recognize that the feedback from overall size which Cooke correctly indicates as necessary is neither an ad hoc addition nor, as Cooke implies, a rather unlikely thing to arise. The essence of the matter lies in appreciating the roles of both reactants and intermediates in controlling the rates of chemical reactions. In accounts of reaction-diffusion theory, the principal theme is usually the unusual feature of rate control by the intermediates X and Y, which confers upon the dynamics their pattern-forming ability. This is, of course, the matter to which attention must first be given in reaction-diffusion theory. It is often concentrated upon at the expense of everything else. Reactants, such as the Brusselator A and B, are pushed into the background by assuming their concentrations constant, and sometimes even rolling them into the values of the rate constants.

Any student who has just taken an elementary course in chemical kinetics, however, if asked what is likely to control the rate of a reaction, is almost certainly going to answer "the reactant concentrations." Throughout my accounts of the Brusselator and its uses in Chapters 9 and 10 I have repeatedly pointed out that one must continually remind oneself whether one is dealing with the A-B pair or the X-Y pair, and that they can be used together to devise quite simple and straightforward models for feedback loops and for hierarchies of processes in series, in which the X of one becomes the A of the next. All of this is in no way a set of ad hoc additions to the Turing model. It is simply part of the chemical kineticist's way of life: putting into a complex reaction mechanism, which may have many steps, just a few steps which have particular kinetic attributes matching something seen experimentally.

In regard to the number of parts in a pattern, or the spacing between repeated parts in a metameric pattern, I have described three types of phenomena: (1) *Acetabularia* measures (i.e., the distance between adjacent hairs in a whorl is constant). This needs no further discussion. It is the one thing that everyone knows a Turing mechanism can do. (2) Metameric pattern, whether in *Drosophila* segmentation or in the somites of vertebrates, counts. The number of parts in the pattern is quite impressively held constant. (3) *Dictyostelium* also maintains a constant number of parts in the pattern, but that number is only two, and the phenomenon is better thought of as regulation of pattern rather than counting of numbers of parts. The regulatory capacity has limits. Very large slugs form a tip at each end and ultimately two fruiting bodies (Hohl and Raper, 1964). But the two-part pattern is certainly regulated within slugs, or artificially severed parts of slugs, covering a length range of a factor between 10 and 20. What do we need in a dynamic model for this kind of regulation, and for the more precise counting in metameric patterns?

The first part of this question was addressed in Lacalli and Harrison (1978), being our first paper on reaction-diffusion theory. We used parameter values in the linear Turing equations which gave pattern growth rate k_g a rather shallow maximum when plotted against wavelength (Figure 7.7a, corresponding to region *B* of parameter space, Figure 7.5). We were somewhat lucky to hit on the appropriate range of values, because we had not yet found our way around the parameter space systematically. In general, one should be very cautious about saying what reaction-diffusion can or cannot do until one has understood that parameter space fairly well.

If one thinks of an elongated organism such as the *Dictyostelium* slug, to which reaction-diffusion theory might reasonably be applied one-dimensionally, the plot of k_g against wavelength needs to be converted into a plot of k_g against length of the slug. Now the conversion factor between wavelength and organism length depends on how many parts there are in the pattern. This leads to a different plot for every possible pattern complexity (Figure 10.13). The message of this diagram is that for the simple Turing model with no additions, the two-part pattern is the fastest-growing and therefore the one that will become established for a range of slug lengths of a factor of about 5. This does not quite match the regulatory capacity of *Dictyostelium* slugs, but it is close, and it is a far greater regulatory capacity than has been indicated by most statements about reaction-diffusion in the biological literature.

To do better than this, one must, as Cooke (1981) pointed out, consider feedback from system size. Once again, feedback is most easily modelled by using the Brusselator reactant *A*. First, an answer to the question posed in the legend to Figure 10.1: How can one devise a mechanism giving a simple

336 *Bringing experiment and theory together*

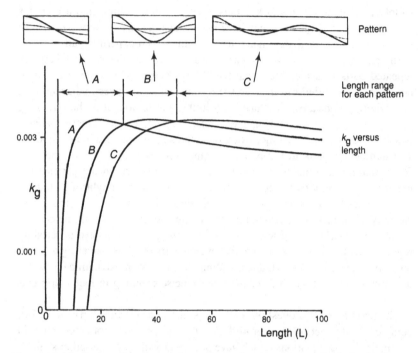

Figure 10.13. Curves of exponential growth rate k_g of pattern amplitude for the linear Turing model with parameters chosen to be in region B of the parameter space in Figure 7.5. For patterns of three different degrees of complexity, A, B, and C, the same curve has to be plotted on three different scales on the axis of abscissae. Pattern A, corresponding to a simple two-part arrangement of differentiation states, as in the *Dictyostelium* slug, has the highest k_g and would therefore become the dominant pattern, for a range of system lengths from 5 to 28 arbitrary length units. Diagram redesigned from the same curves in Lacalli and Harrison (1978).

inverse dependence of wavelength on A? If one writes the first step of the Brusselator as bimolecular in A, and therefore likely to be kinetically second-order, $2A \rightarrow X$ instead of $A \rightarrow X$, then A is replaced by A^2 everywhere that it appears in the Brusselator equations. These include the approximate wavelength expression, equation (10.2). If A^2 is replaced by A^4 in that expression, the required reciprocal relationship is obtained.

With the aid of that reciprocal relationship, it is possible to devise reaction-diffusion models in which the length of a pattern repeat can vary with system length in any way one might want. I first mentioned this in a review (Harrison, 1982) under the heading "pattern repeat *versus* system size: anything goes." For instance, suppose that an organism of elongated form, such as a *Dictyostelium* slug or a *Drosophila* egg, is always the same shape (i.e., all instances of it are similar figures in the strict geometric sense), but that length L varies between individuals. Suppose that substance A is supplied at a rate dependent on surface area (i.e., proportional to L^2), either from the outside if we are thinking of part of a larger organism or by activity at the inner side of the cell membrane. Suppose also that A is used up in a first-order decay everywhere in the volume, proportional to L^3. Then A should settle into a steady-state concentration proportional to $1/L$. Correspondingly, wavelength would be proportional to L, and the number of parts in the pattern to which A is input should be independent of L. Reaction-diffusion can count; it can regulate; it can measure. What more does it need to do?

10.5 Confirmed predictions of kinetic theory

I return finally to the philosophical question of predictions in science, as discussed in Section 10.1.2. The following appear to me to be important verifications of the predictions of kinetic theory:

1. Turing's theory postulated that morphological pattern should be preceded by corresponding prepattern of distribution of chemical substances. Very few such prepatterns were known in 1952. But in the 1980s and early 1990s, this prediction has been repeatedly verified, above all by the enormous and still rapidly increasing information on patterns of gene products in the *Drosophila* blastoderm. There are many other instances, including the prepatterns in *Polysphondylium* (Section 10.3.2), and for chemistry more remote from the genome, my calcium prepatterns in *Acetabularia*.
2. Turing's theory predicts that in the formation of a pattern of repeated parts with a single spatial periodicity, there will be intermediate stages Fourier-analyzable into a mixture of periodicities from which all but one are gradually lost. The *Polysphondylium* data show this (Figure 10.9).
3. Reaction-diffusion theory predicts that increasing the concentrations of inputs to the patterning mechanisms will cause decreased spacing between repeated parts, but that a big enough increase may switch off the patterning effect of the mechanism. The first part of this prediction is verified by my work on calcium in *Acetabularia* whorl patterning. The second part corresponds to what is seen in the relation of the *Drosophila* striped pattern to the maternal gradients closely enough that its status as a verified prediction should be getting a lot of attention from drosophilologists.
4. Reaction-diffusion theory predicts the simultaneous presence in a given tissue of an activator and an inhibitor for a pattern-forming event. This is confirmed in *Hydra* head formation, *Dictyostelium* differentiation, and *Ambystoma* heart formation.

These things are not yet widely accepted as verified predictions, and that is because the kinetic preconception itself is not yet the basis of most developmental biologists' thinking. Why not?

References

Akam, M. E. (1987). The molecular basis for metameric pattern in the *Drosophila* embryo. *Development 101*, 1–22.

Akam, M. E. (1989). Making stripes inelegantly. *Nature 341*, 282–3.

Andronov, A. A., and Pontryagin, L. S. (1937). Coarse systems. *Dokl. Akad. Nauk SSSR 14*, 247–51.

Andronov, A. A., Vitt, A. A., and Khaikin, S. E. (1937). *Theory of Oscillators* (in Russian). English translation, 1966. Oxford: Pergamon Press.

Armstrong, J. B. (1989). A Turing model to explain heart development. *Axolotl Newsletter 18*, 23–5.

Babloyantz, A., and Hiernaux, J. (1975). Models for cell differentiation and generation of polarity in diffusion-governed morphogenetic fields. *Bull. Math. Biol. 37*, 637–57.

Bard, J., and Lauder, I. (1974). How well does Turing's theory of morphogenesis work? *J. Theor. Biol. 45*, 501–31.

Bard, J. B. L. (1981). A model for generating aspects of zebra and other mammalian coat patterns. *J. Theor. Biol. 93*, 363–85.

Barlow, P. W. (1984). Positional controls in root development. In *Positional Controls in Plant Development*, ed. P. W. Barlow and D. J. Carr, pp. 281–318. Cambridge University Press.

Barlow, P. W., and Carr, D. J. (eds) (1984). *Positional Controls in Plant Development*. Cambridge University Press.

Bateson, W. (1894). *Materials for the Study of Variation* (pp. 84–5). London: Macmillan.

Berger, S., de Groot, E. J., Neuhaus, G., and Schweiger, M. (1987). *Acetabularia:* a giant single cell organism with valuable advantages for cell biology. *Eur. J. Cell Biol. 44*, 349–70.

Bernfield, M., Banerjee, S. D., Koda, J. E., and Rapraeger, A. C. (1984). Remodelling of the basement membrane as a mechanism of morphogenetic tissue interaction. In *The Role of Extracellular Matrix in Development*, ed. R. L. Trelstad, pp. 545–71. New York: Alan R. Liss.

Bodenmüller, H., and Schaller, H. C. (1981). Conserved amino acid sequence of a neuropeptide, the head activator, from coelenterates to humans. *Nature 293*, 579–80.

Bonner, J. T. (ed.) (1961). Abridgement of D. W. Thompson, *On Growth and Form* (1917, 1942). Cambridge University Press.

Bonner, J. T. (1967). *The Cellular Slime Molds*, 2nd ed. Princeton: Princeton University Press.

Bopp, M. (1984). Cell pattern and differentiation in bryophytes. In *Positional Controls in Plant Development*, ed. P. W. Barlow and D. J. Carr, pp. 157–91. Cambridge University Press.

340   *References*

Bopp, M., and Feger, F. (1960). Das Grundschema der Blattentwicklung bei Lebermoosen. *Revue Bryologique et Lichénologique 29*, 256–73.

Bragg, W. L., and Nye, J. F. (1947). A dynamical model of a crystal structure. *Proc. R. Soc. Lond. A190*, 474–81.

Brière, C., and Goodwin, B. C. (1988). Geometry and dynamics of tip morphogenesis in *Acetabularia. J. Theor. Biol. 131*, 461–75.

Brown, N. A., and Wolpert, L. (1990). The development of handedness in left/right asymmetry. *Development 109*, 1–9.

Bullough, W. S. (1962). The control of mitotic activity in adult mammalian tissues. *Biol. Rev. 37*, 307–42.

Bullough, W. S. (1983). *The Dynamic Body Tissues.* New York: Van Nostrand Reinhold.

Burton, W. K., Cabrera, N., and Frank, F. C. (1951). The growth of crystals and the equilibrium structure of their surfaces. *Philos. Trans. R. Soc. Lond. A243*, 299–358.

Byrne, G., and Cox, E. C. (1986). Spatial patterning in *Polysphondylium:* monoclonal antibodies specific for whorl prepatterns. *Dev. Biol. 117*, 442–55.

Byrne, G., and Cox, E. C. (1987). Genesis of a spatial pattern in the cellular slime mold *Polysphondylium pallidum. Proc. Natl. Acad. Sci. USA 84*, 4140–4.

Carroll, S. B., and Scott, M. P. (1986). Zygotically active genes that affect the spatial expression of the *fushi tarazu* segmentation gene during early *Drosophila* embryogenesis. *Cell 45*, 113–26.

Castets, V., Dulos, E., Boissonade, J., and de Kepper, P. (1990). Experimental evidence of a sustained standing Turing-type nonequilibrium chemical pattern. *Phys. Rev. Lett. 64*, 2953–6.

Cooke, J. (1975). Control of somite number during development of a vertebrate *Xenopus laevis. Nature 254*, 196–9.

Cooke, J. (1981). The problem of periodic patterns in embryos. *Philos. Trans. R. Soc. Lond. B295*, 509–24.

Cotton, G., and Vanden Driessche, T. (1987). Identification of calmodulin in *Acetabularia:* its distribution and physiological significance. *J. Cell Sci. 87*, 337–47.

Crombie, A. C. (1959). *Medieval and Early Modern Science,* 2nd ed., 2 vols. Garden City, N.Y.: Doubleday.

Crutchfield, J. P., Farmer, J. D., Packard, N. H., and Shaw, R. S. (1986). Chaos. *Sci. Am. 255*, 46–57.

Decker, P. (1973). Evolution in open systems: bistability and the origin of molecular asymmetry. *Nature [New Biol.] 241*, 72–4.

Decker, P. (1974). The origin of molecular asymmetry through the amplification of "stochastic information" (noise) in bioids, open systems which can exist in several steady states. *J. Mol. Evol. 4*, 49–65.

Decker, P. (1979). Spontaneous generation and amplification of molecular asymmetry through kinetical bistability in open systems. In *Origins of Optical Activity in Nature,* ed. D. C. Walker, pp. 109–24. Amsterdam: Elsevier.

Demetrius, L. (1985). The units of selection and measures of fitness. *Proc. R. Soc. Lond. B225*, 147–59.

Demetrius, L., and Ziehe, M. (1984). The measurement of Darwinian fitness in human populations. *Proc. R. Soc. Lond. B222*, 33–50.

Dodd, J. E. (1984). *The Ideas of Particle Physics: An Introduction for Scientists.* Cambridge University Press.

Driever, W., and Nüsslein-Volhard, C. (1988). A gradient of *bicoid* protein in *Drosophila* embryos. The *bicoid* protein determines position in the *Drosophila* embryo in a concentration-dependent manner. *Cell 54*, 83–93, 95–104.

Edelman, G. M. (1984). Cell-adhesion molecules: a molecular basis for animal form. *Sci. Am. 250,* 118–29.

Edelman, G. M. (1988). Topobiology: an introduction to molecular embryology. New York: Basic Books.

Edelstein-Keshet, L. (1988). *Mathematical Models in Biology.* New York: Random House.

Edgar, B. A., Odell, G. M., and Schubiger, G. (1989). A genetic switch, based on negative regulation, sharpens stripes in *Drosophila* embryos. *Dev. Genet. 10,* 124–42.

Eichele, G., and Thaller, C. (1987). Characterization of concentration gradients of a morphogenetically active retinoid in the chick limb bud. *J. Cell Biol. 105,* 1917–23.

Ermentrout, B. (1991). Stripes or spots? Nonlinear effects in bifurcation of reaction-diffusion equations on the square. *Proc. R. Soc. Lond. A434,* 413–17.

Feller, W. (1968). *An Introduction to Probability Theory and Its Applications,* 3rd ed., Vol. 1, Chap. 3. New York: Wiley.

Fenby, D. V. (1981). Einstein and molecular reality. *Chemistry in Britain 17,* 114–18.

Field, R. J., Körös, E., and Noyes, R. M. (1972). Oscillations in chemical systems. II: Thorough analysis of temporal oscillations in the bromate-cerium-malonate system. *J. Am. Chem. Soc. 94,* 8649–64.

Field, R. J., and Noyes, R. M. (1974). A model illustrating amplification of perturbation in an excitable medium. *Symp. Faraday Soc. 9,* 21–7.

Fisher, R. A. (1937). The wave of advance of advantageous genes. *Ann. Eugenics 7,* 355–69.

Frank, F. C. (1953). On spontaneous asymmetric synthesis. *Biochim. Biophys. Acta 11,* 459–63.

Frankel, J. (1989). *Pattern Formation: Ciliate Studies and Models.* Oxford: Oxford University Press.

Frankel, J. (1990). Positional order and cellular handedness. *J. Cell Sci. 97,* 205–11.

French, V., Bryant, P. J., and Bryant, S. V. (1976). Pattern regulation in epimorphic fields. *Science 193,* 969–81.

French, V., Ingham, P., Cooke, J., and Smith, J. (eds.) (1988). Mechanisms of segmentation. *Development [suppl.] 104,* 1–254.

Fritsch, F. E. (1956). *The Structure and Reproduction of the Algae.* Cambridge University Press.

Gardner, M. (1964). *The Ambidextrous Universe.* Harmondsworth, U. K.: Penguin Books.

Gardner, M. (1982). *The Ambidextrous Universe,* 2nd ed. Harmondsworth, U. K.: Penguin Books.

Gierer, A., and Meinhardt, H. (1972). A theory of biological pattern formation. *Kybernetik 12,* 30–9.

Gilbert, S. F. (1988). *Developmental Biology,* 2nd ed. Sunderland, Mass.: Sinauer Associates.

Goel, N. S., and Rogers, G. (1978). Computer simulation of engulfment and other movements of embryonic tissues. *J. Theor. Biol. 71,* 103–40.

Goodwin, B. C., and Kauffman, S. A. (1990). Spatial harmonics and pattern specification in early *Drosophila* development. I: Bifurcation sequences and gene expression. *J. Theor. Biol. 144,* 303–19.

Goodwin, B. C., and Pateromichelakis, S. (1979). The role of electrical fields, ions and the cortex in the morphogenesis of *Acetabularia. Planta 145,* 427–35.

Goodwin, B. C., Skelton, J. L., and Kirk-Bell, S. M. (1983). Control of regeneration and morphogenesis by divalent cations in *Acetabularia mediterranea. Planta 157,* 1–7.

Goodwin, B. C., and Trainor, L. E. H. (1980). A field description of the cleavage process in morphogenesis. *J. Theor. Biol. 86,* 757–70.

Goodwin, B. C., and Trainor, L. E. H. (1985). Tip and whorl morphogenesis in *Acetabularia* by calcium-regulated strain fields. *J. Theor. Biol. 117,* 79–106.

Green, B. R. (1976). Approaches to the genetics of *Acetabularia*. In *The Genetics of Algae*, ed. R. A. Levin, Chap. 12. London: Blackwell.

Green, P. B. (1980). Organogenesis – a biophysical view. *Annu. Rev. Plant Physiol. 31,* 51–82.

Green, P. B., and Poethig, R. S. (1982). Biophysics of the extension and initiation of plant organs. In *Developmental Order: Its Origin and Regulation*, ed. S. Subtelny and P. B. Green, pp. 485–509. New York: Alan R. Liss.

Gross, J. D., Peacey, M. J., and von Strandmann, R. P. (1988). Plasma membrane proton pump inhibition and stalk cell differentiation in *Dictyostelium discoideum*. *Differentiation 38,* 91–8.

Gross, J. D., Town, C. D., Brookman, J. J., Jermyn, K. A., Peacey, M. J., and Kay, R. R. (1981). Cell patterning in *Dictyostelium. Philos. Trans. R. Soc. Lond. B295,* 497–508.

Gunning, B. E. S. (1981). Microtubules and cytomorphogenesis in a developing organ: the root primordium of *Azolla pinnata*. In *Cytomorphogenesis in Plants*, ed. O. Kiermayer, pp. 301–26. Vienna: Springer-Verlag.

Gunning, B. E. S. (1982). The root of the water fern *Azolla*: cellular basis of development and multiple roles for cortical microtubules. In *Developmental Order: Its Origin and Regulation*, ed. S. Subtelny and P. B. Green, pp. 379–421. New York: Alan R. Liss.

Gunning, B. E. S., Hardham, A. R., and Hughes, J. E. (1978a). Pre-prophase bands of microtubules in all categories of formative and proliferative cell division in *Azolla* roots. *Planta 143,* 145–60.

Gunning, B. E. S., Hardham, A. R., and Hughes, J. E. (1978b). Evidence for initiation of microtubules in discrete regions of the cell cortex in *Azolla* root tip cells, and an hypothesis on the development of cortical arrays of microtubules. *Planta 143,* 161–79.

Gunning, B. E. S., Hughes, J. E., and Hardham, A. R. (1978). Formative and proliferative cell divisions, cell differentiation, and developmental changes in the meristem of *Azolla* roots. *Planta 143,* 121–44.

Hallett, J., and Mason, B. J. (1958). The influence of temperature and supersaturation on the habit of ice crystals grown from the vapour. *Proc. R. Soc. Lond. A241,* 440–53.

Harold, F. M. (1990). To shape a cell: an inquiry into the causes of morphogenesis of microorganisms. *Microbiol. Rev. 54,* 381–431.

Harrison, L. G. (1973). Evolution of biochemical systems with specific chiralities: a model involving territorial behaviour. *J. Theor. Biol. 39,* 333–41.

Harrison, L. G. (1974). The possibility of spontaneous resolution of enantiomers on a catalyst surface. *J. Mol. Evol. 4,* 99–111.

Harrison, L. G. (1979). Molecular asymmetry and morphology: big hands from little hands. In *Origins of Optical Activity in Nature*, ed. D. C. Walker, pp. 125–40. Amsterdam: Elsevier.

Harrison, L. G. (1981). Physical chemistry of biological morphogenesis. *Chem. Soc. Rev. 10,* 491–528.

Harrison, L. G. (1982). An overview of kinetic theory in developmental modelling. In *Developmental Order: Its Origin and Regulation*, ed. S. Subtelny and P. B. Green, pp. 3–33. New York: Alan R. Liss.

Harrison, L. G. (1987). What is the status of reaction-diffusion theory thirty-four years after Turing? *J. Theor. Biol. 125,* 369–84.

Harrison, L. G., Graham, K. T., and Lakowski, B. C. (1988). Calcium localization during *Acetabularia* whorl formation: evidence supporting a two-stage hierarchical mechanism. *Development 104*, 255–62.

Harrison, L. G., and Green, B. R. (1988). Kinetically cooperative models: boundary movement in optical resolution, phase transitions, and biological morphogenesis. *Can. J. Chem. 66*, 839–51.

Harrison, L. G., and Hillier, N. A. (1985). Quantitative control of *Acetabularia* morphogenesis by extracellular calcium: a test of kinetic theory. *J. Theor. Biol. 114*, 177–92.

Harrison, L. G., and Kolář, M. (1988). Coupling between reaction-diffusion prepattern and expressed morphogenesis, applied to desmids and dasyclads. *J. Theor. Biol. 130*, 493–515.

Harrison, L. G., and Lacalli, T. C. (1978). Hyperchirality: a mathematically convenient and biochemically possible model for the kinetics of morphogenesis. *Proc. R. Soc. Lond. B202*, 361–97.

Harrison, L. G., and Lyons, M. J. (in press). Stripes and spots and isolated structures: the pattern-forming abilities of diverse non-linearities in reaction-diffusion mechanisms. In *Dynamical Phenomena at Interfaces, Surfaces and Membranes*, ed. D. Beysens and G. Forgacs. Commack, N.Y.: Nova Science Publishers.

Harrison, L. G., Snell, J., and Verdi, R. (1984). Turing's model and pattern adjustment after temperature shock, with application to *Acetabularia* whorls. *J. Theor. Biol. 106*, 59–78.

Harrison, L. G., Snell, J., Verdi, R., Vogt, D. E., Zeiss, G. D., and Green, B. R. (1981). Hair morphogenesis in *Acetabularia mediterranea*: temperature-dependent spacing and models of morphogen waves. *Protoplasma 106*, 211–21.

Harrison, L. G., and Tan, K. Y. (1988). Where may reaction-diffusion mechanisms be operating in metameric patterning of *Drosophila* embryos? *BioEssays 8*, 118–24.

Held, L. I. (1992). *Models for Embryonic Periodicity*. Basel: Karger.

Herak, M., Kochansky-Devidé, V., and Gušić, I. (1977). The development of the dasyclad algae through the ages. In *Fossil Algae: Recent Results and Developments*, ed. E. Flügel, Chap. 15. Berlin: Springer-Verlag.

Herschkowitz-Kaufman, M. (1975). Bifurcation analysis of nonlinear reaction-diffusion equations. II: Steady-state solutions and comparison with numerical simulations. *Bull. Math. Biol. 37*, 589–636.

Hildebrand, J., and Scott, R. L. (1962). *Regular Solutions*. Englewood Cliffs, N.J.: Prentice-Hall.

Hohl, H. R., and Raper, K. B. (1964). Control of sorocarp size in the cellular slime mold *Dictyostelium discoideum*. *Dev. Biol. 9*, 137–53.

Hope, R. A., Hammond, B. J., and Gaze, R. M. (1976). The arrow model: retinotectal specificity and map formation in the goldfish visual system. *Proc. R. Soc. Lond. B194*, 447–66.

Houck, J. C. (ed.) (1976). *Chalones*. Amsterdam: North-Holland.

Hunding, A. (1989). Turing patterns of the second kind simulated on supercomputers in three curvilinear coordinates and time. In *Cell to Cell Signalling: From Experiments to Theoretical Models*, ed. A. Goldbeter, pp. 229–36. London: Academic Press.

Hunding, A., Kauffman, S. A., and Goodwin, B. C. (1990). *Drosophila* segmentation: supercomputer simulations of prepattern hierarchy. *J. Theor. Biol. 145*, 369–84.

Inoué, S. (1982). The role of self-assembly in the generation of biologic form. In *Developmental Order: Its Origin and Regulation*, ed. S. Subtelny and P. B. Green, pp. 35–76. New York: Alan R. Liss.

Inoué, S., and Okazaki, K. (1977). Biocrystals. *Sci. Am. 236*, 82–92.

Irish, V., Lehmann, R., and Akam, M. (1989). The *Drosophila* posterior-group gene *nanos* functions by repressing *hunchback* activity. *Nature 338*, 646–8.

Jacobson, A. G. (1960). Influences of ectoderm and endoderm on heart differentiation in the newt. *Dev. Biol. 2*, 138–54.

Jacobson, A. G., and Duncan, J. T. (1968). Heart induction in salamanders. *J. Exp. Zool. 167*, 79–103.

Jacobson, A. G., and Gordon, R. (1976). Changes in the shape of the developing vertebrate nervous system analyzed experimentally, mathematically and by computer simulation. *J. Exp. Zool. 197*, 191–246.

Jaffe, L. F. (1980). Calcium explosions as triggers of development. *Ann. N.Y. Acad. Sci. 339*, 86–101.

Jaffe, L. F. (1982). Developmental currents, voltages, and gradients. In *Developmental Order: Its Origin and Regulation*, ed. S. Subtelny and P. B. Green, pp. 183–215. New York: Alan R. Liss.

Kalthoff, K. (1976). Specification of the antero-posterior body pattern in insect eggs. In *Insect Development*, ed. P. A. Lawrence, pp. 53–75. Oxford: Blackwell.

Kalthoff, K., and Sander, K. (1968). Der Entwicklungsgang der Missbildung "Doppel-abdomen" im partiell UV-bestrahlten Ei von *Smittia parthenogenetica* (Diptera, Chironomidae). *Wilhelm Roux Arch. EntwMech. Org. 161*, 129–46.

Kauffman, S. A. (1977). Chemical patterns, compartments, and a binary epigenetic code in *Drosophila*. *Am. Zool. 17*, 631–48.

Kauffman, S. A. (1981). Pattern formation in the *Drosophila* embryo. *Philos. Trans. R. Soc. Lond. B295*, 567–94.

Kauffman, S. A., Shymko, R. M., and Trabert, K. (1978). Control of sequential compartment formation in *Drosophila*. *Science 199*, 259–70.

Kepler, J. (1966). *The Six-Cornered Snowflake* (English translation of 1611 Latin text). Oxford: Oxford University Press.

Kiermayer, O. (1970). Causal aspects of cytomorphogenesis in *Micrasterias*. *Ann. N.Y. Acad. Sci. 175*, 686–701.

Kirschner, M., Gerhart, J., Hara, K., and Ubbels, G. (1980). Initiation of the cell cycle and establishment of bilateral symmetry in *Xenopus* eggs. In *The Cell Surface: Mediator of Developmental Processes*, ed. S. Subtelny and N. K. Wessells, pp. 187–215. New York: Academic Press.

Klug, A. (1972). Assembly of tobacco mosaic virus. *Fed. Proc. 31*, 30–42.

Konijn, T. M., van de Meene, J. G. C., Bonner, J. T., and Barkley, D. S. (1967). The acrasin activity of adenosine 3′,5′-cyclic phosphate. *Proc. Natl. Acad. Sci. USA 58*, 1152–4.

Kopell, N., and Howard, L. N. (1973). Horizontal bands in the Belousov reaction. *Science 180*, 1171–3.

Kuhn, T. S. (1962). *The Structure of Scientific Revolutions*. Chicago: University of Chicago Press.

Lacalli, T. C. (1973). Morphogenesis in *Micrasterias*. Ph.D. thesis, University of British Columbia, Canada.

Lacalli, T. C. (1975a). Morphogenesis in *Micrasterias*. I: Tip growth. *J. Embryol. Exp. Morphol. 33*, 95–115.

Lacalli, T. C. (1975b). Morphogenesis in *Micrasterias*. II: Patterns of morphogenesis. *J. Embryol. Exp. Morphol. 33*, 117–26.

Lacalli, T. C. (1976). Morphogenesis in *Micrasterias*. III: The morphogenetic template. *Protoplasma 88*, 133–46.

Lacalli, T. C. (1981). Dissipative structures and morphogenetic pattern in unicellular algae. *Philos. Trans. R. Soc. Lond. B294*, 547–88.

Lacalli, T. C. (1990). Modeling the *Drosophila* pair-rule pattern by reaction-diffusion: gap input and pattern control in a 4-morphogen system. *J. Theor. Biol. 144*, 171–94.

Lacalli, T. C., and Harrison, L. G. (1978). The regulatory capacity of Turing's model for morphogenesis, with application to slime moulds. *J. Theor. Biol. 70*, 273–95.

Lacalli, T. C., and Harrison, L. G. (1979). Turing's conditions and the analysis of morphogenetic models. *J. Theor. Biol. 76*, 419–36.

Lacalli, T. C., and Harrison, L. G. (1987). Turing's model and branching tip growth: relation of time and spatial scales in morphogenesis, with application to *Micrasterias*. *Can. J. Bot. 65*, 1308–19.

Lacalli, T. C., and Harrison, L. G. (1991). From gradients to segments: models for pattern formation in early *Drosophila* embryogenesis. *Seminars in Developmental Biology 2*, 107–17.

Lacalli, T. C., Wilkinson, D. A., and Harrison, L. G. (1988). Theoretical aspects of stripe formation in relation to *Drosophila* segmentation. *Development 104*, 105–13.

Lavenda, B. H. (1985). Brownian motion. *Sci. Am. 252*, 70–85.

Lehninger, A. L. (1975). *Biochemistry*, Chap. 36. New York: Worth Publishers.

Leicester, H. M. (1974). *Development of Biochemical Concepts from Ancient to Modern Times* (pp. 170–1). Cambridge, Mass.: Harvard University Press.

Lindenmayer, A. (1982). Developmental algorithms: lineage *versus* interactive control mechanisms. In *Developmental Order: Its Origin and Regulation*, ed. S. Subtelny and P. B. Green, pp. 219–45. New York: Alan R. Liss.

Lindenmayer, A., and Rozenberg, G. (eds.) (1976). *Automata, Languages, Development*. Amsterdam: North-Holland.

Lintilhac, P. M. (1984). Positional controls in meristem development: a caveat and an alternative. In *Positional Controls in Plant Development*, ed. P. W. Barlow and D. J. Carr, Chap. 4. Cambridge University Press.

Lisman, J. E. (1985). A mechanism for memory storage insensitive to molecular turnover: a bistable autophosphorylating kinase. *Proc. Natl. Acad. Sci. USA 82*, 3055–7.

Locke, M. (1967). The development of patterns in the integument of insects. *Advances in Morphogenesis 6*, 33–88.

Lück, H. B., and Lück, J. (1976). Cell number and cell size in filamentous organisms in relation to ancestrally and positionally dependent generation times. In *Automata, Languages, Development*, ed. A. Lindenmayer and G. Rozenberg, pp. 109–24. Amsterdam: North-Holland.

Lyons, M. J. (1991). *Mathematical Models for Biological Pattern Formation in Two Dimensions*. Ph.D. thesis, University of British Columbia, Canada.

Lyons, M. J., and Harrison, L. G. (1991). A class of reaction-diffusion mechanisms which preferentially select striped patterns. *Chem. Phys. Lett. 183*, 158–64.

Lyons, M. J., Harrison, L. G., Lakowski, B. C., and Lacalli, T. C. (1990). Reaction diffusion modelling of biological pattern formation: application to the embryogenesis of *Drosophila melanogaster*. *Can. J. Physics 68*, 772–7.

McLaughlin, S., and Poo, M.-m. (1981). The role of electro-osmosis in the electric field-induced movement of charged macromolecules on the surfaces of cells. *Biophys. J. 34*, 85–94.

McNally, J. G., and Cox, E. C. (1989). Spots and stripes: the patterning spectrum in the cellular slime mould *Polysphondylium pallidum*. *Development 105*, 323–33.

MacWilliams, H. K. (1982). Numerical simulation of *Hydra* head regeneration using a proportion-regulating version of the Gierer-Meinhardt model. *J. Theor. Biol. 99*, 681–703.

Mandelbrot, B. B. (1977). *Fractals, Form, Chance and Dimension*. San Francisco: Freeman.

Mandelbrot, B. B. (1982). *The Fractal Geometry of Nature*. San Francisco: Freeman.

Marchase, R. B., Barbera, A. J., and Roth, S. (1975). A molecular approach to retinotectal specificity. In *Cell Patterning, CIBA Foundation Symposium 29* (n.s.), pp. 315–41. Amsterdam: Elsevier.

Mason, B. J. (1961). The growth of snow crystals. *Sci. Am. 204*, 120–31. Reprinted 1979 in *The Physics of Everyday Phenomena* (readings from *Sci. Am.*), pp. 4–13. San Francisco: Freeman.

Matela, R. J., and Fletterick, R. J. (1980). Computer simulation of cellular self-sorting: a topological exchange model. *J. Theor. Biol. 84*, 673–90.

Maynard Smith, J. (1968). *Mathematical Ideas in Biology*. Cambridge University Press.

Meinhardt, H. (1977). A model of pattern formation in insect embryogenesis. *J. Cell Sci. 23*, 117–39.

Meinhardt, H. (1982). *Models of Biological Pattern Formation*. London: Academic Press.

Meinhardt, H. (1984). Models of pattern formation and their application to plant development. In *Positional Controls in Plant Development*, ed. P. W. Barlow and D. J. Carr, Chap. 1. Cambridge University Press.

Meinhardt, H. (1986). Hierarchical induction of cell states: a model for segmentation in *Drosophila*. *J. Cell Sci. [Suppl.] 4*, 357–81.

Meinhardt, H. (1988). Models for maternally supplied positional information and the activation of segmentation genes in *Drosophila* embryogenesis. *Development [Suppl.] 104*, 95–110.

Meinhardt, H. (1989). Models for positional signalling with application to the dorso-ventral patterning of insects and segregation into different cell types. *Development [Suppl.]*, 169–80.

Mills, W. H. (1932). Stereochemistry and catalysis. *Chemistry and Industry*, pp. 750–9.

Mitchison, G. J. (1980). A model for vein formation in higher plants. *Proc. R. Soc. Lond. B207*, 79–109.

Mitchison, G. J. (1981). The polar transport of auxin and vein patterns in plants. *Philos. Trans. R. Soc. Lond. B295*, 461–71.

Mitchison, G. J., and Wilcox, M. (1972). Rule governing cell division in *Anabaena*. *Nature 239*, 110–11.

Moore, W. J. (1972). *Physical Chemistry*, 4th ed. Englewood Cliffs, N.J.: Prentice-Hall.

Mörtberg, L. (1974). Celestial asymmetry as a possible cause of small chemical asymmetry and autocatalytic amplification of the latter. In *Generation and Amplification of Asymmetry in Chemical Systems*, ed. W. Thiemann, pp. 109–19. Jul-Conf-13, Kernforschungsanlage Jülich.

Murray, J. D. (1977). *Lectures on Nonlinear-Differential-Equation Models in Biology*. Oxford: Oxford University Press.

Murray, J. D. (1981a). A pre-pattern formation mechanism for animal coat markings. *J. Theor. Biol. 88*, 161–99.

Murray, J. D. (1981b). On pattern formation mechanisms for lepidopteran wing patterns and mammalian coat markings. *Philos. Trans. R. Soc. Lond. B295*, 473–96.

Murray, J. D. (1988). How the leopard gets its spots. *Sci. Am. 258*, 80–7.

Murray, J. D. (1989). *Mathematical Biology*. Berlin: Springer-Verlag.

Nagorcka, B. N. (1988). A pattern formation mechanism to control spatial organization in the embryo of *Drosophila melanogaster*. *J. Theor. Biol. 132*, 277–306.

Nagorcka, B. N. (1989). Wavelike isomorphic prepatterns in development. *J. Theor. Biol. 137*, 127–62.

Nagorcka, B. N., and Mooney, J. R. (1982). The role of a reaction-diffusion system in the formation of hair fibres. *J. Theor. Biol. 98*, 575–607.

Nagorcka, B. N., and Mooney, J. R. (1985). The role of a reaction-diffusion system in the initiation of primary hair follicles. *J. Theor. Biol. 114*, 243–72.

Nicolis, G., and Prigogine, I. (1977). *Self-Organization in Non-Equilibrium Systems*. New York: Wiley.

Odell, G. M., Oster, G., Alberch, P., and Burnside, B. (1981). The mechanical basis of morphogenesis. I: Epithelial folding and invagination. *Dev. Biol. 85*, 446–62.

Okazaki, K., and Inoué, S. (1976). Crystal property of the larval sea urchin spicule. *Dev. Growth Differ. 18*, 413–34.

Orowan, E. (1934). Zur Kristallplastizität. III: Ueber den Mechanismus des Gleitvorganges. *Z. Physik 89*, 634–59.

Oster, G. F. (1983). Mechanochemistry and morphogenesis. In *Biological Structures and Coupled Flows*, ed. A. Oplatka and M. Balaban, pp. 417–43. New York: Academic Press.

Oster, G. F., Murray, J. D., and Harris, A. K. (1983). Mechanical aspects of mesenchymal morphogenesis. *J. Embryol. Exp. Morphol. 78*, 83–125.

Oster, G. F., and Odell, G. M. (1980). Morphogenesis, mechanics and evolution. *Mathematical Problems in the Life Sciences 13*. Providence, R. I.: American Mathematical Society.

Ouyang, Q., and Swinney, H. L. (1991). Transition from a uniform state to hexagonal and striped Turing patterns. *Nature 352*, 610–11.

Pai, H. S., Dehm, P., Schweiger, M., Rahmsdorf, H. J., Ponta, H., Hirsch-Kaufmann, M., and Schweiger, H. G. (1975). Protein kinase of *Acetabularia*. *Protoplasma 85*, 209–18.

Pankratz, M. J., Gaul, U., Hoch, M., Seifert, E., Nauber, U., Gerwin, N., Rothe, M., Bronner, G., Forsbach, V., Goerlich, K., and Jäckle, H. (1990). Overlapping gene activities generate pair-rule stripes and delimit the expression domains of homeotic genes along the longitudinal axis of the *Drosophila* blastoderm embryo. In *Genetics of Pattern Formation and Growth Control*, ed. A. P. Mahowald, pp. 17–29. New York: Wiley-Liss.

Pasteur, L. (1848). Mémoire sur la relation qui peut exister entre la forme cristalline et la composition chimique, et sur la cause de la polarisation rotatoire. *C. R. Acad. Sci. (Paris) 26*, 535–6.

Penrose, R. (1989). *The Emperor's New Mind*. Oxford: Oxford University Press.

Phillips, H. M., and Steinberg, M. S. (1969). Equilibrium measurements of embryonic chick cell adhesiveness. I: Shape equilibrium in centrifugal fields. *Proc. Natl. Acad. Sci. USA 64*, 121–7.

Pickett-Heaps, J. D., and Northcote, D. H. (1966). Organization of microtubules and endoplasmic reticulum during mitosis and cytokinesis in wheat meristems. Cell division in the formation of the stomatal complex of the young leaves of wheat. *J. Cell Sci. 1*, 109–20, 121–8.

Polanyi, M. (1934). Ueber eine Art Gitterstörung, die einen Kristall plastich machen könnte. *Z. Physik 89*, 660–6.

Polanyi, M. (1946). *Science, Faith and Society*. London: Oxford University Press.

Polanyi, M. (1949). Scientific Beliefs. Lecture at the University of Liverpool.

Poole, T. J., and Steinberg, M. S. (1981). Amphibian pronephric duct morphogenesis: segregation, cell rearrangement, and directed migration of the *Ambystoma* duct rudiment. *J. Embryol. Exp. Morphol. 63*, 1–16.

Prigogine, I. (1967). Dissipative structures in chemical systems. In *Fast Reactions and Primary Processes in Chemical Kinetics*, ed. S. Claesson, pp. 371–82. New York: Interscience.

Prigogine, I., and Lefever, R. (1968). Symmetry-breaking instabilities in dissipative systems. II. *J. Chem. Phys. 48*, 1695–700.

Prusinkiewicz, P., and Hanan, J. (1989). *Lindenmayer Systems, Fractals, and Plants.* New York: Springer-Verlag.

Puiseux-Dao, S. (1970). Acetabularia *and Cell Biology*, transl. P. Malpoix-Higgins. New York: Springer.

Ramsay, J. D. F. (1986). Recent developments in the characterization of oxide sols using small angle neutron scattering techniques. *Chem. Soc. Rev. 15*, 335–71.

Rashevsky, N. (1940). An approach to the mathematical biophysics of biological self-regulation and of cell polarity. *Bull. Math. Biophys. 2*, 15–25.

Reiss, H.-D., and Herth, W. (1979). Calcium gradients in tip-growing plant cells visualized by chlorotetracycline fluorescence. *Planta 146*, 615–21.

Rogers, G., and Goel, N. S. (1978). Computer simulation of cellular movements: cell-sorting, cellular migration through a mass of cells and contact inhibition. *J. Theor. Biol. 71*, 141–66.

Russell, M. A. (1985). Positional information in insect segments. *Dev. Biol. 108*, 269–83.

Sachs, T. (1969). Polarity and the induction of organized vascular tissues. *Ann. Bot. 33*, 263–75.

Saito, Y., Goldbeck-Wood, G., and Müller-Krumbhaar, H. (1988). Numerical simulation of dendritic growth. *Phys. Rev. A38*, 2148–57.

Sander, L. M. (1987). Fractal growth. *Sci. Am. 256*, 94–100.

Sater, A. K., and Jacobson, A. G. (1990). The role of the dorsal lip in the induction of heart mesoderm in *Xenopus laevis. Development 108*, 461–70.

Schaller, H. C. (1973). Isolation and characterization of a low-molecular-weight substance activating head and bud formation in *Hydra. J. Embryol. Exp. Morphol. 29*, 27–38.

Schaller, H. C., and Gierer, A. (1973). Distribution of the head-activating substance in *Hydra* and its localization in membranous particles in nerve cells. *J. Embryol. Exp. Morphol. 29*, 39–52.

Schindler, J., and Sussman, M. (1977). Ammonia determines the choice of morphogenetic pathways in *Dictyostelium discoideum. J. Mol. Biol. 116*, 161–9.

Schmid, R., Idziak, E.-M., and Tunnermann, M. (1987). Action spectrum for the blue-light-dependent morphogenesis of hair whorls in *Acetabularia mediterranea. Planta 171*, 96–103.

Seelig, F. F. (1971a). Systems-theoretic model for the spontaneous formation of optical antipodes in strongly asymmetric yield. *J. Theor. Biol. 31*, 355–61.

Seelig, F. F. (1971b). Mono- or bistable behaviour in a weakly or strongly open chemical reaction system. *J. Theor. Biol. 32*, 93–106.

Seelig, F. F. (1972). On spontaneous asymmetric synthesis. *J. Theor. Biol. 34*, 197–8.

Sel'kov, E. E. (1968). Self-oscillations in glycolysis. 1: A simple kinetic model. *Eur. J. Biochem. 4*, 79–86.

Shaw, L. J., and Murray, J. D. (1990). Analysis of a model for complex skin patterns. *S.I.A.M. J. Appl. Math. 50*, 628–48.

Slack, J. M. W. (1983). *From Egg to Embryo.* Cambridge University Press.

Smith, S. C. (1990). Control of heart development in the Mexican axolotl (*Ambystoma mexicanum*). Ph.D. thesis, University of Ottawa, Canada.

Smith, S. C., and Armstrong, J. B. (1990). Heart induction in wild-type and cardiac mutant axolotls (*Ambystoma mexicanum*). *J. Exp. Zool. 254*, 48–54.

Smith, S. C., and Armstrong, J. B. (1991). Heart development in normal and *cardiac-lethal* mutant axolotls: a model for the control of vertebrate cardiogenesis. *Differentiation 47*, 129–34.

Sperry, R. W. (1965). Embryogenesis of behavioral nerve nets. In *Organogenesis*, ed. R. L. DeHaan and H. Ursprung, Chap. 6. New York: Holt, Rinehart & Winston.

Staehelin, A., and Giddings, T. H. (1982). Membrane-mediated control of cell wall microfibrillar order. In *Developmental Order: Its Origin and Regulation*, ed. S. Subtelny and P. B. Green, pp. 133–47. New York: Alan R. Liss.

Starling, E. H. (1905). The chemical correlation of the functions of the body. *Lancet 2*, 339–41.

Steinberg, M. S. (1970). Does differential adhesion govern self-assembly processes in histogenesis? Equilibrium configurations and the emergence of a hierarchy among populations of embryonic cells. *J. Exp. Zool. 173*, 395–434.

Steinberg, M. S. (1975). Adhesion-guided multicellular assembly: a commentary upon the postulates, real and imagined, of the differential adhesion hypothesis, with special attention to computer simulations of cell sorting. *J. Theor. Biol. 55*, 431–43.

Steinberg, M. S., and Poole, T. J. (1981). Strategies for specifying form and pattern: adhesion-guided multicellular assembly. *Philos. Trans. R. Soc. Lond. B295*, 451–60.

Steinberg, M. S., and Wiseman, L. L. (1972). Do morphogenetic tissue movements require active cell movements? *J. Cell Biol. 55*, 606–15.

Tabony, J., and Job, D. (1990). Spatial structures in microtubular solutions requiring a sustained energy source. *Nature 346*, 448–51.

Taylor, G. I. (1934). The mechanism of plastic deformation of crystals. I: Theoretical. *Proc. R. Soc. Lond. A145*, 362–87.

Teilhard de Chardin, P. (1955). *Le Phénomène Humain*. Paris: Editions du Seuil.

Thiemann, W. (ed.) (1974). *Generation and Amplification of Asymmetry in Chemical Systems*. Jul-Conf-13, Kernforschungsanlage Jülich.

Thom, R. (1975). *Structural Stability and Morphogenesis*, trans. D. H. Fowler. Reading, Mass.: W. A. Benjamin.

Thompson, D. W. (1917). *On Growth and Form*. Cambridge University Press.

Thompson, D. W. (1942). *On Growth and Form*, 2nd ed. Cambridge University Press.

Tickle, C., Alberts, B. M., Wolpert, L., and Lee, J. (1982). Local application of retinoic acid to the limb bud mimics the action of the polarizing region. *Nature 296*, 564–5.

Town, C. D., Gross, J. D., and Kay, R. R. (1976). Cell differentiation without morphogenesis in *Dictyostelium discoideum. Nature 262*, 717–19.

Townes, P. S., and Holtfreter, J. (1955). Directed movements and selective adhesion of embryonic amphibian cells. *J. Exp. Zool. 128*, 53–120.

Turing, A. M. (1952). The chemical basis of morphogenesis. *Philos. Trans. R. Soc. Lond. B237*, 37–72.

Tyson, J. J. (1985). A quantitative account of oscillations, bistability, and traveling waves in the Belousov-Zhabotinsky reaction. In *Oscillations and Traveling Waves in Chemical Systems*, ed. R. Field and M. Burger, pp. 93–144. New York: Wiley.

Tyson, J. J., and Kauffman, S. A. (1975). Control of mitosis by a continuous biochemical oscillation. *J. Math. Biol. 1*, 289–310.

Tyson, J. J., and Light, J. C. (1973). Properties of two-component bimolecular and trimolecular chemical reaction systems. *J. Chem. Phys. 59*, 4164–73.

Vanden Driessche, T. (1990). Calcium as a second messenger in *Acetabularia:* calmodulin and signal transduction pathways. In *Calcium as an Intracellular Messenger in Eucaryotic Microbes*, ed. D. H. O'Day, Chap. 16. Washington, D.C.: American Society for Microbiology.

Vöchting, H. (1877). Ueber Theilbarkeit im Pflanzenreich und die Wirkung innerer und äusserer Krafte auf Organbildung an Pflanzentheilen. *Pflüger's Arch. 15*, 153–90.

Vöchting, H. (1878). *Ueber Organbildung im Pflanzenreich*, Vol. 1. Bonn, Germany: Max Cohen & Sohn.

von Ehrenstein, G. (1980). Embryonic cell lineages and development of the nematode *Caenorhabditis elegans;* film shown in a talk with this title, relevant to abstracts D1337, D1338, and D1339. *Eur. J. Cell Biol. 22,* 449–50.

Waddington, C. H. (1956). *Principles of Embryology* (p. 423). London: Allen & Unwin.

Wain, R. L. (1977). Chemicals which control plant growth. *Chem. Soc. Rev. 6,* 261–75.

Wald, G. (1957). The origin of optical activity. *Ann. N.Y. Acad. Sci. 69,* 352–67.

Walker, D. C. (ed.) (1979). *Origins of Optical Activity in Nature.* Amsterdam: Elsevier.

West, W., and West, G. S. (1905). *British Desmidaceae.* London: Adlard & Sons for the Ray Society.

Williams, R. J. P. (1976). Calcium chemistry and its relation to biological function. In *Calcium in Biological Systems,* ed. C. J. Duncan, pp. 1–17. Cambridge University Press.

Williams, R. J. P. (1977). Calcium chemistry and its relation to protein binding. In *Calcium-Binding Proteins and Calcium Function,* ed. R. H. Wasserman, R. A. Carradino, E. Carafoli, R. H. Kretsinger, D. H. MacLennan, and F. L. Siegel, pp. 3–12. New York: North-Holland.

Willshaw, D. J., and von der Malsburg, C. (1976). How patterned neural connections can be set up by self-organization. *Proc. R. Soc. Lond. B194,* 431–45.

Willshaw, D. J., and von der Malsburg, C. (1979). A marker induction mechanism for the establishment of ordered neural mappings: its application to the retinotectal problem. *Philos. Trans. R. Soc. Lond. B287,* 203–43.

Winfree, A. T. (1987). Editor's preface to translation of V. S. Zykov, *Simulation of Wave Processes in Excitable Media,* pp. x–xxvi. Manchester: Manchester University Press.

Wolfe, S. L. (1981). *Biology of the Cell,* 2nd ed. Belmont, Calif.: Wadsworth.

Wolfram, S. (1984). Cellular automata as models of complexity. *Nature 311,* 419–23.

Wolk, C. P., Austin, S., Bortins, J., and Galonsky, A. (1974). Autoradiographic localization of ^{13}N after fixation of ^{13}N-labeled nitrogen gas by a heterocyst-forming blue-green alga. *J. Cell Biol. 61,* 440–53.

Wolk, C. P., and Quine, M. P. (1975). Formation of one-dimensional patterns by stochastic processes and by filamentous blue-green algae. *Dev. Biol. 46,* 370–82.

Wolpert, L. (1970). Positional information and pattern formation. In *Towards a Theoretical Biology,* ed. C. H. Waddington, Vol. 3, pp. 198–230. Edinburgh: Edinburgh University Press.

Wolpert, L. (1981). Positional information and pattern formation. *Philos. Trans. R. Soc. Lond. B295,* 441–50.

Zeeman, E. C. (1974). Primary and secondary waves in developmental biology. In *Lectures on Mathematics in the Life Sciences, 7: Some Mathematical Questions in Biology VI,* pp. 69–161. Providence, R.I.: American Mathematical Society.

Index

Acetabularia, xvi, 7, 15, 19, 24, 62, 238, 299–309, 312, 325, 327–8, 334–7
action at a distance, 31, 97
activator (*see also* morphogen), 37–45, 268, 323
Acton, A. B., xv
adhesion: differential, 113, 118–28; heterophilic, 122–3; homophilic, 122–3
aging, cell surface, 312–14
Ambystoma, see axolotl
ammonia, as morphogen, 324–6
Amphioxus, 95
amplifier, 24–7, 37, 134, 154–6, 167–9, 321–3
Anabaena, 68–71, 74, 281
Arabidopsis, 300
Aristotle, 28, 89, 93, 146, 173
asymmetry, 167–72
atomic theory, 21–2; Dalton's, 298
autocatalysis, 38, 58–9, 134, 169, 257, 296, 323–4
automata, cellular, 28, 68–9, 153
auxin, 296–7
Avogadro's Law, 14
axolotl, 115, 119, 124, 323, 328–31
Azolla, 72–80, 109

Belousov–Zhabotinski reaction, xvii, 332
Berthollet, 298
bicoid, 316, 318, 320
bifurcation, 201–3, 207
binaphthyl, xiii
blastoderm, 15, 61, 83, 129, 281, 315–22
blastula, 146–8
Boltzmann, L., 8, 22–4, 299
boundary conditions: Dirichlet, 62, 69, 78; no-flux (Neumann), 61–2, 78
brusselator, 53, 82, 186–7, 191, 198, 234, 243–6, 248, 258–9, 265–83, 309, 312–14, 323, 328
bubble raft, 133

Caenorhabditis, 94–5, 119
calcite, 106–8

calcium, 7, 96, 134–6, 254, 303–7, 325–6
calculus, necessity for, 9, 173–5, 209
cAMP, cyclic adenosine monophosphate, 324–5
Cannizzaro, S., 14
cardiac-lethal, axolotl, mutant, 330–1
catastrophe theory, 205–8
causes, 155–6; efficient, 89, 93; etiology, 90, 93
cell: division, 67–80, 108–11; as molecule, in theories, 21, 26, 110–19, 136, 141, 255
cellular automata, *see* automata, cellular
Chaetomorpha, 69–71
Chain reactions, balanced, 265–7
chalones, 46–7
chaos, 24–7, 148
chemotaxis, 119
Cheshire Cat, sine wave pattern fading by diffusion, 175–7, 181–2, 225–6
chick: cells, embryonic, sorting, 112–13, 120–2; feather follicles (*see also* hexagonal patterns), 112
chirality (*see also* hyperchirality; optical resolution), xiii, 154–72, 261–3, 283–91
chlorotetracycline, 304, 307
classifications: of developmental theories, 252–6; of reaction-diffusion models, 256–63
collagen, 116, 149
concanavalin A, 289, 306
contractile processes, 96
Cosmarium, 311
Cotton effect, 170
counter-intuitive properties, 186, 229, 323–4
Crick, F., 23
crystal growth, 101–8, 210–11
cubic equations, 206–8
Cymopolia, 308
cytoskeleton (*see also* microtubules), 129, 306

dead reckoning, in navigation and surveying, *see* traverse, open

"death dominates," importance of removal processes, 197
depletion models, *see also* brusselator, 258
desmids, *see also Micrasterias,* 310–14
dichotomous branching, 310–14
Dictyostelium, 15, 61, 113–14, 119, 238, 240, 262, 323–5, 334–7
differential equations: introduction to, 175–83; partial, 178–9
diffusion: in Belousov–Zhabotinski reaction, 332; of a boundary, 163; of cells, 113, 136; as communication, xviii, 40, 136, 318–22; diffusion-limited aggregation, 27; diffusivity, 94, 179; diffusivity ratio, 235–43; equations, Fick's Law, 178–9, 212; first use in equations for pattern formation, 253; in nonlinear models, 264–97; in reaction-diffusion models, classification, 252–63; in reaction-diffusion models, introduction, 188–90; in Turing's model, 223–9; and wavelength, 36, 50–3, 232–3, 302
DIF-1, as morphogen, 324–6
dislocations in crystals, 94–5, 133
dissipative structure, 196, 332
DNA, 117, 322
double-abdomen deformity, 82–4
Drosophila: gene-classes, 315–17; imaginal discs, 238; segmentation, 11, 15, 19–20, 56, 262, 279, 281–2, 293, 302, 314–23, 334–7

Echinoderms, 95, 106–8
Einstein, A., 22, 156, 170
electro-osmosis, 99–100
enantiomers (*see also* optical resolution), 155, 158, 167, 285
endoderm, 328–31
engrailed, 282, 317
engulfment experiments, tissues *in vitro,* 118–23
entities, esp. a pattern as an entity, 19, 181–3, 225, 229, 279, 301
entropy, 23, 26, 94, 122, 148, 169, 198–201
epithelia: axolotl embryo, 328–31; development, in classification of theories, 254; in glandular development, 116–17; mechanochemical model for deformation of, 134–7; vertebrate skin, 294–5
equilibrium, 91–2, 118–23, 255; approach to *versus* departure from, 98, 101, 113, 118–23; shapes of crystals, 102–4
errors, 152–3, 166
Euastrum, 311
even-skipped, 316–7
exponential growth, 189–90, 209, 231–2, 241, 252

feedback, esp. positive feedback loops, 37, 169, 257–8, 307–9, 312–14, 317, 323

Fick's Law, 179
field: diversity of meaning, 96–7; physical, e.g., electrical, 74, 96–7; of science, xviii–xix; theory, 97, 206
filamentary organisms, 67–74
filopodium, 130, 133
flatland, 288–9
fluctuations, *see also* noise, 160–1, 225
flux, autocatalytic, 296–7
Flying Dutchman, opera, as an analogy, 30
fractals, 27
Frank, F. C., 104, 166
free energy, 35–6, 92, 97, 103–4, 108–9, 121, 130, 255
fushi tarazu, 316–17

gambling, 161–3
gap junctions, 281
gastrulation: echinoderms, 115, 119, 134–7; nematodes, 94–5, 112, 115, 119
Gedanken experiment, 170–2
genetic code, 167
Gierer–Meinhardt model, 32, 38, 50, 58–9, 63, 67, 71, 80–8, 183, 259–60, 279–81, 291–4, 325, 330–1
gland, mouse submandibular, salivary, 116–17, 119
governor, on engine, analogy, 227–9
gradients, 20, 33–7, 71, 81, 84–8, 115, 150, 272–5, 277, 315, 317–23; of adhesion, 115, 123–8
Green, B. R., xvi
Green, P. B., xvi, 8, 11

hairy, 317, 321
Harvey, W., 17
heart formation, axolotl, 328–31
Heisenberg, W., 23, 25
hexagonal patterns, 56–7, 149, 279–81
hierarchies, 118–21, 307–9, 314–20
homeosis, 317
homunculus, 37
hormone, 33, 46
Hoyle, F., 6
hunchback, 316, 319–20
Hydra, 15, 58, 71, 84–8, 292, 323
hyperchirality, 55, 261–3, 283–91, 312, 322

IAA, indole-3-yl acetic acid, *see* auxin
induction: endoderm-to-mesoderm, 328–31; homeogenetic, 262
inhibitor (*see also* morphogen; nonlinearity; Turing, A. M.), 46–55, 256, 268, 292, 330
int-1, 317
invagination, 254

Joule, J. P., 14

Kepler, J., essay on the snowflake, 102
kinase, protein, 307

kinetic theory of pattern: advances in, 323; in classification of developmental theories, 254; definition, 92–3, 100–1; opposition to, 315; scope, 252–5
knirps, 316, 321
Krüppel, 316, 319–21

Lacalli, T. C., xiii–xv, 9–10
Lachner, F., conductor, in musical analogy, 30
Lavoisier, A. L., 18–19
Liebig, J. von, first to recognize proteins, 322
linearization, 184–6, 258, 270–1
liverwort, 152–3
localization of pattern-forming region, 272–5
Lophocolea, 153

Mach, Ernst, 8, 22
macroscopics, 1, 8
Malthus, J., 90
mathematics in biology, 3–4, 247–9
Maxwell distribution, 25
Mayer, J. R., 14
measuring *versus* counting, 56–7, 334–7
mechanical force, 92, 94–6, 114, 128, 133–7, 314
mechanochemistry, 94–6, 115, 133–7, 254, 331
mesenchyme, 96, 108, 112, 116–19, 136, 138
mesoderm, 328–31
Michaelis–Menten equation, 295, 306
Micrasterias, xiii, xv, 15, 62, 149–50, 287, 310–14
microfilaments, 129
microtubules, 72–3, 110–11, 129, 333
Mills, W. H., mechanism for optical resolution, 39, 85, 155, 157–8, 166, 191, 264
model, meaning of the word, 247–9
Moelwyn-Hughes, E. A., 13
Moore, Walter J., 14, 22
morphogen, 31–55, 189, 305–7; precursors, 268–9, 275–8, 305–7, 334; type I (Wolpert), 32–7, 315, 318; type II (Turing), 32, 47–55, 188–9, 214, 223–46, 315, 318; variables A, H, U, V, X, Y, 58–9, 223–4, 268–9
moss, 109
motion, as ultimate reality, 28–30, 89, 172
Murray, J. D., 294–5

nanos, 318
Needham, J., 9
noise (*see also* fluctuations), 43–4, 320–1
nonlinearity, 50, 58, 63–4, 185–6, 248, 252–6, 259, 264–97

Ockham, William of (and Ockham's razor), 31, 38, 90, 276

ocular dominance, 290–1
optical resolution, xiii, 38–45, 143, 154–72, 187–8, 192–203, 254
oregonator, 254, 332
oscillations, *see* waves
oskar, 318

paradigms (*see also* preconceptions), xiv, 7, 12–13, 15, 17–18, 21, 28–30; paradigm shift, xvi
Paramecium, 284
parameter space, 213–14, 235–45, 273–5
parity, 155
Pasteur, L., 155, 169
Perrin, 22
phase plane, 213–14
phlogiston theory, 18–19
phosphorylation, 307
phylogeny, 308
Pincock, R. E., xiii–xiv
polar furrow, 109, 130
polarity, 37, 61, 69–70, 168–9
Polysphondylium, 15, 325–7
positional information, 15, 32–7, 212, 254, 315
positivism, 8, 22, 299
preconceptions (*see also* paradigms), xiv, 7, 12–13, 15, 17–18, 21, 107, 248, 298
prediction, in scientific method, 302–5, 337
preprophase band, 72–4, 110–11

range of activation, 83
receptors, 286, 306
regulation, 334–7
retinotectal specificity, 37, 125–8, 169
Rhabdoporella, 308
Rhodnius, 317
root development, 72–80
Rumford, Count, 14
runt, 317

saturation, 295
scales of distance (*see also* wavelength), 36–7, 46–7
Schrödinger equation, 16
scientific method, xiv–xv, 3–7, 12–18, 28, 89–90, 247–51, 322
sea urchin (*see also* Echinoderms), 106–8
segmentation, *see Drosophila*
self-assembly, 26, 91, 106–8, 143
self-electrophoresis, 98–101
self-enhancement (*see also* autocatalysis), 101, 169
self-organization, 91, 101, 118–19, 143, 148, 154
sequences of patterns, 275–8, 307–9, 315–22
simulation, 162
sine waves, 41–5, 60–2, 65–6, 175–8, 225–6, 229–33, 325–6

Smittia, 83, 318
Snider, R. F., 8
snowflake, 27, 93, 101–2, 104–6
sorting-out of cells, 112–36
source, localized (Saunders), 33–7, 318
species-specific shapes, 153
spicule, echinoderm, 106–8
spline, 146–54, 289
steady-state approximation, in kinetics, 266–7
Stokes' Law, 29
stripes, 20–1, 55, 260–1, 290–1, 294–5, 315–16, 318, 320–2, 332–3
structural: stability (dynamic equations), 10, 203–4, 214; theory of morphogenesis (*see also* self-assembly), 90–1, 254
surface tension, 121–3
switching, 281–2
symmetry, 1, 145–72; bilateral, 145–6, 284; breaking and making, 41, 136, 143, 145–72, 284; elements, 101; of kinetics, 261–3, 285–7; spherical, 146–7

Taricha, 330–1
tectum, optic, *see* retinotectal specificity
teratogenesis, 80–4, 281, 292
termolecular step, 191
Tetrahymena, 284
thermodynamic driving force, 194
Thompson, Benjamin, 14
threshold, 192–201, 268–9
tip growth, 62, 276, 300–14

transitive property, 118, 120
traverse, open (surveying), 151, 166
Turing, A. M., and Turing's model (*see also* morphogen), xiii, xix, 32, 38, 47–9, 71, 84, 115, 146, 148, 167, 206, 214, 223–46, 293, 323
Turing's conditions (*see also* parameter space), 233–46, 274
Turing's constants *versus* rate constants of mechanism, 270–2, 293

Ulothrix, 69, 71
uncertainty principle, 23, 25

vesicles, intracellular, 134, 312, 325–6
visco-elasticity, 128

Wagner, R., in musical analogy to biological dynamics, 30
Watson, J. D., 23
Watt, James, 197
wavelength, 43–5, 49–53, 62–6, 81–3, 86, 188–90, 224–7, 256, 278–82, 326–8
waves, travelling or stationary, xvii, 237–40, 332
whorls, xvi, 15, 19, 238, 448–61
wingless, 317
Wolpert signaller, *see* morphogen, type I
Wulff's theorem, 102–4, 210

Xanthidium, 311
Xenopus, 168, 330